"The book is highly relevant, and its publication comes at the perfect time to stimulate reflection on the benefits and challenges of integrating artificial intelligence and emerging technologies into evaluation. The various authors, all renowned experts in the field of program evaluation, draw on their know-how to offer cutting-edge analyses and prospective solutions for integrating AI technologies into evaluation. In short, this book is an enriching and enlightening read, carried by renowned authors, designed to enlighten minds on a subject of crucial importance in our time."

Steve Jacob, *Professor of Political Science, Laval University*

"AI is poised to change the field of evaluation in a multitude of ways, yet the evidence base is still slim. This book is a welcome addition to the literature on emerging technologies and their implications for evaluation. Key reasons that evaluators were not taking up big data approaches included a lack of relevant use cases for the public and not-for-profit sectors and a mistrust of big data approaches due to ethical concerns. This book provides a rich set of potential use cases, documentation and learning from those examples, and an exploration of ethics and equity themes related to data science and evaluation, and offers an important contribution to the literature base for the use of AI in evaluation. The work comes at an important time for the Evaluation field, when interest in AI is quite high, yet, in many cases, capacity and knowledge need to be enhanced and supplemented. It offers a unique perspective on ways that AI can be integrated into Evaluation, drawing on the work of well-respected evaluation professionals and academics who have long-standing experience to share."

Linda Raftree, *Founder, MERL Tech*

Artificial Intelligence and Evaluation

Artificial Intelligence and Evaluation: Emerging Technologies and Their Implications for Evaluation is a groundbreaking exploration of how the landscape of program evaluation will be redefined by artificial intelligence and other emerging digital technologies.

In an era where digital technologies and artificial intelligence (AI) are rapidly evolving, this book presents a pivotal resource for evaluators navigating the transformative intersection of their practice and cutting-edge technology. Addressing the dual dimensions of how evaluations are conducted and what is evaluated, a roster of distinguished contributors illuminate the impact of AI on program evaluation methodologies. Offering a discerning overview of various digital technologies, their promises and perils, they carefully dissect the implications for evaluative processes and debate how evaluators must be equipped with the requisite skills to harness the full potential of AI tools. Further, the book includes a number of compelling use cases, demonstrating the tangible applications of AI in diverse evaluation scenarios. The use cases range from the application of GIS data to advanced text analytics. As such, this book provides evaluators with inspirational cases on how to apply AI in their practice as well as what pitfalls one must look out for.

Artificial Intelligence and Evaluation is an indispensable guide for evaluators seeking to not only adapt to but thrive in the dynamic landscape of evaluation practices reshaped by the advent of artificial intelligence.

Steffen Bohni Nielsen (PhD) is Director General at the Danish National Research Centre for the Working Environment, and a member of the Danish National Research and Innovation Council. He is a member of the International Evaluation Research Group (INTEVAL) and has published extensively in the field of evaluation.

Francesco Mazzeo Rinaldi (PhD) is Professor at the University of Catania, Italy, and serves as the Director of the University's Research Center, LAPOSS. His main research interests include program and policy evaluation, Big Data analytics, and Artificial Intelligence. He is a member of the International Evaluation Research Group (INTEVAL).

Gustav Jakob Petersson (PhD) is Senior Analyst at the Swedish Research Council. He is a member of the International Evaluation Research Group (INTEVAL) and has co-edited one of its volumes on cyber society, Big Data, and evaluation in 2017.

Comparative Policy Evaluation
Edited by Ray C. Rist

The Comparative Policy Evaluation series is an interdisciplinary and internationally focused set of books that embodies within it a strong emphasis on comparative analyses of governance issues—drawing from all continents and many different nation states. The lens through which these policy initiatives are viewed and reviewed is that of evaluation. These evaluation assessments are done mainly from the perspectives of sociology, anthropology, economics, policy science, auditing, law, and human rights. The books also provide a strong longitudinal perspective on the evolution of the policy issues being analyzed.

Long Term Perspectives in Evaluation
Edited by Kim Forss, Ida Lindkvist, Mark McGillivray

The Realpolitik of Evaluation
Edited by Markus Palenberg, Arne Paulson

Changing Bureaucracies
Edited by Burt Perrin, Tony Tyrrell

Ethics for Evaluation
Edited by Rob D. van den Berg, Penny Hawkins, Nicoletta Stame

Towards Sustainable Futures
The Role of Evaluation
Edited by Per Øyvind Bastøe, Kim Forss, Ida Lindkvist

Evaluation in the Post Truth World
Edited by Mita Marra, Karol Olejniczak, Arne Paulson

Theories of Change in Reality: Strengths, Limitations and Future Directions
Edited by Andrew Koleros, Marie-Hélène Adrien, Tony Tyrrell

Artificial Intelligence and Evaluation
Emerging Technologies and Their Implications for Evaluation
Edited by Steffen Bohni Nielsen, Francesco Mazzeo Rinaldi, and Gustav Jakob Petersson

Artificial Intelligence and Evaluation

Emerging Technologies and Their Implications for Evaluation

Edited by Steffen Bohni Nielsen, Francesco Mazzeo Rinaldi and Gustav Jakob Petersson

Routledge
Taylor & Francis Group
NEW YORK AND LONDON

First published 2025
by Routledge
605 Third Avenue, New York, NY 10158

and by Routledge
4 Park Square, Milton Park, Abingdon, Oxon, OX14 4RN

Routledge is an imprint of the Taylor & Francis Group, an informa business

The Open Access for this title has been funded by: Bixal Solutions, Inc; Independent Evaluation Group, World Bank; Bench S.r.l., University of Catania; The Global Environment Facility, Independent Evaluation Office; NFA - National Research Centre for the Working Environment.

Library of Congress Cataloging-in-Publication Data
Names: Nielsen, Steffen Bohni, editor. | Mazzeo Rinaldi, Francesco, editor. | Petersson, Gustav Jakob, editor. Title: Artificial intelligence and evaluation : emerging technologies and their implications for evaluation / edited by Steffen Bohni Nielsen, Francesco Mazzeo Rinaldi, Gustav Jakob Petersson. Description: New York, NY : Routledge, 2024. | Series: Comparative policy evaluation | Includes bibliographical references and index. | Identifiers: LCCN 2024030964 (print) | LCCN 2024030965 (ebook) | ISBN 9781032843896 (hardback) | ISBN 9781032856803 (paperback) | ISBN 9781003512493 (ebook) Subjects: LCSH: Evaluation research (Social action programs)--Technological innovations. | Evaluation--Methodology--Technological innovations. | Artificial intelligence. Classification: LCC H62 .A6693 2024 (print) | LCC H62 (ebook) | DDC 001.4--dc23/eng/20240725 LC record available at https://lccn.loc.gov/2024030964LC ebook record available at https://lccn.loc.gov/2024030965

ISBN: 978-1-032-84389-6 (hbk)
ISBN: 978-1-032-85680-3 (pbk)
ISBN: 978-1-003-51249-3 (ebk)

DOI: 10.4324/ 9781003512493

Typeset in Times New Roman
by Deanta Global Publishing Services, Chennai, India

Contents

Acknowledgments

Writing a book is a journey that requires the support of many people. We are deeply grateful to each and every person who has played a part in its creation.

First and foremost, we are profoundly thankful to the talented people who contributed their expertise to the creation of this work. Collaborating with such brilliant authors has been a privilege, and we are grateful for the opportunity to bring together a diverse range of voices and experiences in this collective effort. Each author played a crucial role in shaping the content, and their commitment to excellence has undoubtedly elevated the quality of this book.

We want to thank the sponsors who sincerely believed in this project. Your generosity has been instrumental in ensuring the widespread availability of this work. Thank you for making this open-access publication a reality. Special mention and heartfelt thanks to:

- Bixal
- Global Environment Facility, Independent Evaluation Office
- Danish National Research Centre for the Working Environment
- University of Catania
- University of Lausanne
- World Bank, Independent Evaluation Group

We are sincerely grateful to the reviewers for their time, expertise, and the invaluable role they have played in shaping this publication.

To our families, thank you for your unwavering support throughout this endeavor. Your patience, understanding, and encouragement have sustained us during the long hours of writing and revision.

We are grateful to the dedicated professionals who assisted in the production of this book. Your commitment to excellence has helped transform our ideas into a polished and cohesive manuscript.

Heartfelt acknowledgment goes to Ray C. Rist, the editor of this excellent book series, a great friend and guide for all of us.

Finally, to the readers who embark on this journey with us, thank you for your time and attention. We hope the words within these pages resonate with you, contributing to a better understanding of AI's challenges to evaluation....

With heartfelt gratitude,
Steffen Bohni Nielsen
Francesco Mazzeo Rinaldi
Gustav Jakob Petersson

Contributors

Anupam Anand, PhD (Geospatial Science) is Senior Evaluation Officer at the Global Environmental Facility Independent Evaluation Office (GEF IEO). He has over 15 years of combined experience in international development, evaluation, and academia. At the GEF-IEO, he leads evaluations on biodiversity, land degradation, and crosscutting themes such as conflict and fragility, as well as nature-based solutions.

Harsh Anuj works on applications of data science to Independent Evaluation Group of the World Bank's evaluative and synthesis products and supports data science capacity building. Previously, he led the Global Delivery Initiative's DeCODE and has provided data science support to management, operations, and corporate teams across World Bank Group.

Michael Bamberger, Ph.D. (Sociology), has 50 years of experience in international development and evaluation. Following 10 years in urban community development, he worked with the World Bank and as an independent evaluation consult with UN agencies, international banks, and development agencies. He is a Senior Research Fellow at the International Initiative for Impact Evaluation (3ie), and is currently working on complexity-responsive evaluation and integrating data science and evaluation.

Geeta Batra, Ph.D. has over 25 years of experience in international development including 15 years of experience in program evaluation. At the Independent Evaluation Office of the Global Environment Facility (GEF IEO), she manages a team of evaluation professionals and provides oversight on the design, implementation, and quality of evaluations.

Kerry Bruce, DrPH is a public health practitioner and monitoring and evaluation expert. She was an early adopter of the use of technology in international development. When she's not plotting how to use technology to improve programming, you'll find her hiking, skiing, or reading science fiction.

Pirmin Bundi is Associate Professor of Public Policy and Evaluation at the Swiss Graduate School of Public Administration at the University of Lausanne. His main interests are policy evaluation, public policy, public administration, and political behavior.

Elvira Celardi, Assistant Professor of Sociology at the University of Catania, specializes in social research methodology and evaluation. Her research focuses on complex social phenomena such as social inclusion, social change, globalization, and poverty. Currently, she's integrating AI and machine learning tools to analyze these intricate social phenomena.

Sung-Woo Cho, Ph.D., is Associate Vice Provost for Academic Data Analytics at the University of Oregon. Reporting to the Provost, he promotes the use of artificial intelligence for predictive, evaluative, and generative purposes at the university. He holds a B.A. from Stanford University and a Ph.D. from Columbia University.

Lars Skage Engebretsen is a Special Advisor at the Office of the Auditor General of Norway (OAGN) and specializes in performance auditing methods. He has a master's degree in economics and has extensive experience from performance auditing.

Valentine J. Gandhi, Ph.D., wears multi-disciplinary hats as a Policy Advisor, Senior Evaluation expert, Tech4Good specialist, Cybersecurity and Knowledge Manager. He has worked in over 40 countries including conflict zones. He currently is the Chief of Party of a 14 country regional USAID MEL Platform in the Indo pacific. He is the founder of the think tank The Development CAFE, and the host of the EvalEdge Podcast for European Evaluation Society.

Laura Gatto is a Postdoctoral Researcher at the Swiss Graduate School of Public Administration at the University of Lausanne. Her research focuses on policy evaluation, public policy, and interest groups. Trained in both quantitative and qualitative methods, Laura is particularly interested in computational text analysis.

Nathan Greenstein is Assistant Director of Machine Learning in the Office of the Provost at the University of Oregon. He facilitates data science projects and focuses on how machine learning can be used ethically and equitably. He holds a B.A. in Cognitive Science from Dartmouth College.

Ariya Hagh works for the methods advisory function of the Independent Evaluation Group (IEG) of the World Bank, where he provides support on evaluation design, data science applications, econometrics, and qualitative data analysis. He completed his Ph.D. at Georgetown University in 2020 and has a background in international relations and research methods.

Jon Holm Ph.D. is Special Advisor at the Research Council of Norway (RCN) where he is responsible for the development of national research assessments administered by RCN. He is also involved in developing methods and processes for research analysis and evaluation at the Department for data management and analysis.

Helge Holtermann is Senior Advisor specializing in methods for performance auditing at the Office of the Auditor General of Norway (OAGN). He holds a Ph.D. in political science and has a background within peace and conflict research.

Frans L. Leeuw (sociologist) is Professor-Emeritus at Maastricht University for social science research of public policy and law. Earlier he was Director at Netherlands National Institute of Justice Research, Professor of Evaluation Studies at Utrecht University, Dean, Humanities Faculty, Netherlands Open University, and Director, Performance Audit & Evaluation department of the Netherlands National Audit Office. He was also the co-founder of the European Evaluation Society and author of over 10 books and 150 (academic) publications.

Vincenzo Miracula, Ph.D. candidate in Complex Systems at the University of Catania, is an expert in computational social sciences and AI. His interests lie in network theory and the spread of fake news. He conducted research as a visiting fellow at the Universidad Complutense de Madrid and has been an Associate Member of the Italian Association for Artificial Intelligence and the Italian Association of Sociology since 2022.

Mari Mjaaland is Senior Advisor at the Office of the Auditor General of Norway (OAGN). She is an experienced performance auditor and a former journalist with a master's degree in sociology.

Tom Næss is Special Advisor and Performance Auditor with the Office of the Auditor General of Norway (OAGN) and an experienced team leader of performance audits. He holds a master's degree in international relations.

Denis Newman-Griffis Ph.D. is Lecturer in Data Science, at the University of Sheffield, and Research Fellow of the Research on Research Institute, where they lead the GRAIL project on responsible AI in research funding and evaluation. Their research investigates principles and practices of responsible AI in science, health, and disability.

Steffen Bohni Nielsen Ph.D. is Director General at the Danish National Research Centre for the Working Environment and a member of the Danish National Research and Innovation Council. He is a member of the International Evaluation Research Group (INTEVAL) and has published extensively in the field of evaluation.

Gustav Jakob Petersson Ph.D. is Senior Analyst at the Swedish Research Council. He is a member of the International Evaluation Research Group (INTEVAL) and has co-edited one of its volumes on cyber society, Big Data and evaluation in 2017.

Antonio Picone, Ph.D. candidate in Complex Systems at the University of Catania, actively participates in natural language processing research. His focus lies in leveraging AI to accurately identify sentiments and emotions in texts, demonstrating a profound dedication to advancing our comprehension of intricate linguistic interactions.

Carolin Prabhu is Senior Advisor and part of the Office of the Auditor General of Norway (OAGN) Innovation Lab. She holds a Ph.D. in physics and has a background within data analytics and machine learning.

Estelle Raimondo Ph.D. is Program Manager of Independent Evaluation Group (IEG) of the the World Bank's Methods Advisory Unit. She is a specialist in evaluation methods and advises teams across IEG on a wide range of research designs.

Francesco Mazzeo Rinaldi Ph.D. is Professor at the University of Catania, Italy, and serves as the Director of the University's Research Center, LAPOSS. His main research interests include program and policy evaluation, Big Data analytics, and Artificial Intelligence. He is a member of the International Evaluation Research Group (INTEVAL).

Juha I. Uitto, Ph.D., served as Director of the Independent Evaluation Office (IEO) of the Global Environment Facility (GEF) from 2014 to 2024. Prior to that, he worked as evaluator in GEF and UNDP since 1999. He spent the 1990s with the UN University as environmental research and training coordinator.

Jos Vaessen Ph.D. is Evaluation Adviser in IEG. His work focuses on evaluation capacity development and evaluation policy. He is a member of the International Evaluation Research Group (INTEVAL).

Joris Vandelanotte is Medical Doctor and Specialist in Public Health Medicine with more than 25 years of experience in public health, health systems strengthening, and monitoring and evaluation, especially in Africa. Joris has a special interest in applying text analytics to evaluation work, and in bicycle maintenance and repair.

Pete York is Principal and Chief Data Scientist at BCT Partners, a consulting firm specializing in data analytics for equitable social impact. He has over 25 years of experience in evaluation and research, and has authored several publications on using machine learning and big data for social change.

Virginia Ziulu is a data scientist in Independen Evaluation Group of the World Bank's Methods Advisory unit. She specializes in complex data science applications including machine learning, natural language processing, remote sensing, and computer vision. She has a background in data science with postgraduate studies completed at the University of Oxford and the University of Edinburgh.

1 Evaluation in the Era of Artificial Intelligence

Steffen Bohni Nielsen, Francesco Mazzeo Rinaldi, and Gustav Jakob Petersson

Introduction

In recent years, combat drones have been a presence in the skies over conflict zones such as Afghanistan, Nagorno-Karabakh, Libya, and Ukraine. These drones, unmanned aerial vehicles (UAVs), provide real-time data for reconnaissance, surveillance, and deadly aerial weaponry. The use of UAVs has moved from intelligence gathering to counter-terrorism or counter-insurgency warfare into full-scale conventional combat. Today, more than 100 nation-states' armed forces possess these capabilities. Particularly, the Ukraine conflict provides a pointer to UAVs' importance in future warfare as the technology becomes ever more sophisticated as it is linked to artificial intelligence (AI) (Marcus, 2022).

When recruiting for positions in the Swedish municipality of Upplands-Bro, candidates face an unusual interviewer. Since 2019, the interviewer has been a physical robot powered by artificial intelligence. The robot will ask the applicant questions and assess the candidates by analyzing their behaviors, problem-solving capacities, and other skills. The aim is to make the recruitment process less biased than traditional interview practices. The robot's interviews are subsequently analyzed and combined with competency scores from the initial application. The robot ranks promising candidates prior to recruiters conducting the final interview with candidates for the position (Misuraca & Van Noordt, 2020).

During the COVID-19 pandemic, many European countries' Centers for Disease Control built applications (apps) for cellular phones that would use GPS technology to track the whereabouts of individuals who downloaded the app. The app would track the whereabouts of the population using geospatial data. If these individuals had been in contact with other individuals with confirmed or potential exposure to COVID-19, they would automatically receive instructions for preventive actions such as self-isolation, testing, and notifying close contacts.

In the fall of 2022, ChatGPT was launched, a generative AI model capable of providing sophisticated answers to a myriad of questions. Its capability enables high performance on a variety of standardized tests for college admission. Its capacity to summarize complex texts and quantitative data led to widespread adoption and posed new challenges to higher education, entire

DOI: 10.4324/9781003512493-1

knowledge-intensive professions, and the labor market at large (Eloundou, Manning, Mishkin & Rock, 2023).

In the field of science, the use of AI is burgeoning and has contributed to collecting, analyzing, and reporting on various studies (Cotton, Cotton & Shipway, 2023). Across scientific disciplines, AI-powered solutions are being developed to support a number of different tasks, such as text screening, coding, translation, transcription, and quantitative and qualitative analysis.

These examples point to the presence and multiplicity of use of digital technologies, particularly in various domains such as human resources, public health, national security, education, and science.

Implications for Professions

Thus, the mention of AI is by no means coincidental. We are witnessing an exponential growth in globally generated data (Nielsen, Ejler & Schretzmann, 2017; Petersson and Breul, 2017). The rapid evolution of new technologies and dramatically decreasing costs of storage have enabled innovation in techniques that instantaneously capture, analyze, and visualize these huge data repositories (Kiron, Kirk Prentice & Boucher Ferguson, 2014; Mazzeo Rinaldi et al., 2017) . AI and similar techniques that process structured and unstructured, quantitative and qualitative data are central to these developments, and massive sums are currently being invested in such technologies.

Largely, these developments concern new sources for *data capture*, such as online searches, social media platforms, satellites, drones, internet of things (IoT; i.e., sensors), mobile phones, telecom records, and administrative registries (structured and unstructured data). Also, new sources for *data storage and management*, such as cloud computing, Digital Ledger Technologies (DLT) (blockchain), and edge computing, have emerged. Finally, *sources for data processing*, such as text analytics, AI/ML, and other quantitative approaches and visualization, have rapidly expanded (Mazzeo Rinaldi et al., 2024). We are in the midst of the fourth (information) revolution: the digital era.

While digital technological innovation has been driven by the private sector, the public sector has also moved to exploit the potentials of digital technologies (Heeks & Bailur, 2007). This is supported by a report from the European Commission documenting that artificial intelligence (AI) finds nascent applications in the public sector in Europe at national, regional, and local levels, and that there is growing interest in using AI to support and improve policy-making and service delivery (Misuraca & van Noordt, 2020). National governments are moving toward new measures to balance the promises and perils of artificial intelligence.

Globally, the public sector is now aware that the amount of data is growing tremendously, and this data deluge is simply unstoppable. Many public agencies and organizations have now realized that the flood of data that comes from the

Internet, smartphones, sensors, satellites, and digital transformation initiatives has great potential when combined with AI and machine learning (ML) algorithms capable of finding valuable insights in the data.

Parallel to these developments, different professions have also responded differently. Digital technologies are pervasive and affect most industries. Some tasks are automated or augmented by digitally driven emerging technologies (ET).

In a comprehensive analysis, management consulting firm, McKinsey concluded:

> Our analysis of more than 2000 work activities across more than 800 occupations shows that certain categories of activities are more easily automatable than others. They include physical activities in highly predictable and structured environments, as well as data collection and data processing. These account for roughly half of the activities that people do across all sectors. The least susceptible categories include managing others, providing expertise, and interfacing with stakeholders.
>
> (2018, p. 2)

In other words, no profession, or industry, will be left unaffected by digital technologies. Focusing solely on Large Language Models (such as ChatGPT), a recent study analyzed more than 1000 occupations at the job task and daily work activity level. The authors concluded that knowledge-intensive industries were among those to become most affected by AI. The technology will have a profound impact on higher-wage occupations with routine cognitive tasks (Eloundou, Manning, Mishkin & Rock, 2023).

These include knowledge-intensive occupations such as accounting, auditing, law, medicine, and research. Evaluation as a (para)profession drawing from social scientific research methods will also be affected.

According to McKinsey, the public and social sectors are among those that will be affected the most (2018). These are the typical evaluands of program evaluation. In other words, what evaluators evaluate and how they evaluate it are likely to undergo significant changes over the next few years.

Scope of the Book

The rapid development of emerging technologies, particularly within generative AI, beckons that we take a renewed look at the interlacing between evaluation and emerging technologies. This book seeks to answer three overarching questions:

1. What are the emerging digital technologies?
2. What requisite skills do evaluators need?

3. What contribution can evaluation make to AI and vice versa?

Let us, therefore, expand on the context and understanding of the salience of these questions.

Proliferation of Emerging Technologies in Evaluation

Social scientists at large have responded to the new opportunities that emerging technologies offer, and forerunners seem to be embracing opportunities in computational social science and collaboration with data science. Its challenge to traditional social science is considered real and impending (Burrows and Savage, 2014), and its potentials and perils are therefore scrutinized intensely (Grossmann et al., 2023).

Yet evaluators seem to have been slow to respond to this new development (Picciotto, 2020; Raftree & Bamberger, 2014). Consider these observations:

- AEA Connect (Evaltalk) (as of July 2022) contains only 253 entries on artificial intelligence and no user tags on Big Data. No topical discussions have been posted on the issue. This is the "chatforum" for the world's largest evaluation community.
- A Google Scholar search with the same search terms yielded between 5-1,480 results when combining the search term with "evaluation". Scanning across these documents, the majority referred to the evaluation of the predictive performance of AI/ML and not integration with evaluation practice.
- Professional development workshops at national evaluation conferences offer limited, if any, training in Big Data analytics.
- AEA Core competencies in program evaluation do not contain explicit reference to Big Data analytics (American Evaluation Association, 2018).
- Professional development curriculum in evaluation training offers limited training opportunities. Leading providers such as The Evaluators Institute's 2023 program offered one module on machine learning. IPDET's 2023 program offered one module on machine learning.
- Curriculum in university-based programs does not explicitly mention BD and seems not to be adapting to the new skills required by BD/AI analytics (Lavelle, 2020);

Rathinam and colleagues (2021) created an evidence gap map of the use of Big Data in international development impact evaluations. This domain appears as leading in applying Big Data in the field of evaluation (Raftree & Bamberger, 2014). They identified 48 impact evaluations using Big Data, with satellite data used in over 80% of the observed cases. To this day, no reviews of national evaluation markets have been carried out (Nielsen, 2023).

Currently, relatively few peer-reviewed articles on digital technologies and their implications on evaluation practice have been published. A search of research articles in nine major evaluation journals: *American Journal of Evaluation, Canadian Journal of Program Evaluation, Educational Evaluation and Policy Analysis, Evaluation, Evaluation and the Health Professions, Evaluation and Program Planning, Evaluation Journal of Australasia, Evaluation Review, Journal of Multidisciplinary Evaluation, and New Directions for Evaluation,* from 2013 to 2023 (May) identified 18 distinct articles with "Big Data," "artificial intelligence," "machine learning," and "text analytics," or "Internet of Things" as the title, keyword, or in the abstract (see also Nielsen, 2023).

In 2023, a *New Directions for Evaluation* issue dedicated to AI and evaluation will be published. This marks the first concerted academic effort to analyze the implications for evaluation practice (Montrosse-Moorhead & Mason, 2023) and may indicate an increasing interest in the application of digital technologies and evaluation most recently.

Examples of the use of AI are starting to be published in the peer-reviewed literature (Bonfiglio, Camaioni, Carta & Cristiano, 2023; Cintron & Montrosse-Moorhead, 2022; Roy & Rambo-Hernandez, 2021). Protagonists call for further cooperation and integration with data science (Bruce, Gandhi & Vandelonotte, 2020; Hejnowicz & Chaplowe, 2021; Raftree, 2020; York & Bamberger, 2020).

Among the first books to focus on the interlacing between digital technologies, specifically BD, and evaluation was the anthology edited by Petersson and Breul (2017). Herein, a survey (using convenience sampling with a remarkably low response rate) among self-reported evaluators in the mid-2010s documented that about ten percent had experience with Big Data (Højlund, Olejniczak & Petersson, 2017). We have identified no other survey of the demand or supply side. Since, York and Bamberger have echoed these findings in a separate publication (2020).

Put bluntly, evaluators are still largely unfamiliar with Big Data. In 2012, John Gargani, perhaps wishfully, predicted that in ten years "evaluations will abandon data collection in favor of data mining … [because] … tremendous amounts of data are being collected in our day-to-day lives and stored digitally. It will become routine for evaluators to access and integrate these data" (Gargani, 2012).

We are still far from this scenario. In the intermittent years, the challenge to evaluation has become even more pertinent as AI, in particular, has evolved rapidly. Perhaps, Peter Daboll's predictions that BD outperform evaluation and increasingly make evaluation irrelevant and obsoleteare becoming real (2013).

BD and data science have come to stay and they are growing. Evaluation is now the little brother in the field of knowledge production.

There are therefore several reasons why there is a need for further integration between evaluation and BD/AI. Petersson, Leeuw, and Olejniczak (2017) identified four challenges for evaluators:

(1) a new role in the policy-making process,
(2) explore designs and tools applied in the BD field,
(3) obtain new competencies to manage data, and
(4) seek collaboration with data scientists.

Similarly, Nielsen, Ejler, and Schretzmann (2017) identified four challenges for evaluation:

(1) Rival,
(2) complementary,
(3) utilization, and
(4) competency challenges.

When *what* we evaluate is changing, *how* we evaluate must inevitably change too. At worst, rival knowledge producing services will outcompete evaluation's offerings. As indicated by the (slow) emergence of publications focusing on evaluation and digital technologies, changes *are* occurring.

According to Raftree (2020), who wrote in the context of international development evaluation, digital technologies have begun to proliferate in this domain in three distinct waves.

Essentially, the first wave would allow evaluation practitioners to keep doing what they did, but augmented by new sources for data capture (geo-spatial data, large administrative registries, and mobile phones).

The second wave focused on new forms of data capture, such as the internet of things (IoT), satellites, and drones, and an escalating focus on AI and ML. This was evidenced by Rathinam and colleagues' evidence map, wherein satellite images were frequently applied (2021).

The third wave came in close or in parallel with the second wave and explored new technologies for data capture, storage, and data processing.

Importantly, Raftree observes, "new disciplines (such as software development and data science) are entering the MERL field, bringing new ideas and ways of working" (2020, p. 15).

It remains to be seen whether Raftree's notion of waves is an appropriate metaphor for adopting digital technologies in evaluation practices at large. As mentioned, only tangential empirical evidence exists about how digital technologies have spread across domain segments in the evaluation industry (see Nielsen, 2023). The recent surge in peer-reviewed articles suggests that new ways of data processing such as texts and photographic images are part of the third wave (Cintron & Montrosse-Moorhead, 2022; York & Bamberger, 2020).

In other words, Big Data applications are slowly, but increasingly, becoming part of the evaluand (what is to be evaluated) *and* a tool for evaluators. The latter tools are under rapid development and at various stages of commodification.

Today, a host of AI-powered technologies tailored to social scientific research already exists. Such commodification may ease access to building requisite capabilities.

Eloundou and colleagues argue that AI can either *displace* or *augment* tasks in various occupations (2023). Arguably, a more nuanced way to understand how digital technologies affect evaluator practice is needed. To appropriately assess emerging technologies' consequences for evaluation practice, one will need to break down its constituent tasks and activities.

- *Displacing tasks*. Emerging technology replaces human tasks, such as translation, transcription, standardized reporting, and abstract screening.
- *Facilitating tasks*. Emerging technology enables human tasks, such as virtual platforms for interviews with difficult-to-reach populations.
- *Augmenting tasks*. Emerging technology enhances human tasks, such as enabling a bigger population scope, merging data, data mining, optimizing analyses, and auto-coding.
- *Generating new tasks*. Emerging technology generates human tasks, such as satellite imagery analysis, drone operations, and prompting.

In different ways, tasks may be solved more *expeditiously*, *efficiently*, or *effectively* using emerging technologies. Implications in terms of *equity and ethics when* using emerging technologies also remain a concern.

We are facing a future, wherein the reality is, if you are to evaluate interventions driven by emerging technologies, your evaluation team will need to possess requisite technological competencies on par with subject matter experts. This anthology offers a number of pertinent examples of such collaborations.

These observations suggest that the evaluation community have yet not adapted to the challenges and opportunities presented by emerging technologies. Echoing Petersson, Leeuw, and Olejniczak (2017), we believe that evaluators should work with and use emerging technologies. Otherwise, the evaluation community will experience an encroachment from more innovative analytical professions that represent rival forms of knowledge production that may promise insights more expediently, more efficiently, and more usefully than what is offered by evaluators.

Structure of the Book

We have structured the book into three sections: In the *first section*, Kerry Bruce, Valentine J. Gandhi and Joris Vandelanotte (Chapter 2) (2025) introduce the emerging technologies supporting design, data collection, storage, and analysis. Therein, they introduce the potentials and perils of their application in evaluation. As such, the chapter provides a point of reference for subsequent chapters in the volume. In the next chapter (Chapter 3) (2025), Pete York and Michael

Bamberger analyze and exemplify how emerging technologies for design, data collection, and analysis can be applied in evaluation and present forms of collaboration between data science and evaluation (2025a). They also discuss what and how evaluation tools can be integrated with data science – and why it is important for evaluators to further this integration. Nathan Greenstein and Sung-Woo Cho (Chapter 4) go on to discuss ethics and equity issues in data science for evaluators and why these issues need special consideration as the emerging technologies hold immense promise, but can also intentionally, or unintentionally, do harm (2025).

In the *second section*, we provide a number of case studies that describe and analyze how analytical tools from data science have been integrated in, and ameliorated, evaluative work through different kinds of benefits. In the first chapter of the section, Virginia Ziulu, Harsh Anuj, Ariya Hagh, Estelle Raimondo, and Jos Vaessen (Chapter 5) analyze how text analytics was used as part of a meta evaluation in the World Bank and, as such, also holds potential for broader application in knowledge management within, and across, organizations (2025). In a similar vein, Tom Næss, Carolin Prabhu, Mari Mjaaland, Helge Holtermann, and Lars Skage Engebretsen then present a case of applying text mining in performance auditing of the Norwegian Police's work on cybercrime in Norway (Chapter 6) (2025). Jon Holm, Denis Newman-Griffis, and Gustav Petersson present a case of applying text analytics, specifically Natural Language Processing (NLP), in evaluating the impact of research and research policies (Chapter 7) (2025). Laura Gatto and Pirmin Bundi then present a case of applying text mining and quantitative text analytics on topical models used to evaluate policy-making processes in Switzerland (Chapter 8) (2025). In the next chapter, Francesco Mazzeo Rinaldi, Elvira Celardi, Antonio Picone, and Vincenzo Miracula describe and discuss the potential of using ML and text analysis tools by presenting a case study on the Ukraine conflict. They illustrate how these technologies facilitate digital contextual analysis, offering policymakers valuable evaluative insights (Chapter 9) (2025). Next, Anupam Anand, Geeta Batra, and Juha I. Uitto analyze the use of geospatial data to harness evaluative evidence in the context of environmental interventions (Chapter 10) (2025). Pete York (Chapter 11) demonstrates how new analytical techniques can be leveraged to inform decision-making, thus augmenting the use of existing data. In the final chapter of the section, Frans Leeuw analyzes the potentials and predicaments in adjudication when artificial and human intelligence are integrated and offers ways in which such interventions should be evaluated (Chapter 12) (2025).

In the *final section*, Steffen Bohni Nielsen (2025) analyzes the implications of emerging technologies for evaluation when the practice is considered an industry. He argues that some tasks and service lines are likely to be displaced, while expert evaluative knowledge will remain important (Chapter 13). In the final chapter, Steffen Bohni Nielsen, Francesco Mazzeo Rinaldi, and Gustav Petersson (Chapter 14) use a cross-case analysis to dig into organizational and competence challenges in the further integration of evaluation and data science. They discuss functions, roles, skills, and technologies (2025).

References

American Evaluation Association (2018). AEA Evaluator Competencies. Retrieved from https://www.eval.org/Portals/0/Self%20assessment%201oct.pdf

Anand, A., Batra, G. & Uitto, J.I. (2025). Harnessing Geospatial Approaches to Strengthen Evaluative Evidence. In S.B. Nielsen, F. Mazzeo Rinaldi and G.J. Petersson (eds.). *Artificial Intelligence and Evaluation. Emerging Technologies and Their Implications for Evaluation* (pp. 196–218). London: Routledge. https://doi.org/10.4324/9781003512493

Bonfiglio, A., Camaioni, B., Carta, V. & Cristiano, S. (2023). Estimating the Common Agricultural Policy milestones and targets by neural networks. *Evaluation and Program Planning*. https://doi.org/10.1016/j.evalprogplan.2023.102296

Bruce K., Gandhi V.J., & Vandelanotte, J. (2020). *Emerging Technologies and Approaches in Monitoring, Evaluation, Research, and Learning for International Development Programs*. MERL Tech Report # 4. Retrieved from https://merltech.org/wp-content/uploads/2020/07/4_MERL_Emerging-Tech_FINAL_7.19.2020.pdf

Bruce, K., Gandhi, V.J. & Vandelanotte, J. (2025). Emerging Technology and Evaluation in International Development. In S.B. Nielsen, F. Mazzeo Rinaldi and G.J. Petersson (eds.). *Artificial Intelligence and Evaluation. Emerging Technologies and their Implications for Evaluation* (pp. 13–36). London: Routledge. https://doi.org/10.4324/9781003512493

Burrows, R., & Savage, M. (2014). After the crisis? Big Data and the methodological challenges of empirical sociology. *Big Data & Society*, *1*(1). https://doi.org/10.1177/2053951714540280

Cintron, D. W., & Montrosse-Moorhead, B. (2022). Integrating Big Data Into Evaluation: R Code for Topic Identification and Modeling. *American Journal of Evaluation*, *43*(3), 412–436. https://doi.org/10.1177/10982140211031640

Cotton, D., Cotton, P.A., & Shipway, J.R. (2023). Chatting and Cheating. Ensuring academic integrity in the era of ChatGPT. *EdArXiv*. https://doi.org/10.1080/14703297.2023.2190148

Daboll, P. (2013). Five reasons why Big data will crush big research. *Forbes*. 2013/12/03. Retrieved from https://www.forbes.com/sites/onmarketing/2013/12/03/5-reasons-why-big-data-will-crush-big-research/?sh=6a0e44555d0f

Eloundou, T., Manning, S. Mishkin, P. & Rock, D. (2023). GPTs are GPTs: An Early Look at the Labor Market Impact Potential of Large Language Models. *arXiv:2303.10130* [econ.GN]. https://doi.org/10.48550/arXiv.2303.10130

Gargani, J. (2012). The Future of Evaluation: 10 Predictions. EVAL BLOG. John Gargani's blog about program design and evaluation. Retrieved from http://evalblog.com/2012/01/30/the-future-of-evaluation-10-predictions/

Gatto, L. & Bundi, P. (2025). The Use of Quantitative Text Analysis in Evaluations. In S.B. Nielsen, F. Mazzeo Rinaldi and G.J. Petersson (eds.). *Artificial Intelligence and Evaluation. Emerging Technologies and their Implications for Evaluation* (pp. 144–167). London: Routledge. https://doi.org/10.4324/9781003512493

Greenstein, N. & Cho, S.-W. (2025). Ethics & Equity in Data Science for Evaluators. In S.B. Nielsen, F. Mazzeo Rinaldi and G.J. Petersson (eds.). *Artificial Intelligence and Evaluation. Emerging Technologies and Their Implications for Evaluation* (pp. 56–77). London: Routledge. https://doi.org/10.4324/9781003512493

Grossmann, I., Feinberg, M., Parker, D.C., Christakis, N.A., Tetlock, P.E. & Cunningham, W.A. (2023). AI and the transformation of social science research. *Science*, 380(6650), 1108–1109. https://doi.org/10.1126/science.adi1778

Heeks, R. & Bailur, S. (2007). Analyzing e-government research: Perspectives, philosophies, theories, methods, and practice. *Government Information Quarterly*, 24(2), 243–265. https://doi.org/10.1016/j.giq.2006.06.005

Hejnowicz, A. & Chaplowe, S. (2021). Catching the wave: Harnessing data science to support evaluation's capacity for making a transformational contribution to sustainable development. *Canadian Journal of Program Evaluation,* 36(2), 162–186. https://doi.org/10.3138/cjpe.71527

Holm, J., & Newman-Griffis, D. & Petersson, G.J. (2025). Big data for big investments: Making responsible and effective use of data science and AI in research councils. In S.B. Nielsen, F. Mazzeo Rinaldi and G.J. Petersson (eds.). *Artificial Intelligence and Evaluation. Emerging Technologies and Their Implications for Evaluation* (pp. 120–143). London: Routledge. https://doi.org/10.4324/9781003512493

Højlund, S., Olejniczak, K. & Petersson, G.J.(2017). The Use of Big Data in Evaluation. In: edited G.J. Petersson & J.Breul. *Cyber Society, Big Data and Evaluation* (pp. 35–60). London: Routledge

Kiron, D., Kirk Prentice, P. & Boucher Ferguson, R. (2014). The Analytics Mandate. Findings from the 2014 Data & Analytics Global Executive Study and Research Report. MIT Sloan Management Review, Research Report. Retrieved from https://sloanreview.mit.edu/projects/analytics-mandate/

LaVelle, J. M. (2020). Educating evaluators 1976–2017: An expanded analysis of university-based evaluation education programs. *American Journal of Evaluation*, 41(4), 494–509. https://doi.org/10.1177/1098214019860914

Leeuw, F.L. (2025). The Algorithmization of Policy and Society: The Need for a Realist Evaluation Approach. In S.B. Nielsen, F. Mazzeo Rinaldi and G.J. Petersson (eds.). *Artificial Intelligence and Evaluation. Emerging Technologies and Their Implications for Evaluation* (pp. 242–265). London: Routledge. https://doi.org/10.4324/9781003512493

Marcus, J. (2022). Combat Drones: We Are in a New Era of Warfare – Here's Why. (2022-02-04). Retrieved from https://www.bbc.com/news/world-60047328

Mazzeo Rinaldi, F., Celardi, E., Miracula, V. & Picone, A. (2025). Artificial Intelligence and Text Analysis in Evaluating Complex Social Phenomena. The Russia-Ukraine Conflict. In S.B. Nielsen, F. Mazzeo Rinaldi and G.J. Petersson (eds.). *Artificial Intelligence and Evaluation. Emerging Technologies and Their Implications for Evaluation* (pp. 168–195). London: Routledge. https://doi.org/10.4324/9781003512493

Mazzeo Rinaldi, F., Giuffrida, G., Nicotra, S., & Dispinseri F. (2024). A Classification Algorithm to Link Official Documents to Sustainable Development Goals. In I. K. Lindkvist., K. Forss & P. Øyvind Bastøe (eds.). *Towards Sustainable Futures - A role for evaluations?* (pp. 166–185). London: Routledge.

Mazzeo Rinaldi, F., Giuffrida, G., & Negrete, T. (2017). Real-time monitoring and evaluation - Emerging news as predictive process using Big Data based approach.

In G. J. Petersson & J. D. Breul (eds.). *Cyber Society, Big Data and Evaluation* (pp. 191–214). New York, NY: Routledge.

McKinsey Global Institute (2018). *Ai, Automation, and the Future of Work: Ten Things to Solve for*. Briefing Note Prepared for the Tech4good Summit. Retrieved from https://www.mckinsey.com/featured-insights/future-of-work/ai-automation-and-the-future-of-work-ten-things-to-solve-for

Misuraca, G. & van Noordt, C. (2020). *Overview of the Use and Impact of AI in Public Services in the EU, EUR 30255 EN*. Luxembourg: Publications Office of the European Union. ISBN 978-92-76-19540-5. https://doi.org/10.2760/039619, JRC120399

Montrosse-Moorhead, B. & Mason, S. (eds.) (2023). Editors' note. *New Directions for Evaluation*, 178–179. 1–6. https://doi.org/10.1002/ev.20554

Nielsen, S.B. (2023). Disrupting evaluation? Emerging technologies and their implications for the evaluation industry. *New Directions for Evaluation*, 178–179, 47–57. https://doi.org/10.1002/ev.20558

Nielsen, S.B. (2025). The Evaluation Industry and Emerging Technologies. In S.B. Nielsen, F. Mazzeo Rinaldi and G.J. Petersson (eds.). *Artificial Intelligence and Evaluation. Emerging Technologies and Their Implications for Evaluation* (pp. 266–286). London: Routledge. https://doi.org/10.4324/9781003512493

Nielsen, S.B., Mazzeo Rinaldi, F.M. & Petersson, G. (2025). Digital Era Evaluation. Future Perspectives. In S.B. Nielsen, F. Mazzeo Rinaldi and G.J. Petersson (eds.). *Artificial Intelligence and Evaluation. Emerging Technologies and Their Implications for Evaluation* (pp. XX–YY). London: Routledge.

Nielsen, S.B., Ejler, N. & Schretzmann, M. (2017). Exploring Big (Data) opportunities: The Case of the Center for Innovation through Data Intelligence, New York City. In G.J. Petersson and J.D. Breul (eds.). *Cyber Society, Big Data and Evaluation* (pp. 147–170). New York, NY: Routledge.

Næss, T., Prabhu, C., Mjaaland, M., Holtermann, H. & Engebretsen, L.S. (2025). Text Mining and Machine Learning in an Evaluation of Police Handling of Cybercrime in Norway. In S.B. Nielsen, F. Mazzeo Rinaldi and G.J. Petersson (eds.). *Artificial Intelligence and Evaluation. Emerging Technologies and Their Implications for Evaluation* (pp. 103–119). London: Routledge. https://doi.org/10.4324/9781003512493

Petersson, G.J. & Breul, J. (eds.). (2017). *Cyber Society, Big Data, and Evaluation*. New York, NY: Routledge

Petersson, G.J. Leeuw, F.L. & Olejniczak, K. (2017). Cyber Society, Big Data and Evaluation: A Future Perspective. In G.J. Petersson and J. Breul (eds). *Cyber Society, Big Data, and Evaluation* (pp. 237–254). New York, NY: Routledge.

Picciotto, R. (2020). Evaluation and the big data challenge. *American Journal of Evaluation*, 41(2), 166–181. https://doi.org/10.1177/1098214019850334

Raftree, L. (2020). *MERL Tech State of the Field. The Evolution of MERL Tech*. MERL Tech Report # 1. Retrieved from https://merltech.org/resources/merl-tech-state-of-the-field-the-evolution-of-merl-tech/

Raftree, L. & Bamberger, M. (2014). *Emerging Opportunities: Monitoring and Evaluation in a Tech-Enabled World*. Retrieved from https://www.rockefellerfoundation.org/report/emerging-opportunities-monitoring/

Rathinam, F., Khatua, S., Siddiqui, Z., Malik, M., Duggal, P., Watson, S. & Vollenweider, X. (2021). Using big data for evaluating development outcomes: A systematic map. *Campbell Systematic Review*, 17(3), e1149. https://doi.org/10.1002/cl2.1149

Roy, A. & Rambo-Hernandez, K. E. (2021). There's So Much to Do and Not Enough Time to Do It! A Case for Sentiment Analysis to Derive Meaning From Open Text Using Student Reflections of Engineering Activities. *American Journal of Evaluation*, 42(4), 559–576. https://doi.org/10.1177/1098214020962576

York, P. (2025). The Future of Evaluation Analytics: Case Studies of Structural Causal Modeling in Action. In S.B. Nielsen, F. Mazzeo Rinaldi and G.J. Petersson (eds.). *Artificial Intelligence and Evaluation. Emerging Technologies and Their Implications for Evaluation* (pp. 219–241). London: Routledge. https://doi.org/10.4324/9781003512493

York, P. & Bamberger, M. (2020). Measuring results and impact in the age of big data: The nexus of evaluation, analytics, and digital technology. The Rockefeller Foundation. Retrieved from: https://www.rockefellerfoundation.org/report/measuring-results-and-impact-in-the-age-of-big-data-the-nexus-of-evaluation-analytics-and-digital-technology/

York, P. & Bamberger, M. (2025). The Applications of Big Data to Strengthen Evaluation. In S.B. Nielsen, F. Mazzeo Rinaldi and G.J. Petersson (eds.). *Artificial Intelligence and Evaluation. Emerging Technologies and Their Implications for Evaluation* (pp. 37–55). London: Routledge. https://doi.org/10.4324/9781003512493

Vaessen, J., Lemire, S. & Befani, B. (2020). *Evaluation of International Development Interventions. An Overview of Approaches and Methods.* Washington: World Bank IEG.

Vigoda-Gadot. E. & Dana R. Vashdi (eds.). (2020). *Handbook of Research Methods in Public Administration, Management and Policy*. Elgar Handbooks in Public Administration and Management, Cheltenham, 2020. https://doi.org/10.4337/9781789903485

Ziulu, V., Anuj, H., Hagh, A., Raimondo, E. & Vaessen, J. (2025). Extracting Meaning from Textual Data for Evaluation. Lessons from Recent Practice at the Independent Evaluation Group of the World Bank. In S.B. Nielsen, F. Mazzeo Rinaldi and G.J. Petersson (eds.). *Artificial Intelligence and Evaluation. Emerging Technologies and Their Implications for Evaluation* (pp. 78–102). London: Routledge. https://doi.org/10.4324/9781003512493

2 Emerging Technology and Evaluation in International Development

Kerry Bruce, Valentine J. Gandhi, and Joris Vandelanotte

Introduction

Since the first ICT4D conference (CRS and ICT4D Conference, n.d.) in 2010 and the first MERLTech conference in 2014 (MERLTech, (n.d.), the international development community, including monitoring and evaluation professionals, has been coming together both to discuss emerging technology and to better understand how to use it for evaluation (Bamberger, 2022). Technology has evolved from adapting existing technology for international development to doing development in a digital world (Department for International Development, 2018).

This chapter will discuss the promises and pitfalls of using emerging technology in the evaluation space in the international development arena. Examples of each technology from practice (where available) and the promises and perils of these technologies are then discussed. In this chapter, we discuss the evidence around two key areas with regard to evaluation: data collection and data analysis. The technologies we discuss have the potential to increase the efficiency, cost-effectiveness, and impact of both international development programming and how we monitor and evaluate programs (Gandhi, 2022).

However, all the technologies discussed present challenges that evaluators need to keep in mind. All the emerging technologies reviewed are viewed through the lens of how they can be used by evaluators to understand how to improve international development programming. We have specifically looked at individual technologies rather than broader fields in which they exist. For example, rather than discussing artificial intelligence (AI), we examined technologies within the field such as machine learning or chatbots. Due to the broader scope that could deserve its own chapter, we have not included discussions of topics like cybersecurity or data storage, both of which have relevance to evaluation but are adjacent to our main discussion on how technologies can be used by evaluators in their work.

DOI: 10.4324/9781003512493-2

Table 2.1 Key emerging technologies in international development evaluation: promises and pitfalls

Emerging technology	*Promises*	*Pitfalls*
Data collection and capture		
Mobile data, administrative registries/geographic data	Rapid collection to use cycle, ability to map outcomes, ability to improve the quality of data collected	Privacy of data (especially location), administrative data may lack context
Social media data and online data	Increased availability of information, more direct access to community opinion (potentially without government interference), low cost to collect	Privacy issues, inadvertent inclusion of fake news or manipulated data, censorship, potential lack of equity in representation, requires extensive data cleaning
Satellites, drones, IoT	Complete and accurate picture of program impacts and outcomes, objective, real-time monitoring possible, possible cost efficiency in data collection	Lack of contextual information, concerns about security, privacy, consent, safety, data quality, and reliability. May require skill to implement
Chatbots, Virtual Agents	Automates the process of data collection in a low bandwidth setting potentially reducing costs, easy to update or change data that are collected, always available	Not sensitive to responses and may annoy respondents, requires skill to deploy, still nascent
Virtual Reality	Ability to aid with formative evaluation and support program design, may reduce need for travel, simulates real life scenarios	Expensive to establish, capacity to implement is high and may not be appropriate for all situations
Data analysis		
Data Analytics (using Big Data)	Increased speed of analysis, can use a wide span of data and can re-use data	May exclude those with less data to contribute, data preparation and analysis require a high level of skill

(*Continued*)

Table 2.1 Continued

Emerging technology	*Promises*	*Pitfalls*
Natural Language Processing and Generative AI (ex: ChatGPT)	Automation, speed and ability to include more information in analyses, and reveal hidden insights, can collaborate with it to support writing and coding	Algorithmic bias possible, extensive data cleaning is needed for a specific corpus, need to train the model on a corpus, black box on how it derives its conclusions, questions of acknowledgment of authorship
Machine Learning	Optimization of documents, potential to discover hidden insights, can assign labels/groupings based on word frequency, with reduced bias, can be automatically updated as new data arrive, getting easier to use	Algorithmic bias may include or exclude key groups, may include irrelevant data due to poor data preparation, difficult to prepare data and have sufficient data to train models

Promises and Pitfalls of Emerging Technology for Monitoring and Evaluation

Table 2.1 presents the key emerging technologies that we should consider in international development evaluation, grouped by the key themes of data collection and data analysis.

More details about each of these technologies are provided below, including a discussion on the promises, perils, and some examples from practice.

Data Collection and Capture

Data collection and capture using technology have the highest rate of adoption in international development. The use of advanced data collection technologies has progressed rapidly over the last decade. Where mobile data collection and online surveys were relatively new in 2010, they are now commonplace today. There are five main technologies that we will discuss in more detail below.

Mobile Data Collection and GPS Location Data

The international development community has widely adopted mobile and online technologies for data collection, often with global positioning system (GPS) location data features (Blumenstock, Cadamuro, & On, 2015). Many mobile data collection platforms were designed for the international development community and range from free or "freemium"[1] type models (such as Kobo Toolbox and Magpi) to paid subscription services that can be used anywhere in the world and in most languages. Mobile platforms are primarily useful for collecting quantitative data with some limited qualitative input. Online platforms (such as SurveyMonkey, Google Forms, or Qualtrics) have expanded substantially. A good resource for understanding the wide array and uses of mobile data collection can be found on ICTWorks (Vota, 2018) and Development CAFE Blogs (The Development Café, 2018).

Promises of Mobile and GPS Data Collection

Mobile data collection can help to standardize the data that is collected, using skip patterns, checks, ranges, and both required and non-required questions. GPS location data features enable improved mapping and disaggregation of data by geography. If data is being collected where there is good internet connectivity, it may be possible to record and transcribe what participants are saying in real-time. Data collection from mobile applications (such as applications for health workers, agricultural extension workers, or teachers) can be a rich source of monitoring data about how a program is (or is not) reaching its target outputs. Advanced mobile data collection systems can access existing databases that allow data collectors to access previous records from the same respondent, easing data collection for cohort surveys. When these database systems are paired with biometrics (Benston et al., 2020), embedded radio frequency identification (RFID) tags, or barcode scanners, the accuracy of identifying a respondent or a specific data collection point increases.

Perils of Mobile and GPS Data Collection

High-quality data collection that is standardized and rigorous still requires trained enumerators, and while mobile data collection has sped up the process of accessing the data, it has not significantly reduced the costs of human labor. Online data collection, while efficient, still suffers from low response rates and uncertain sampling frames, which compromise the quality and possibly the accuracy of the data. GPS data can be too accurate and potentially reveal too much information about a respondent and a higher level of analytical skill is required to mask data (although the best platforms now allow for masking at the point of collection). For all data collected, privacy and the ability to identify individuals can be a risk if data is not properly stored and if protocols for de-identification and re-use of data are not clearly defined. As with all data collection, the primarily quantitative nature of this data can mean that the context and meaning behind the data are lost or difficult to interpret.

Box 1: SMA is an interdisciplinary research area that is concerned with developing, adapting, and extending informatics tools, frameworks, and methods to track, collect, and analyze a large amount of structured, semi-structured, and unstructured social media data to extract useful patterns and information (Wikipedia, n.d).

Social Media Data and Online Data

Social media are interactive online tools that provide space for user-created content (creation, discussion, exchange). Facebook, YouTube, WhatsApp, Instagram, TikTok, Messenger, WeChat, LinkedIn, and Telegram have the most users (Wikipedia (n.d.). Beyond social media, internet websites form a vast repository of data and information that may be of use to evaluators.

Social Media Analysis (SMA) (Zachlod, Samuel, Ochsner & Werthmüller, 2022) is an interesting methodological tool for evaluators. Traditionally used by marketing and sales companies, they have developed several commercially available analytics tools such as Tweettracker (Kumar, Barbier, Abbasi & Liu, 2021), Hootsuite, Planable, and Loomly (Marvin & Sevilla, 2019).

Promises of Social Media Data and Online Data

There are many evaluation fields where SMA has the potential to contribute toward broader and better evaluation approaches and results. For example, programs that aim for large-scale social transformation, public policy transformation (Fabra-Mata & Mygind, 2019), or programs to influence public opinion could potentially be assessed using SMA (Mazzeo, Celardi, Miracula & Picone, 2025). Programs that aim to change knowledge, attitudes, and practices in a variety of fields (e.g., agriculture, education, health, disasters, politics, economics) and evaluation of the quality of (public) services can benefit from SMA.

Both SMA and analyses of publicly available online data can contribute to a better understanding of societal phenomena and changes in public opinion over time, and provide more direct access to community opinion. They can provide evaluators with a broader and larger sample of opinions and information, beyond traditional evaluation methods (surveys or key informant interviews). Findings from these sources may support the discovery of additional hypotheses, reveal unintended consequences of programming, and alternative opinions and information, and can be particularly useful to validate or triangulate other findings.

Analytic approaches include sentiment analysis, word frequency analysis, topic modeling, and content analysis. Time trends can demonstrate changes in opinion, knowledge, attitudes, and practices. GIS information can further enhance SMA to add a spatial analysis component; social network analysis of social interaction and exchange can enrich SMA.

The mere availability of social media or online data doesn't warrant its use in evaluation, though. It remains important to use these types of online data appropriately within an evaluation framework. Some key questions to ask prior to using these sources may include:

1. What are the specific evaluation questions that will be answered?
2. What are the most appropriate social media platforms or search terms that can provide (supportive) evidence to answer these questions?
3. What indicators or metrics will be measured using social media data or online data sources?
4. How do SMA and online data complement the other (more traditional) evaluation methods, and are the costs and time involved worth it?

Perils Online Data and Social Media Data

SMA in evaluation comes with a number of challenges. Social media data and online data often need extensive data cleaning and preparation for use. Data needs to be extracted from the chosen social media site, digitized, vectorized, tokenized, and made ready for analysis. These are often the skills of a computer scientist rather than a social scientist evaluator.

Accessing and preparing data on a social media platform or from specific websites require the use of application programming interfaces (APIs), which is not a traditional social scientist's skill. Access to data may come at a financial cost.

The choice of data source (social media platform or online source) needs to be carefully considered, as online platforms and sites serve different purposes and uses. Specific communities or stakeholder groups might be excluded from certain social media platforms or online sources. This exclusion of certain groups may negatively influence the representativeness of the evaluation findings, exclude specific populations, or make the findings difficult to interpret.

Ethical considerations, including privacy and consent to use the data for evaluation purposes, need to be addressed (Greenstein & Cho, 2025). General Data Protection Regulations (GDPR) in Europe have stringent privacy and consent requirements, which may make useful data inaccessible.

The data quality from any online source needs to be carefully assessed to avoid bias. Think of data generated by bots, fake accounts providing information that could create a false impression of a specific outcome. On the opposite side, censorship could remove actual sentiment and user content, which would also change the outcome. Understanding how these possibilities affect an online sourced dataset will be key to interpreting the final findings.

Best practice to mitigate the impact of bots or misinformation could be to use industry estimates of the problem for a given social media platform. However, these estimates are difficult to obtain for online data.

Examples from Practice: Social Media and Online Data

The Evaluation Department at the Norwegian Agency for Development Cooperation used Twitter data alongside other data sources to evaluate Norway's role in the peace process between the Government of Colombia and the FARC.[2] The Twitter analysis aimed to: determine critical points in the peace process; expand the sample to include Colombian society and actors not included in direct data collection; identify relevant stakeholders; and yield new insights. The analysis involved using keywords, content analysis, sentiment analysis, trend and time series analysis, and social network analysis to extract useful information from tweets and Twitter accounts. The Twitter analysis proved useful in corroborating evidence of an increase in trust during the peace process and providing additional insights that supported the facilitation team. The authors conclude that social media analysis can add value to evaluations of large-scale social transformation or public policy development or implementation, but basic methodological precautions remain necessary (Fabra-Mata & Mygind, 2019).

Satellites, Drones, and the Internet of Things

Satellites, drones, and the Internet of Things (IoT) devices have been in use for data collection for a number of years (Anand, Batra & Uitto, 2025). Each of these devices uses a number of sensors to create thousands of specific, objective data points, including data like geo-references, temperature, moisture, flow, and color. These data points can then be assimilated to give an evaluator a unique perspective and view of the item under surveillance.

In the field of evaluation, using remote collection devices like these can be a great way to collect geospatial data, data about ongoing functions, track the progress of projects in areas ranging from crop cover in agriculture, to temperature change, to the movement of migrant workers, or teacher absenteeism at schools. Combined with machine learning, the use of the data from these devices may be easier to analyze. Here we discuss some of the possibilities that these technologies can bring to the field of monitoring, evaluation, and learning.

Promises of Satellites, Drones, and IoT

Satellites, drones, and IoT devices can provide high-quality, near-real-time data on a wide range of indicators, including health supply chain (Dubin, Greve & Triche, n.d), environmental and climate data (Gong, Geng & Chen, 2015), agriculture (Petkovic, Petkovic & Petkovic, 2017), and infrastructure conditions. The data can provide a more complete and accurate picture of program impacts and outcomes. Satellites, drones, and IoT devices can provide real-time data on program impacts and outcomes, allowing for immediate feedback and course correction. More immediately available data can help to ensure that program interventions are responsive to changing circumstances and that they are achieving the desired results.

Satellites, drones, and IoT devices can provide precise and accurate data, particularly when combined with advanced analytics and machine learning techniques (Geospatial Commission, 2019). The types of data collected can help to identify patterns and trends in program impacts and outcomes that may be difficult to detect using traditional monitoring and evaluation approaches. These technologies can reduce the need for manual data collection and analysis, potentially saving time and resources. Using satellites, drones, or IoT devices to automate and routinize the collection of data can allow program staff to focus on program implementation and improvement, rather than on data collection and analysis (World Bank, 2017).

Importantly, these technologies can provide a more complete and accurate record of program implementation and impacts, improving accountability and transparency. This improved information can help to build trust between program stakeholders, including beneficiaries, donors, and implementing partners (UNEP, 2021).

Perils of Satellites, Drones, and IoT

The use of satellites, drones, and IoT devices can be costly and complex, particularly in terms of data processing and analysis. While the SERVIR program[3] has democratized access to satellite data for analysis, if you need data from a very specific geography or timeframe, it may still be expensive. Drones have reduced in cost, but given they've been used for military purposes, using them for program monitoring may run into national laws that do not allow drone use or a perception of an association with the military. However, experts have argued that with increased availability of 5G and portable drones, their use will continue to expand (Marchese, Moheddine & Patrone, 2019).

One of the main problems with the data from these devices is the volume of information (which is often more than needed) and how to make sense of exactly what is needed to understand a program's performance. The use of satellites, drones, and IoT devices requires technical expertise and capacity, in terms of both hardware and software. The need for specialized skills can make it difficult for some programs or organizations to adopt these technologies and make effective use of the data collected.

The quality and reliability of data collected through satellites, drones, and IoT devices can be affected by a range of factors, including weather conditions, technical malfunctions, and data transmission errors. These errors and gaps can make it difficult to ensure the accuracy and completeness of data (Penn State University, n.d.). While data from these devices is objective, gaps in the data can mean losing understanding of program performance, where traditional methods may have provided a more complete picture.

The use of satellites, drones, and IoT devices can raise privacy and security concerns, particularly in relation to data collection and storage. Without adequate monitoring, unauthorized access could occur, which could jeopardize data quality and the privacy of information (Greenstein & Cho, 2025). IoT devices are particularly vulnerable because a weakness in any one portion of an interlinked system could possibly lead to compromised data in the entire system (Tawalbeh, Muheidat, Tawalbeh & Quwaider, 2020). A break in the system could make it difficult to ensure the confidentiality and integrity of data. Data protection is a key factor that monitoring and evaluation professionals need to consider. For example, while satellite data can help with climate monitoring, to prevent deforestation, it could also be used to help poachers find wildlife (Skrabania, 2021). For evaluators, understanding stakeholders' concerns with regard to data protection and possible misuse of technology will be key.

The use of satellites, drones, and IoT devices can raise questions of data ownership and control, particularly in cases where data is collected from multiple sources. In the case of satellite or drone data, where data is collected on private land, the practice can raise questions of privacy (Fitzpatrick, 2021). Questions of privacy can make it difficult to ensure that data is used appropriately and fairly.

Similar to previous data collection examples (mobile data collection and social media/online data), drones, satellites and the IoT may be able to provide objective data, but they do not provide any context. The situation in which the data was collected remains opaque and cannot help an evaluator to understand how and why something is happening.

Examples from Practice: Satellites, Drones, and IoT

Drones have been effectively used, especially for monitoring community forest programs, where access can be difficult due to the area of interest, and the main outcome is ensuring the forest canopy remains intact (FAO, 2018). Drones have been used to monitor agricultural programs to detect plant diseases (Abbas, Zhang, Zheng et al., 2023), to track soil organic carbon index, vegetation indices, soil moisture, and soil erosion (MasterCard Foundation, Mercy Corps and AGRIFIN Accelerate, 2019). No specific examples of evaluations using these technologies were found in the literature, although evaluations may have made use of monitoring data as a source of information.

Chatbots and Virtual Agents

A virtual agent, also known as a virtual assistant or chatbot, is a computer program that uses artificial intelligence (AI) and Natural Language Processing (NLP) to simulate human conversation and assist users in performing tasks or answering questions (Janssen, Passlick, Cardona & Breitner, 2020). For the purposes of monitoring and evaluation, what could be interesting about these tools is how they could be used to collect information. To date, they've mostly been used in international development for the provision of information and resources, training, broad support services, and reporting (Carrington, 2022).

Promises of Chatbots and Virtual Agents

Chatbots generally operate with relatively low bandwidth, meaning they can be accessed in low connectivity areas. For the purposes of reporting information, they can be available when the respondent wants to use them. Chatbots can ensure consistency and accuracy in the data collected since they use predefined scripts and prompts to gather information from users. A chatbot-based data collection approach can reduce the risk of errors or inconsistencies that can occur when data is collected manually. Chatbots can handle a large volume of requests and data collection tasks simultaneously, making them a scalable solution for organizations that need to collect data from a large number of users or across multiple channels. Carefully designed chatbots can even be used for clinical research purposes (Chaix, Bibault, Romain et al., 2022), and one research paper found virtual agents aided socially anxious respondents to interact more and disclose more than with human data collectors (Kang & Gratch, 2010).

Perils of Chatbots and Virtual Agents

In order for chatbots to be effective, they need to be well designed, using a full range of user experience (language, look, bandwidth, etc.). One possible problem is that the research to date is very fragmented across disciplines (from technology to sectoral) and across application domains (Følstad, Araujo, Law et al., 2021).

Chatbots are largely dependent on the respondent reaching out to interact with them, which will affect the representativeness of the data collected and may be influenced by a population's comfort with technology. Chatbots use predefined scripts and prompts to collect data, which can limit their flexibility in handling complex or unexpected situations. Users may find it difficult to communicate their needs or concerns to the chatbot if the available response options are limited. Chatbots may have technical limitations, such as language understanding and processing, which can affect the accuracy and completeness of the data collected. In addition, chatbots may have difficulty handling multiple languages, accents, and dialects, which can limit their applicability in diverse settings.

Chatbots may not be able to provide the same level of empathy, understanding, and support that a human enumerator can provide, which can be especially important in sensitive or emotional situations and could impact the quality of data collection. Chatbots may collect personal information from users, which can raise privacy and security concerns. Users may be hesitant to share personal information with a chatbot, especially if they are uncertain about who is collecting the information and how it will be used (Hasal, Nowaková, Saghair, et al., 2021).

Examples from Practice: Chatbots and Virtual Agents

CivicTech developed a chatbot deployed through Facebook to monitor public works in their area of Madagascar through a World Bank-funded project. CivicTech's chatbot both provided information on public works projects in their area to be monitored (e.g., type of work, cost, and timeline) and allowed participants to submit information about the status of public works and anonymously report potential irregularities (Rakotomalala, Peixoto & Kumagai, 2020).

Virtual Reality

Virtual Reality (VR) technology has become increasingly popular in recent years, providing new opportunities for monitoring, evaluation, and learning (MEL) processes. VR can create a digital environment that simulates real-life situations, providing an immersive experience that can enhance participant engagement and improve data quality. Despite the promises of VR for MEL, it also poses several perils, such as ethical concerns, technical challenges, and limited generalizability of findings.

Promises of VR

VR can simulate real-life scenarios and provide a controlled environment that allows researchers to observe participant behavior and collect data in a standardized way. For example, in the field of education, VR can simulate a virtual classroom, allowing researchers to observe and collect data on student behavior, interactions, and learning outcomes. VR can enhance data quality by providing a more objective measurement of behavior, reducing the potential for observer bias because the observer is less visible (Liu, Wang, Lei, et al., 2020).

VR technology can be a cost-effective method for MEL processes, using simulations to reduce the need for expensive and time-consuming on-site data collection. Traditional MEL methods often require extensive travel, accommodation, and equipment expenses, which can be challenging to manage, especially in low-resource settings. VR can overcome these limitations by providing a virtual environment that mimics real-life situations. Using a VR approach may save time, reduce costs, and improve accuracy and privacy concerns, particularly when collecting sensitive data. These ethical considerations must be carefully addressed, and participants' safety and well-being should be prioritized.

Perils of VR

While VR may be cost-effective (Farra, Gneuhs, Hodgson et al., 2019) in the long run, the initial costs of implementing VR can be high. VR equipment and software are expensive, and the technical expertise required to set up, create simulations, and operate the system will be high. VR requires specialized equipment, software, and technical expertise, which can be a barrier to implementation, particularly in low-resource settings. Technical challenges can lead to data loss, inaccuracies, and errors, undermining the validity and reliability of the evaluation. Cost and technical expertise can be significant barriers to implementation, particularly for small projects or organizations with limited resources. VR technology is relatively new, and technical challenges can arise during implementation.

Potential ethical implications of VR include physiological and cognitive impacts and behavioral and social dynamics. Identifying and managing procedures to address emerging ethical issues will happen not only through regulations and laws (e.g., government and institutional approval) but also through ethics-in-practice (respect, care, morals, and education) (Kenwright, 2019).

Examples from Practice: Virtual Reality

The International Committee of the Red Cross has used virtual reality to increase the efficacy of teaching complex and variable subject matter, like International Humanitarian Law (ICRC, n.d.). UNICEF has invested in companies that are using VR to conduct reading assessments (Kitheka & Szymczak, 2022). Both ICRC and UNICEF note that VR and augmented reality are on the frontier of development practice but could have practical applications. VR technology has been used in the education sector to enhance learning outcomes and improve teacher training. The World Bank found that VR training was equal or more conducive to improving learning outcomes than traditional training methods (Angel-Urdinola, Castillo & Hoyos, 2021).

Data Processing/Analysis

Unlike a decade ago when the best emerging technologies were in the data collection arena, today the most interesting and potentially useful technologies are in the data analysis and processing realm, providing more potential to use more and different types of data in analysis.

Data Analytics

Data analytics using Big Data is discussed extensively in the chapter by York and Bamberger (2025). Here we discuss the promises and perils for use in monitoring and evaluation in the international context (Arockia, Varneekha & Veneshia, 2017).

Promises of Data Analytics

One of the key advantages of using data analytics to parse Big Data for monitoring and evaluation may be the speed at which insights can be generated.

Traditional evaluation methods can be time-consuming and expensive, while Big Data analytics can provide near real-time insights, using large volumes of data (van der Vink, Carlson, Phillips, et al., 2023). These types of analyses can enable organizations to respond more quickly to emerging challenges and adjust their programs and interventions accordingly once their data is in a format that can readily be used and re-used.

The other main advantage of data analytics is that one can use a much wider variety of data sources than would be possible for a human to analyze and derive insights from these multiple sources in ways that were not previously possible. It is possible to reuse big datasets and return to them with new and different questions, potentially increasing the insights available from the data.

Perils of Data Analytics

The ability to access and process Big Data using data analytics will likely be affected by the digital divide, where international and "northern" consultants have more and better computing power than many evaluands. At the same time, though, there are fewer "Big Data" sources in the global south, which may mean any analyses that are conducted miss the point. Specifically, if only existing secondary data is used, without specifically targeting key groups, vulnerable and marginalized populations may be further marginalized.

Evaluators focus on the quality of data collection and make efforts to ensure that there is inclusivity and representativeness, and that respondents are protected from harm. These facets of data collection may have less prominence if Big Data is exclusively used for analysis. Similarly, there may be a tendency to be overconfident in the reliability and validity of data analytics that need to be examined and considered in each scenario. Finally, many have expressed concerns with what is and is not included and considered in automated algorithms.

As with many of these technologies, the extraction, transformation, and analyses of these datasets require specialized training and expertise. The divide between the evaluators and the computer scientists who have these skills is substantial, and there is still a need for better understanding between these two professions in order for data analytics to be meaningful for evaluation.

Natural Language Processing (NLP) and Generative AI

NLP is a machine technology that gives computers the ability to interpret, manipulate, and comprehend human language. Generative AI refers to a category of artificial intelligence (AI) algorithms that generate new outputs based on the data they have been trained on. In the context of language (as opposed to images), NLP offers a way to process large volumes of text (which humans cannot do in a time-effective manner) and then, using generative AI, queries of a body of documents (or corpus) are made.

Promises of NLP and Generative AI

In 2023, the advent of widely available Generative AI models (like ChatGPT, Bard, and Bing among many other examples) is a game changer. They are trained on a large corpus, and then fine-tuned in a human-supervised approach (transfer learning). A few specific use cases for evaluation could include:

1. Developing and drafting proposals, log frames, theories of change, evaluation methodologies, and reports – as a thought partner
2. Reviewing and summarizing large volumes of structured or unstructured data (including text, video, audio, or even images)
3. Enhancing and simplifying data collection through automated translation, transcription, and speech-to-text tools.
4. Analyzing data through topic modeling, named entity recognition, sentiment analysis, and content analysis.
5. Generating the code for qualitative or quantitative data analysis.

Generative AI might support more nuanced understanding of situations and suggest possible ethical concerns to consider. By combining the strengths of AI with the unique perspective of evaluators, we can create more effective and personalized interactions between users and AI systems.

Part of this book chapter was written in collaboration with ChatGPT and reviewed with Hemmingway.AI. We used ChatGPT as a thought partner to think through things we may have missed. We used it to improve our language and summarize sections that were too long.

Perils of NLP and Generative AI

There are still clear limitations with NLP and Generative AI technology that need to be considered. Generative AI models like ChatGPT seem very insightful, fast, and incredibly accurate. Yet, they can't think. At their core, they are probability machines that recombine words into coherent text, based on statistical probabilities and a vast corpus of text and information. Generative AIs sometimes get things wrong, convincingly and eloquently, which might fool a naïve operator. It's important for evaluators to keep in mind that Generative AI is capable of being highly persuasive, even if their response is misleading or may not be truthful or accurate. They lack empathy, intuition, creativity, and judgment and are essentially statistical predictors of the "next word."

Some Generative AIs do not reference articles or sources to corroborate their output. When asked to do so, they oblige, but the references are developed in the same way as the text: through a probability-based recombination of words. The references look good, but they are actually complete fabrications. For example, we asked ChatGPT, "What are the benefits of using chatbots for data collection?

Please provide a citation too." It responded with an answer and provided this citation: "Simões, L. M., de Lima, E. F. F., & Neves, M. C. (2020). Chatbots in data collection and monitoring: An overview. Journal of Business Research, 116, 13-25. https://doi.org/10.1016/j.jbusres.2020.04.022." The authors of this chapter were excited to read an article they hadn't come across yet – but it turns out that this citation ***does not exist***. You still have to do your own research – ChatGPT will not do that part for you.

Building language models like ChatGPT requires large amounts of data, is costly in terms of time, computing resources, and energy, and can have an environmental impact that may harm marginalized communities disproportionately. Sharing language models, so that research teams and evaluators can fine-tune them for their purposes, will be key to saving resources and reducing environmental costs.

One of the biggest challenges is that large language model developers have scraped large amounts of data from the internet, taking both the best and the worst of what is available, to develop their models. The models are shaped by the worldviews and ideologies present in these corpuses. Biases in the corpuses will be perpetuated and reinforced in the language models and can result in harmful outcomes: perpetuating stereotypes, discrimination, and offensive or culturally inappropriate language. It will be important to be transparent about the data sources used to develop the language models. Human fine-tuning and the use of good judgment in reviewing AI output can mitigate these challenges.

AI and generative AI are self-reinforcing language systems: they can produce eloquent log frames, persuasive proposals, and glowing reports, all based on developmental and evaluation jargon used in previous evaluation documents. These will feed into more of the same type of reports and strengthen existing and accepted models. They do not necessarily reflect reality. Evaluators should guard against this kind of confirmation bias through critical thinking and through the inclusion of alternative epistemologies, perspectives, and insights.

Machine Learning for Evaluation

Machine learning – specifically supervised machine learning (SML) and unsupervised machine learning (UML) – holds great potential for the field of evaluation. SML is an approach that teaches the machine to understand and code content in a similar way. A human shows the computer or adjudicates a computer-assigned code for groups of text. Once a sufficient number of examples have been provided, the machine can code segments on its own. UML generally uses a topic modeling approach to group related concepts into groups. Humans need to interpret the resulting topics and assign meaning to them, but UML can be a useful way to process large volumes of text and derive meaning quickly (Gatto & Bundi, 2025; Mazzeo, Celardi, Miracula & Picone, 2025; Næss, Prabhu, Mjaaland, Holtermann & Engebretsen, 2025).

Promises of Machine Learning for Evaluation

Both SML and UML could potentially help evaluators to rapidly process and use much larger textual datasets than humans could manually process in a time and cost-efficient manner. By automating the analysis and textual "coding" process through machine learning, evaluators could ensure they are considering a much wider corpus of information in their evaluations and taking a systematic approach to coding this corpus. If evaluation commissioners (donors, governments, multilateral institutions, and implementers) would invest in automated document section extraction routines, like the World Bank has done (Toetzke, Banholzer & Feuerriegel, 2022; Ziulu, Anuj, Hagh, Raimondo & Vaessen, 2025), this will speed the extraction of relevant evidence sections from the text and speed the process of data cleaning and allow us to use targeted data more quickly.

What UML may help us to do is to increase the speed of systematic literature review for evaluations. Researchers have developed a methodology and openly available code to speed the review of tens of thousands of articles (Thiabaud, Triulzi, Orel et al., 2020). They used topic modeling (UML) to understand the major themes in the data. However, they noted this method was only available for databases that provide free APIs for open access to full-text articles and only worked for machine-readable text. They found that the topic modeling approach they used would be negatively affected by smaller corpora (such as those that might be used for an evaluation), yielding possibly uninterpretable findings.

Perils of Machine Learning for Evaluation

The main drawback of both SML and UML for evaluation is the need to examine a specific "corpus" of evidence or documents. While large language models like ChatGPT are a fascinating resource, they use a much broader range of information than we would like to consider for most evaluations, and those sources are difficult to parse. Further, ChatGPT will sometimes "hallucinate" results (Zuccon, Koopman & Shaik, 2023) in its attempt to answer our prompts.

In order to create a specific corpus for an evaluation to which we could apply SML or UML to support the analysis, the process of preparing the data is complicated and time-intensive. For example, in interview transcripts, we would need to remove all the questions that the interviewer repeatedly asks, lest the machine consider these extraneous components to be part of the dataset. In documents that we'd like the machine to include for consideration as part of the evidence for an evaluation, such as implementer or government reports, it would be helpful to remove any "confounding" or repetitive information such as proforma titles and section headers, so they would not mislead the machine, which generally uses word frequency and word proximity to make meaning of text. While preprocessing the data is becoming easier, it is not easy yet, and putting too much or the wrong data into a model will skew the results.

There is a high level of skill needed to prepare the text for analysis. Both the World Bank and Toetzke and colleagues (Toetzke, Banholzer & Feuerriegel, 2022) followed a similar process to prepare their data for analysis. Broadly, they followed these steps.

1. Translate (into English) and preprocess (select and make machine-readable) the textual descriptions to be included in the dataset.
2. Vectorize the text (turning the text into numerical representations – most often frequencies).
3. Apply SML or UML techniques to the data to code or create topic "clusters" of meaning.
4. Validate and make meaning of the findings (through review, visualization of the results).

While these steps may seem simple, few evaluators are trained in the data preparation techniques to extract, tokenize, vectorize, and prepare data for machine learning. While generative AI like ChatGPT can support this process, there is still a need for subject matter expertise in Python or Java to debug the code.[4] Evaluators are not trained in the process of selecting a model (e.g., SML or UML, or selecting a sentence-level or paragraph-level vectoring model) and the pros and cons of different approaches and how they could affect the final outcome of an analysis.

The World Bank found that machine learning can bias the analysis when "the data were highly imbalanced by class." This class imbalance occurs when some labels are used frequently (more than 200 inputs per label) and others are rarely used (fewer than 10 inputs per label). As a result, there are many types of labels to predict but few examples of less frequent labels to learn from (Franzen, Cuong, Schweizer et al., 2022). How authors write, what words are used, and how frequently they are used will influence a machine learning analysis.

There is still a need for subject matter experts to weigh in on "topics" or "domains" to ensure that the computer got it right and to assign meaning to the vague "topics" that topic modeling generates (Chang, Boyd-Graber, Wang et al., 2009). An excellent example of the complexities of topic modeling interpretation can be found in Chakrabarti and Frye (2017) in their seminal work describing the analysis of over 4000 handwritten journals on HIV, which show the need for human interpretation (see especially Table 2.1 in their paper).

Examples from Practice: Machine Learning

USAID used AI/machine learning to review program documents for evidence of outcomes (positive or negative) related to equity for marginalized racial and ethnic groups. Specifically, they identified publicly available evaluation reports and activity final reports for USAID programs from the past 10 years, extracted text from these PDF files and converted them to machine-readable files, and developed a lexicon of general and country-specific racial and ethnic terms. They then developed an NLP algorithm coded in Python and R languages to identify instances in the documents of USAID programming actions and outcomes that included the racial and ethnic terminology, reviewed the algorithm output for relevance and false positives, and analyzed the results for patterns over time, across countries/regions, and across programming sectors. But the results were not all that helpful. USAID's hypothesis on why the analysis was not helpful posited that there was a lack of data science knowledge on the USAID side, unrealistic expectations for what machine learning could accomplish, and a lack of international development, racial/ethnic equity, & social science research knowledge on the part of the machine learning contractors doing the work. The authors noted the high level of risk aversion and unclear USAID policies and guidance related to AI and machine learning at USAID (Roen & Gallager, 2022).

As was referenced above, the World Bank used both SML and UML to analyze a corpus of its evaluations to label content and understand if SML was a feasible method for future analyses. They used UML to identify factors affecting intervention success. They found both methods useful, but not seamless.

Conclusion

Emerging technology holds great promise for enabling both monitoring and evaluation to make better use of a much wider range of data, both quantitative and qualitative, and thereby improve the breadth of the evidence base used to make program improvement decisions.

Data collection and analysis remain largely an "extractive" industry. The technology to share rapid feedback on data to visualize data in highly inventive ways exists but is too sparsely used. This lack of rapid feedback on data collected is both because the commissioners/owners of the data do not want to share their data and because there are too few incentives to do so. Closing the feedback loop and using emerging technologies to help us do this will be key, but providing data in machine-readable, API-accessible formats will be needed to improve the use of data.

Just because technology may make data collection more efficient and faster does not mean that our understanding of how and what the data show will be improved. Many of these technologies fail to assist monitoring and evaluation specialists in understanding why something is working and the conditions that must exist for it to continue.

Data privacy and re-identification of individuals through Big Data and machine learning techniques remain a problem. While individuals are assured of the confidentiality of information at the point of data collection, once their data becomes part of a larger dataset and can be triangulated against other datasets, it may be easy to identify the individual. The re-use of data, intended for one purpose, where consent was provided for that purpose, which is then reused for another purpose, violates the consent that was provided. And yet, the data if it could be de-identified, could be useful and could avoid wasting further data collection scarce resources.

There is still a wide capacity gap in the use of advanced data analysis technologies. This capacity gap existed for mobile data collection a decade ago but is now largely closed. As data analysis becomes more automated and the use of the technologies we describe becomes more widespread, this gap should lessen. Today, using these technologies requires intensive data preparation, cleaning, and coding, which is beyond the skill set of many evaluators. Human interpretation and support to train models are still required and the expertise of evaluators will be necessary to support quality and transparent analyses.

While there is substantial training available online in machine learning, sentiment analysis, and other relevant fields, there are precious few training opportunities targeted for evaluators. Those that are available are offered by a narrow range of evaluators. As mobile data collection and computer-aided qualitative data analysis courses have made their way into the mainstream of subject matter and generalist evaluation training courses, it is time for data science and machine learning basics to appear. These courses will make evaluators – both commissioners and practitioners – more informed on what is and is not possible within the constraints of evaluation contracts.

Because most evaluations are contracts, until clients start requesting evaluations that expect the use of these technologies, evaluators and evaluation companies are not being driven to develop these skills. The US government with the Foundations for Evidence-Based Policymaking Act of 2018 (US Congress, 2018) may be starting to move in this direction, but it has a lot of ground to cover to ensure comprehensive monitoring and evaluation within the US Federal government before it gets to more advanced topics. Promisingly, Executive Order 13960 from the White House (US White House, 2022) requires all federal agencies to inventory their AI use cases and share their inventories with other government agencies and the public. This order may drive commissioners in the US government at least toward using more advanced techniques. Whatever happens on the policy front, both practitioners and commissioners need to have a better grasp of what lies behind emerging technology techniques so they can accurately assess the risks, biases, and advantages these technologies may bring.

There is still a gap between the skills needed for evaluation and the skills needed to be able to employ many of the technologies described in this chapter. There is a need for evaluators to stay abreast of the wider use of technology in program implementation in order to be able to evaluate its relevance, effectiveness, efficiency, sustainability, coherence, and impact. But there is a growing need to include training in these emerging technologies, NLP, Big Data, and machine learning into courses on evaluation, so that the next generation of practitioners knows how to make use of this technology and work with their technology colleagues to use evidence to make decisions. Training in these domains is an opportunity to work together in shaping these technologies to avoid the pitfalls and enhance the promises by applying evaluation methods and skills in the development of emerging technologies (Gandhi, 2023).

Notes

1 The best example of a "freemium" model is Gmail – for most users, it is free, but if you want to use it for a lot of data or its advanced features, you need to pay.
2 Fuerzas Armadas Revolucionarias de Colombia-Ejército del Pueblo (The Revolutionary Armed Forces of Colombia – People's Army, known as the FARC-EP, or simply FARC).
3 This is a joint program between USAID, NASA, and other stakeholders to make use of geo-spatial data. See https://www.servirglobal.net.
4 An excellent illustration of the complexity of using APIs to import a specific corpus of information into ChatGPT to analyze themes is outlined in this blog: https://beebom.com/how-train-ai-chatbot-custom-knowledge-base-chatgpt-api/.

References

Abbas, A., Zhang, Z., Zheng, H., et al. (2023). Drones in plant disease assessment, efficient monitoring, and detection: A way forward to smart agriculture. *Agronomy*, 13(6):1524. https://doi.org/10.3390/agronomy13061524

Anand A., Batra, G., & Uitto, J.I. (2025). Harnessing Geospatial Approaches to Strengthen Evaluative Evidence. In S.B. Nielsen, F. Mazzeo Rinaldi and G.J. Petersson (eds.). *Artificial Intelligence and Evaluation. Emerging Technologies and Their Implications for Evaluation* (pp. 196–218). London: Routledge. https://doi.org/10.4324/9781003512493

Angel-Urdinola, D., Castillo, C., & Hoyos, A. (2021, April 27). *Can Virtual Reality Simulators Develop Students' Skills?* World Bank Blogs. Retrieved from https://blogs.worldbank.org/developmenttalk/can-virtual-reality-simulators-develop-students-skills

Arockia, S.P., Varneekha, S.S., & Veneshia, K.A. (2017). The 17 V's of big data. *International Research Journal of Engineering and Technology (IRJET)*, 4(9). Retrieved from https://www.irjet.net/archives/V4/i9/IRJET-V4I957.pdf

Bamberger, M. (2022, August 17). *Using Digital Data to Strengthen the Evaluation of Complex Programs*. AEA354. Retrieved from https://aea365.org/blog/aeas-digital-data-tech-working-group-week-using-digital-data-to-strengthen-the-evaluation-of-complex-programs-by-michael-bamberger/

Benston, A.M., Kumwenda, W., Lurie, M., et al. (2020). Improving monitoring of engagement in HIV care for women in option B+: A pilot test of biometric fingerprint scanning in Lilongwe, Malawi. *AIDS Behavior*, 24(2):551–559. https://doi.org/10.1007/s10461-019-02748-6

Blumenstock, J., Cadamuro, G., & On, R. (2015). Predicting poverty and wealth from mobile phone metadata. *Science*, 350:1073–1076.

Carrington, C. (2022, December 14). *Chatbots for the International Development and Humanitarian Sectors: What Works?* DAI. Retreved November 12, 2023, from https://dai-global-digital.com/chatbots-for-the-international-development-and-humanitarian-sectors-what-works.html?utm_source=daidotcom

Chaix, B., Bibault, J.-E., Romain, R., et al. (2022). Assessing the performances of a chatbot to collect real-life data of patients suffering from primary headache disorders. *Digit Health,* 2022:8. https://doi.org/10.1177/20552076221097783

Chakrabarti, P., & Frye, M. (2017). A mixed-methods framework for analyzing text data: Integrating computational techniques with qualitative methods in demography.*Demographic Research,* 37(42):1351–1382. https://doi.org/10.4054/DemRes.2017.37.42

Chang, J., Boyd-Graber, J., Wang, C., et al. (2009). Reading tea leaves: How humans interpret topic models. *Advances in Neural Information Processing Systems,* 32:288–296.

CRS and ICT4D Conference. (n.d.). *Past ICT4D Conferences, CRS and ICT4D.* Retrieved from https://www.ict4dconference.org/past-events/

Department for International Development. (2018, January 23). *Doing Development in a Digital World.* Retrieved from https://www.gov.uk/government/publications/dfid-digital-strategy-2018-to-2020-doing-development-in-a-digital-world/dfid-digital-strategy-2018-to-2020-doing-development-in-a-digital-world#Chapter2

Dubin, S., Greve, A., & Triche, R. (n.d.). *Drones in International Development, Innovating the Supply Chain to Reach Patients in Remote Areas.* PEPFAR and USAID.Retrieved November 12, 2023, from https://www.updwg.org/wp-content/uploads/2020/11/Drones_in_International_Development_Innovating_the_Supply_Chain_to_Reach_Patients_in_Remote_Areas_2_1.pdf

Fabra-Mata, J., & Mygind, J. (2019). Big data in evaluation: Experiences from using Twitter analysis to evaluate Norway's contribution to the peace process in Colombia. *Evaluation,* 25(1):6–22. https://doi.org/10.1177/1356389018804259

FAO. (2018, October 14). *e-Agriculture, Promising Practice Drones for community monitoring of forests.* FAO. Retrieved from https://www.fao.org/3/I8760EN/i8760en.pdf

Farra, S.L., Gneuhs, M., Hodgson, E., et al. (2019). Comparative cost of virtual reality training and live exercises for training hospital workers for evacuation. *Computer Informatics Nursing,* 37(9):446–454. https://doi.org/10.1097/CIN.0000000000000540

Fitzpatrick, N. (2021, March 2). *Who owns Geospatial Data?* TechUK. Retrieved from https://www.techuk.org/resource/who-owns-geospatial-data.html

Følstad, A., Araujo, T., Law, E.L.C., et al. (2021). Future directions for chatbot research: an interdisciplinary research agenda. *Computing,* 103:2915–2942. https://doi.org/10.1007/s00607-021-01016-7

Franzen, S., Cuong, Q., Schweizer, A., et al. (2022). *Advanced Content Analysis: Can Artificial Intelligence Accelerate Theory-Driven Complex Program Evaluation?* IEG Methods and Evaluation Capacity Development Working Paper Series. Independent Evaluation Group. World Bank. Retrieved from https://ieg.worldbankgroup.org/methods-resource/advanced-content-analysis-can-artificial-intelligence-accelerate-theory-driven-complex

Gandhi, V. J. (2022, August 20). *'Lucky' Luke and Doing Evaluation in a Digital World.* AEA365. Retrieved from https://aea365.org/blog/aeas-digital-data-technology

-working-group-week-lucky-luke-and-doing-evaluation-in-a-digital-world-by-valentine-j-gandhi/

Gandhi, V. J. (2023, October 5). AI for MEL and MEL for AI, MEL Community of Practice lecture series - Bixal. Retrieved from https://www.youtube.com/watch?v=r9xM7IzSDHs

Gatto, L., & Bundi, P. (2025). The Use of Quantitative Text Analysis in Evaluations. In S.B. Nielsen, F. Mazzeo Rinaldi and G.J. Petersson (eds.). *Artificial Intelligence and Evaluation. Emerging Technologies and Their Implications for Evaluation* (pp. 144–167). London: Routledge. https://doi.org/10.4324/9781003512493

Geospatial Commission. (2019, August 27). *Future Technologies Review*. Gov.UK. Retrieved from https://www.gov.uk/government/publications/future-technologies-review/future-technologies-review

Gong, J., Geng, J., & Chen, Z. (2015). Real-time GIS data model and sensor web service platform for environmental data management. *International Journal of Health Geographics*, 14:2. https://doi.org/10.1186/1476-072X-14-2

Greenstein, N., & Cho, S.-W. (2025). Ethics & Equity in Data Science for Evaluators. In S.B. Nielsen, F. Mazzeo Rinaldi and G.J. Petersson (eds.). *Artificial Intelligence and Evaluation. Emerging Technologies and their Implications for Evaluation* (pp. 56–77). London: Routledge. https://doi.org/10.4324/9781003512493

Hasal, M., Nowaková, J., Saghair, K.A., et al. (2021). Chatbots: Security, privacy, data protection, and social aspects. *Concurrency Computation: Practice and Experience,* 33:e6426. https://doi.org/10.1002/cpe.6426

ICRC. (n.d.). *Virtual Reality and Innovation.* ICRC. Retrieved from https://www.icrc.org/en/what-we-do/virtual-reality

Janssen, A., Passlick, J., Cardona, D., & Breitner, M. (2020). Virtual assistance in any context - A taxonomy of design elements for domain-specific chatbots. *Business & Information Systems Engineering*, 62:211–225). https://doi.org/10.1007/s12599-020-00644-1

Kang, S.-H., & Gratch, J. (2010). Virtual humans elicit socially anxious interactants' verbal self-disclosure. *Computer Animation and Virtual Worlds*, 21:473–482. https://doi.org/10.1002/cav.345

Kenwright, B. (2019, January 14). *Virtual Reality: Ethical Challenges and Dangers*, Technology and Society. IEEE. Retrieved from https://technologyandsociety.org/virtual-reality-ethical-challenges-and-dangers/

Kitheka, K., & Szymczak, C. (2022). *Interacting with AR and VR*. UNICEF. Retrieved from https://www.unicef.org/innovation/stories/interacting-ar-and-vr

Kumar, S., Barbier, G., Abbasi, M., & Liu, H. (2021). TweetTracker: An analysis tool for humanitarian and disaster relief. *Proceedings of the International AAAI Conference on Web and Social Media*, 5(1), 661–662. https://doi.org/10.1609/icwsm.v5i1.14079

Liu, R., Wang, L., Lei, J., et al. (2020). Effects of an immersive virtual reality-based classroom on students' learning performance in science lessons. *British Journal of Educational Technology*, 51:2034–2049. https://doi.org/10.1111/bjet.13028

Marchese, M., Moheddine, A., & Patrone, F. (2019). IoT and UAV integration in 5G Hybrid Terrestrial- Satellite Networks. *Sensors (Basel)*, 19(17):3704. https://doi.org/10.3390/s19173704

Marvin, G., & Sevilla, G. (2019, August 23). *The Best Social Media Management and Analytics Tools*. PCMag. Retrieved from https://www.pcmag.com/picks/the-best-social-media-management-analytics-tools

MasterCard Foundation, Mercy Corps and AGRIFIN Accelerate. (2019, December). *AgriFin Accelerate Five Year Overview*. Retrieved from https://www.mercycorpsagrifin.org/wp-content/uploads/2020/03/AgriFinALE4_OpeningPlenaryAllSlides.pdf

Mazzeo Rinaldi, F., Celardi, E., Miracula, V., & Picone, A. (2025). Artificial Intelligence and Text Analysis in Evaluating Complex Social Phenomena. The Russia-Ukraine Conflict. In S.B. Nielsen, F. Mazzeo Rinaldi and G.J. Petersson (eds.). *Artificial Intelligence and Evaluation. Emerging Technologies and Their Implications for Evaluation* (pp. 168–195). London: Routledge. https://doi.org/10.4324/9781003512493

MERLTech. (n.d.). *Past MERLTech Conferences*. MERLTech. Retrieved from https://merltech.org/events/past-merl-tech-conferences/

Næss, T., Prabhu, C., Mjaaland, M., Holtermann, H., & Engebretsen, L.S. (2025). Text Mining and Machine Learning in an Evaluation of Police Handling of Cybercrime in Norway. In S.B. Nielsen, F. Mazzeo Rinaldi and G.J. Petersson (eds.). *Artificial Intelligence and Evaluation. Emerging Technologies and Their Implications for Evaluation* (pp. 103–119). London: Routledge. https://doi.org/10.4324/9781003512493

Penn State University, Department of Geography. (n.d.). *Geog 160 – Mapping Our Changing World, Geospatial Data Quality, Validity, Accuracy and Precision*. Retrieved from https://www.e-education.psu.edu/geog160/node/1922

Petkovic, S., Petkovic, D., & Petkovic, A. (2017). IoT devices VS. drones for data collection in agriculture. *DAAAM International Scientific Book*, 16:63–80.

Rakotomalala, O., Peixoto, T., & Kumagai, S. (2020). *Chatbots for Third-Party Monitoring: CivicTech Pilot in Madagascar*. Governance Notes; No. 23. World Bank. Retrieved from https://openknowledge.worldbank.org/handle/10986/34076

Roen, E., & Gallager, J. (2022). *Can Artificial Intelligence Identify Evaluation Findings That Advance Equity? USAID's Mixed Experience*. Presentation at the American Evaluation Association Meeting in New Orleans, November 2022 (included with permission from the authors).

Skrabania, L. (2021, March 18). *Interview: How Problematic Are Satellites and Drones in Terms of Data Protection?* Reset Digital for Good. Retrieved from https://en.reset.org/interview-how-problematic-are-satellites-and-drones-terms-data-protection-03072021/

Tawalbeh, L., Muheidat, F., Tawalbeh, M., & Quwaider, M. (2020). IoT privacy and security: Challenges and solutions. *Applied Sciences*, 10(12):4102. https://doi.org/10.3390/app10124102

The Development Cafe. (2018). *Blog*. The Development Café. Retrieved from https://www.dev-cafe.org/blog/

Thiabaud, A., Triulzi, I., & Orel, E., et al. (2020). Social, behavioral, and cultural factors of HIV in Malawi: Semi-automated systematic review. *Journal of Medical Internet Research*, 22(8):e18747. https://doi.org/10.2196/18747

Toetzke, M., Banholzer, N., & Feuerriegel, S. (2022). Monitoring global development aid with machine learning, nature sustainability. *Nature Sustainability*, 5:533–541. https://doi.org/10.1038/s41893-022-00874-z

UNEP. (2021, February 18). *Making peace with nature: A scientific blueprint to tackle the climate, biodiversity and pollution emergencies.* UNEP. Retrieved from https://www.unep.org/resources/making-peace-nature

U.S. Congress. (2018). *Foundations for evidence-based policymaking act of 2018 5 USC301(2019).* Public Law No: 115-435 (01/14/2019). Retrieved from https://www.congress.gov/115/plaws/publ435/PLAW-115publ435.pdf

U.S. Whitehouse. (2022). *Agency Inventories of AI use Cases, Executive Order 13960, 2022.* Retrieved from https://www.ai.gov/ai-use-case-inventories/

van der Vink, G.E., Carlson, K.N., Phillips, E., et al. (2023). Identifying vulnerability to human trafficking in Bangladesh: An ecosystem approach using weak-signal analysis. *Journal of International Development*, 1–17. https://doi.org/10.1002/jid.3824

Vota, W. (2018, November 21). *The Definitive Guide to Mobile Data Collection in International Development.* ICTWorks. Retrieved from https://www.ictworks.org/guide-mobile-data-collection/

Wikipedia. (n.d.). *List of Social Platforms with at Least 100 Million Active Users.* Wikipedia. Retrieved from https://en.wikipedia.org/wiki/List_of_social_platforms_with_at_least_100_million_active_users

World Bank. (2017, April 25). *Tapping the potential of drones for development.* World Bank. Retrieved from https://www.worldbank.org/en/topic/transport/brief/drones-for-development

York, P., & Bamberger, M. (2025). The Applications of Big Data to Strengthen Evaluation. In S.B. Nielsen, F. Mazzeo Rinaldi and G.J. Petersson (eds.). *Artificial Intelligence and Evaluation. Emerging Technologies and Their Implications for Evaluation* (pp. 37–55). London: Routledge. https://doi.org/10.4324/9781003512493

Zachlod, C., Samuel, O., Ochsner, A., & Werthmüller, S. (2022). Analytics of social media data – State of characteristics and application. *Journal of Business Research*, 144:1064–1076. https://doi.org/10.1016/j.jbusres.2022.02.016

Ziulu, V., Anuj, H., Hagh, A., Raimondo, E., & Vaessen, J. (2025). Extracting Meaning from Textual Data for Evaluation. – Lessons from Recent Practice at the Independent Evaluation Group of the World Bank. In S.B. Nielsen, F. Mazzeo Rinaldi and G.J. Petersson (eds.). *Artificial Intelligence and Evaluation. Emerging Technologies and their Implications for Evaluation* (pp. 78–102). London: Routledge. https://doi.org/10.4324/9781003512493

Zuccon, G., Koopman, B., & Shaik, R. (2023, September 17). *ChatGPT hallucinates when attributing answers.* Cornell University, arXiv.org. Retrieved from https://www.proquest.com/working-papers/chatgpt-hallucinates-when-attributing-answers/docview/2866249831/se-2

3 The Applications of Big Data to Strengthen Evaluation

Pete York and Michael Bamberger

Introduction

In Chapter 2, Kerry Bruce, Valentine J. Gandhi, and Joris Vandelanotte (2025) provide an overview and assessment of the emerging digital technologies for the monitoring and evaluation of development projects. They discuss digital technologies in terms of data capture, data storage, and data processing and analysis, and consider the promises and perils of each technology, as well as providing examples from practice. They show that these powerful new technologies are already transforming the nature and applications of Monitoring and Evaluation (M&E) and that the process of change is accelerating.

The focus of their chapter is on describing and assessing these technologies and discussing how they can be used for monitoring and evaluation of development programs.

The purpose of the present chapter is to build on this analysis to explore how these new technologies are transforming and broadening program evaluation into a set of tools providing real-time analysis of complex interventions, making evaluation a dynamic management and policy tool to improve the performance of ongoing programs, and to provide a broader vision of the goals and impacts of large and complex programs than was previously possible. We take the examples of causal modeling (precision analytics) and the application of geospatial analysis to illustrate how these advances are applied in programs to promote social equity. In Chapter 11, York (2025) describes these applications in more detail.

These tools, particularly machine learning (ML) and artificial intelligence (AI), can also take over many of the time-consuming routine tasks of data collection and analysis, so that evaluators can devote much more time to evaluative thinking around issues such as constructing theory-based models, causal and predictive analysis, and organizing real-time findings to improve the ongoing implementation of service delivery. As Bruce, Gandhi, and Vandelanotte (2025) also show, the speed with which the new technologies can find patterns in huge volumes of data also provides evaluators with important new insights into complex patterns and associations that previously could never have been detected.

DOI: 10.4324/9781003512493-3

In the present chapter, we explore further the practical applications of these technological advances, specifically their role in promoting social equity through sophisticated program evaluations.

This chapter is structured to first define and contextualize the emerging concepts and terminologies within the domain of big data and data science. It then elucidates the synergistic relationship between these disciplines and evaluation, building upon the technological competencies previously discussed. Finally, we will address the conceptual contributions of evaluation to this dialogue, particularly through the lens of causal modeling and precision analytics, before presenting tangible examples that showcase the burgeoning potential of big data in the evaluation of complex programs. The first section provides a brief overview and definition of big data, data analytics, and data science. The second section then discusses how big data and data science can strengthen development evaluation. We build on the discussion of the strengths and limitations of big data in the previous chapter (Bruce, Gandhi & Vandelanotte, 2025), and illustrate how the different technologies have been applied in development evaluation. We discuss how the different applications of artificial intelligence and machine learning are starting to transform the nature of evaluation, including how this is making it possible to harvest and repurpose the huge volumes of administrative data that can now be analyzed (see Table 2.1 in the previous chapter (Bruce, Gandhi & Vandelanotte, 2025) and Table 3.1 in the present chapter). The third section then discusses what evaluation can bring to the table, focusing on causal modeling (and precision analytics). The fourth section then provides examples of the exciting new applications of big data, including a greater ability to evaluate complex development programs.

Box 3.1: Defining big data, data analytics, and data science.

Big data: the different types of digital data (described in Table 3.1)

Data analytics: the new analytical tools and methods for the analysis of big data (and also data that is not big).

Data science: the research processes that combine digital data collection (big data) and the analysis of big data (data analytics).

Note: Data analytics can also be used for the analysis of smaller data sets.

The term "Big Data" is used to describe both digital methods for data collection and also the new statistical techniques for the analysis of digital data. For greater clarity, we use the terms *big data* and *data analytics* to refer respectively to the sources of digital data and the tools for the processing and analysis of the data, and *data science* to describe the overall process of data collection and analysis (Box 3.1).

Table 3.1 Examples of data analytics tools

Analytical tool/approach	*Explanation*
1. A/B testing[1] and experimental on-line designs	Experimental and quasi-experimental designs that match test groups with controls to assess whether there is evidence of change or improvement
2. Integrating different data sets	Different data sets from within an agency or data from different agencies are combined into a common metric. This makes it possible to assess the influence of a much wider range of variables, in particular contextual variables, on program outcomes
3. Data mining	Combining tools from machine learning and statistics to extract patterns from large data sets
4. Machine learning (ML)	Machines are taught to search for key terms or images (pattern recognition) in large data sets. ML is also used to automate routine data collection and analysis activities
5. Natural language processing (NLP), text analytics, and image recognition	Algorithms are used to recognize human language and unstructured data such as faces and images (e.g., X-rays). NLP can also be used to conduct qualitative topic/thematic modeling
6. Analysis of topics discussed in radio call-in programs	This is a variant of text analytics used by humanitarian agencies to identify potential signs of social conflict, particularly as they affect refugee populations (see Chapter 9)
7. Large Language Transformer models, like BARTLarge, GPT-4, etc.	The model is trained on large data sets and can then compose text and images drawing on this database. These large language transformer models can also be used for automating qualitative analysis (thematic analysis/topic modeling)
8. Statistical analysis	Conventional and big data-specific statistical tools and data visualization are available in a wide range of apps such as R, Python, and Knime (among many others)
9. Geospatial analysis and Geographic Information Systems (GIS)	Maps are generated that locate physical objects (houses, roads, rivers) or activities (crime or accident sites) by their geographic coordinates on a map. Layers are created where non-geographic information such as poverty hot-spots, disease incidence, nutritional levels, or crop yield are defined for each physical object. The layers can be combined to permit many different kinds of analysis

(*Continued*)

Table 3.1 Continued

Analytical tool/approach	*Explanation*
10. Analysis of phone records	This is a variant of GIS analysis where phone records are used to track mobility and the attributes of different groups of identifiable phone users (e.g., refugees)
11. Decision-making algorithms	Algorithms are widely used to guide decisions on selection of university applications, mortgage and loan approval, identifying fraud, and planning police patrols. The level of human input into the decisions can vary from little or no human input to significant input

Big data is often defined in terms of the "3 Vs" originally – Velocity, Volume, and Variety, originally proposed by Doug Laney in 2001. However, many other characteristics have subsequently been proposed (Box 3.2). While these characteristics are *descriptive,* Ashfaque (2020) has proposed 10 criteria that can be used both to describe and to **evaluate** the quality of big data.[2]

Box 3.2: Some of the multiple characteristics proposed for defining big data.

Laney's 3 Vs:

Velocity: Generated very fast – often in real-time
Volume: The volume of data is too large to analyze on a single computer
Variety: Can combine numerical, audio, and visual data

Other characteristics:

Veracity: Big data sources vary in terms of their quality, accuracy, and validity
Networked: Connected through computers and systems
Always on: (constant stream of data)
Non-reactive: The data was usually collected for a different purpose, so that accessing the data does not change it

Complexity: Big data is often representative of complex relationships between different data sets. It is also used to model dynamic relationships within systems

Source: Salganik (2016), York and Bamberger (2020)

Big data can be classified into three groups according to whether the unit of analysis is:

- **Individuals and social groups** (social media, purchases/card swipes, blogs, podcasts, Google searches, personal Internet of Things (IoT) like smart watches, census data, household surveys, crowdsourcing).
- **Organizations and systems** (human resource data, program administrative data, electronic health records, customer relations management information systems).
- **Geographic area** (satellites, community video monitors, phone call data records [telecom]).

These three categories vary in terms of the unit of analysis, how the data is used, who inputs the data, who analyzes the data, how the findings are used, and whether the data source is aware that data is being collected on them and how it will be used (see York & Bamberger, 2020, Section 2.2).

Data Analytics

Data analytics is the process of examining data sets (in text, audio, and video format) and drawing conclusions and inferences using a wide range of software. Table 3.1 describes the authors' opinions on some of the most common analytical tools, most of which are discussed in this publication.

Having introduced some of the basic big data concepts, the next section discusses how these tools and techniques can strengthen evaluation practice.

How Big Data and Data Science Can Strengthen Evaluation

The Benefits and Limitations of Data Science for Evaluation Practice

Lazer et al. (2021) argue that data science is completely transforming the social sciences by "making the unmeasurable-measurable," so that a vastly expanded range of quantitative and qualitative data can now be measured and analyzed. The significant reduction in the cost and time required for the collection and analysis of data makes it possible to work with much larger samples, to conduct more disaggregated analyses, and to assess program impacts on different minorities and vulnerable groups. Another benefit is that geospatial analysis makes it possible to generate longitudinal data over longer periods of time before a project begins and after it has ended. This is particularly valuable for assessing project sustainability. Big data also makes it possible to evaluate programs that operate in complex contexts and to capture data on processes and behavioral change in program environments with multiple actors (Bamberger & Mabry, 2020, Chapter 16).

Table 2.1 in the previous chapter (Bruce, Gandhi & Vandelanotte, 2025) illustrates the multiple ways that big data techniques are already being used in international development research and evaluation. For example, satellites and drones are widely used to track the movements of refugees and to project the growth of the population of refugee camps and the corresponding demand for food, medicine, and construction materials.

It is, however, important to recognize sources of bias and other limitations of data science for program evaluation (see following section). Lazer et al. (2021) argue that despite its many benefits, the findings and recommendations from data science should always be suspect and carefully reviewed before being used (see Table 3.2). Most big data were collected for a different purpose and may

Table 3.2 Some limitations of big data for management and evaluation

a.	Digital data must be critically assessed before use in program evaluation because it was collected for a different purpose, and the available data is often not well suited for the purpose for which it is used. Issues of construct validity
b.	Inaccessible to many potential users due to cost or administrative (and sometimes political) control on access
c.	Selectivity bias – only covers users of a particular app
d.	Data collection and analysis often reflect cultural biases
e.	Drifting – users change over time
f.	Proprietary algorithms – users often do not know how algorithms were constructed
g.	Relies on indirect measures – data is usually collected for a different purpose and may not be a good measure of the variables of interest (issues of construct validity)
h.	Concerns about data quality and the limited use of triangulation and ground-truthing
i.	Issues around remote data collection (decontextualized, potential issues of social exclusion, unable to interpret the context in which data generated)

Source: Adapted from Salganik (2019), York and Bamberger (2020), and Lazer et al. (2021).

not be appropriate for the purposes of a particular evaluation, and it is often difficult to know exactly how the data was collected and processed. Also, the data, particularly for social media analysis, often comes from a biased sample (e.g., only people who use a particular app or who have access to a smartphone), and many administrative data sets may exclude certain vulnerable or difficult-to-reach groups such as the homeless, refugees, the undocumented, or certain ethnic groups. There are also many issues relating to *reflexivity*, the fact that people's behavior often changes when they know they are being observed.[3]

Table 3.2 summarizes some of the limitations of big data. The limitations, as well as the benefits (discussed in the following sections), vary according to the unit of analysis (household, community, district, organization, individual, etc.) as well as the particular tool being used (social media analysis, satellite images, phone call data records, etc.).

It is also important to understand the complex and often not very transparent algorithms used by different platforms. Many platforms restrict access to their information, sometimes depending on their subscription plans, limiting access to sensitive information, and because of how they filter information to different users. These rules must be understood by researchers trying to understand how access to information affects behavior (Lazer et al., 2021).

Finally, evaluators and other researchers must challenge the myth that big data is more objective and unbiased than conventional survey data because it avoids human bias. D'Ignacio and Klein's (2020) *Data Feminism* provides a detailed analysis of how these biases result in different treatment of women and men in much social research, and the under-representation of gender-related issues in many government and private sector reports and socio-economic research.

Sources of bias in big data and data analytics, and ways to address them, are discussed in the following section.

Applications of Artificial Intelligence (AI) and Machine Learning in Evaluation

Data science also offers powerful tools for the integration and analysis of huge, multicomponent data sets. These techniques include the following:

- Machine learning can automate many routine data analysis tasks related to large-scale surveys – freeing researchers from these time-consuming tasks so that they have more time and resources to devote to the true tasks of evaluation.
- Creating integrated data platforms that merge data from different sources and agencies. These permit an understanding of the multidimensional nature of most socio-economic problems, which was not possible with the separate analysis of each data set.

- Scanning large data sets comprising multiple variables to identify naturally occurring associations between variables used to construct *natural experiments*.
- Predictive analytics: testing recommendations from natural experiments.
- Text analytics: originally used to find patterns and trends in large volumes of PDF files, but now the applications have greatly expanded through *natural language processing* (NLP).
- Simulations and *digital twins*: creating digital images of programs, cities, and systems to model alternative scenarios.

The applications of AI are discussed more fully in the previous and next sections of this chapter and in Chapter 11 (York, 2025). AI is also widely used in humanitarian programs, for example, to predict movements of refugees and to project demand for food, medicine, and construction materials for displaced populations. Detecting fake news and hate speech is a third example.

The increasing use of program administrative data combined with machine learning algorithms now makes it possible to more precisely tailor program designs to the unique needs of different population subgroups. Data science can collect and analyze data much more rapidly and cost-effectively, disaggregating populations with matching backgrounds and contexts to provide real-time insights about what works for whom and under what conditions.

The ability to work with large data platforms that integrate a wide range of input, output, and outcome indicators enables evaluators to compare multiple short-term program outcomes. Examples include worker or program staff reports on outcomes for individuals or groups (such as troubled youth, patients with different kinds of behavior problems, small businesses receiving microloans) who have received different combinations of program treatments or services and have different combinations of attributes (sex, age, type of business, family history). These types of data can be used to predict outcomes for different subgroups when interventions are varied. Predictive modeling provides managers with a whole new range of implementation and diagnostic tools to understand factors determining program outcomes and to adapt the range or intensity of the interventions.

In the first two sections, and referring to the previous chapter, we have discussed the increasing number of big data information sources and analytical tools that can broaden the range of topics that evaluation can address, and the powerful new analytical methods offered by AI and the rapidly evolving generative AI tools (e.g., ChatGPT). In the next section, we will show how these can be used to strengthen both the evaluation of international development programs and evaluation practice more generally (Figure 3.1).

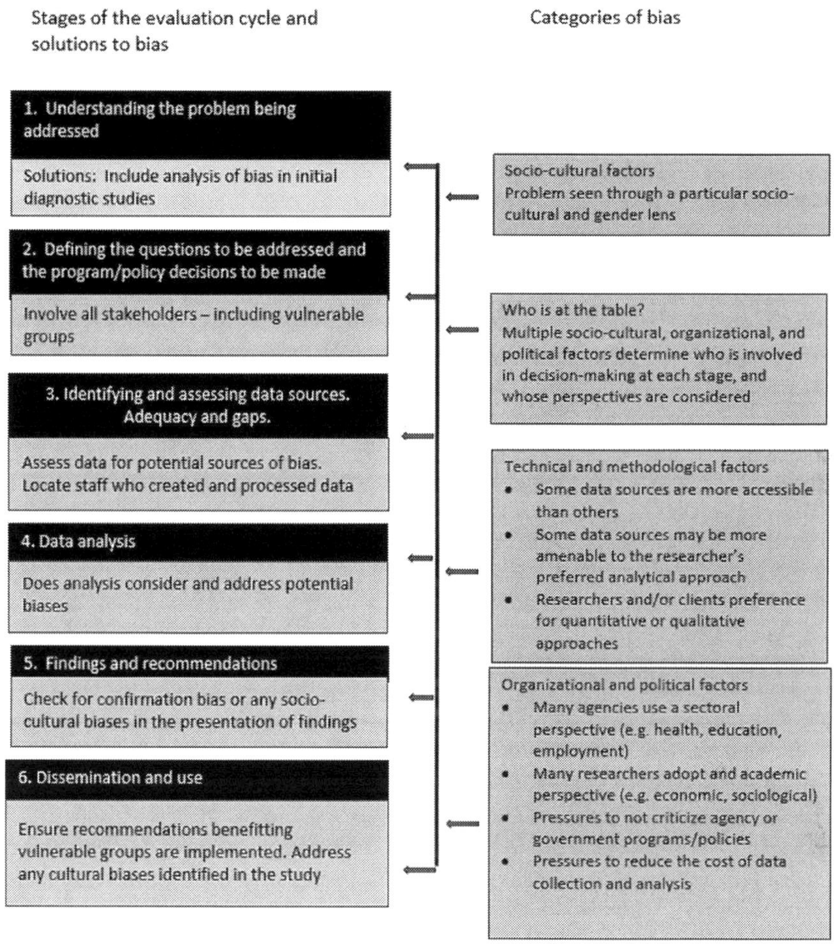

Figure 3.1 Stages of the evaluation cycle and solutions to bias.

Incorporating Values (Equity, Racial Equality, Gender, etc).

We argued previously that all researchers and evaluators, including those using big data, frame their research through a particular set of socio-cultural, professional, and organizational lenses. This is inevitable, and values help define the purpose and focus of research.

However, it is important that values and perspectives are made explicit and that they are recognized and understood by researchers and clients. This is not always done, either because researchers do not recognize the assumptions and values underpinning their work, or because they may not wish to inform the client of some of the values they bring to the table. For example, a researcher who has a strong personal commitment to issues such as social or economic justice, gender equality, or the rights of particular ethnic groups may not wish to make these explicit for fear of prejudicing their chances of winning the contract.[4]

The Power of Big Data-Driven Evaluations

Evaluation of Large, Multi-component Programs

While development agencies continue to support many stand-alone projects, there is a move in both developing and industrial nations toward large, integrated, multicomponent policies and programs (Sarker, 2021). Big data and data science are rapidly developing the tools and techniques to generate and analyze the huge volumes of data required to model and analyze these multisectoral programs. In Chapter 11, Pete York (2025) presents case studies illustrating applications of these new analytical techniques in support of government agencies, corporations, and non-profit organizations in the United States. All these techniques are starting to become applicable in developing countries as their capacity to generate and manage large data sets increases. The techniques described in Chapter 11 are as follows:

- ***Precision Analytics*** (PA) is a causal analytic method that combines subject matter experts, existing big data sets, and machine learning algorithms to build highly accurate, valid, and reliable assessment, evaluation, and decision-making tools. The PA approach to evaluation trains machine learning algorithms to build predictive, prescriptive, and evaluative models that determine what causes the desired outcome for each target population segment, such as individuals, groups, organizations, or communities. This is achieved by conducting quasi-experimental observational studies using historical big and/or program administration data. The subject matter experts train machine learning algorithms to find naturally occurring experiments in history to determine what interventions have been tried by and for similar groups in the past and which efforts produced the most significant positive results over time.
- ***Equitable Impact Platform (EquIP)*** is a geospatial big data platform that assesses and evaluates the nonprofit sector's contribution to equitable community improvement. EquIP combines data from IRS 990 tax forms and the Census Bureau's American Community Survey (ACS) with BCT's Precision Analytics modeling approach. This platform helps funders and donors identify communities in greatest need, prioritize marginalized communities, find the most accessible nonprofits that can serve these communities best, and receive assessment, predictive, and prescriptive insights about the types of

financial and capacity-building support these organizations need to make a difference.

Transforming Evaluation from Ex-post Accountability to a Dynamic Management Tool

In the experience of the present authors, monitoring and evaluation have been considered by many (but certainly not all) funding and implementing agencies as tools for accountability to ensure programs are complying with the objectives defined in their results-based management (RBM) framework. RBMs are used by most regional development banks, and many bilateral and UN agencies. Computer-based M&E systems made it possible to efficiently collect large amounts of information and to present this in the form of progress reports. However, for many agencies, data was reported separately for each project or office, and they did not have the capacity to integrate different data sets or to use the data to improve program performance.

However, new analytical tools are becoming available so that data sets from different projects, departments, or agencies can be merged into an integrated data platform. These integrated data sets are transforming evaluation into dynamic management tools that can use AI and machine learning to apply analytical techniques such as the construction of natural experiments, to create dynamic management tools that learn from ongoing program activities to provide rapid feedback, suggesting ways to improve performance.

Chapter 11 (York, 2025) presents a case study illustrating how precision analytics was combined with causal modeling to improve the performance of a multi-program social service agency that provides behavioral health, education, and prevention services to children and families experiencing emotional and behavioral difficulties (York, 2021).

From Natural Language Processing (NLP) to Natural Language Understanding (NLU) (PY)

As program evaluation continues to evolve, there is a growing need for more efficient and effective methods of analyzing data. One approach that has emerged as promising for program evaluation is natural language understanding (NLU), which utilizes advanced algorithms to automate the process of analyzing large amounts of unstructured qualitative text data, such as program reports, documents, interview or focus group notes, open-ended surveys, and case notes.

In this chapter, we will explore the potential benefits and challenges of using NLU for program evaluation, as well as its potential to transform the field. Recent advancements in natural language processing (NLP) and its subfield, NLU, have enabled rapid progress in the automation and improved efficiency, reliability, and validity of the qualitative data analysis process. Large language models, which are computer programs that utilize deep learning algorithms to process vast amounts of textual data to understand, generate, and manipulate human language, are at the forefront of these advancements. In particular, transformer models, like BERT (Bidirectional Encoder Representations from Transformers), GPT (Generative Pre-Trained Transformer), and RoBERTa (Robustly Optimized BERT Pre-Training Approach), are a type of large language model used in natural language processing (NLP) that have recently emerged as a major advancement in NLU. These models use self-attention mechanisms to capture the relationships between words in a sentence or document, which allows them to understand the meaning and context of words in a more sophisticated way.

The emergence of transformer models in the field of natural language understanding (NLU) has greatly advanced the capabilities of natural language processing (NLP), allowing for rapid progress in the automation and improved efficiency, reliability, and validity of qualitative data analysis.

For program evaluations, mixed methods, combining both qualitative and quantitative data collection, analysis, synthesis, and triangulation, represent an ideal approach. By leveraging the strengths of both methods and addressing their limitations, a more comprehensive understanding of program implementation and outcomes can be achieved.

With the assistance of NLP and large language models, program evaluators can analyze large volumes of qualitative data more accurately and efficiently, in a more streamlined, reliable, and valid process of qualitative data analysis in program evaluation, which can more effectively combine with quantitative data analysis in service to the advantages that will be realized with a mixed methods approach.

To demonstrate the potential of NLU and large language models in program evaluation, we will provide an example from a big data science for evaluation project conducted by one of the authors. A major metropolitan public transportation system in the United States sought to analyze a customer database containing over 100,000 comments submitted through emails, phone calls, and social media posts. To develop evidence-informed solutions and engage in rigorous research and evaluation, the transportation system wanted to use NLU techniques to conduct a more in-depth analysis of the data. Specifically, the system sought to utilize transformer models to understand both explicit and latent themes in the data, related to their theory of change, with the goal of guiding improved operations, informing service planning, supporting safety, identifying meaningful outcomes related to the system's services and products, and identifying emerging topics or themes that supported their ongoing evaluative learning. Through this approach, they sought to automate the extraction of valuable insights efficiently and effectively from customer comments, to provide more accurate decision-support insights to key system leaders and managers, with the ultimate goal of improving the overall customer experience.

The research team leveraged recent advancements in transformer models and machine learning algorithms to develop an automated process for conducting accurate and reliable qualitative analysis of text data, which we called "Automated Topic Modeling" (ATM). This approach was developed and tested through a number of projects that analyzed a large corpus of proposal and report document data for the National Science Foundation, to evaluate their investments in science and the broadening participation of underrepresented minorities in science.

The ATM process is a human-in-the-loop process, which involves collaboration between data science researchers and subject matter experts. The humans train large language models by developing and refining a coding schema that is applied to every block of text or paragraph for every document. The human-in-the-loop process is critical for fine-tuning, validating, and ensuring that biased results are addressed. ATM is conducted always in close consultation with project stakeholders, including organizational and programmatic leaders, managers and staff, as well as internal researchers and evaluators.

More specifically, the NLU for evaluation approach involves several key steps to prepare and analyze text data for program evaluation. Firstly, the data are cleaned and pre-processed, which includes structuring the text data and integrating comment context data. Next, NLP and NLU algorithms are trained to tag each comment with the logic model components to which it is most associated. Additionally, experts, in collaboration with the large language algorithms, iteratively develop a coding schema for each construct within the broader logic model components (e.g., inputs, strategies, outcomes). This coding schema is then applied to tag and score the similarity of every comment against each selected keyword and phrase in the coding schema. Finally, descriptive, evaluative, and prescriptive analyses are conducted on the structured data to produce answers to research questions. These analyses can include the use of causal precision modeling, a quasi-experimental method that trains Machine learning algorithms to find and evaluate natural experiments. Throughout the process, a human-in-the-loop approach is maintained, involving project stakeholders in multiple iterations to ensure the accuracy and validity of the results. The goal of this approach is to automate the production of accurate and reliable insights for evidence-based decision-making.

There are multiple iterations involved in the process of training the algorithms, and each of these steps, usually three or more iterations, involves the engagement of key project stakeholders to solicit their review, validation, input, and provide refinements to the coding schema to ensure that the results from the training are accurate and valid. Predictive machine learning algorithms, like random forests, are also used in the fine-tuning process, whereby human experts have labeled when NLU algorithms were correct or incorrect with respect to coding qualitative evidence as representing logic model constructs. All of this human-in-the-loop training process ensures that the modeling results reflect the contextual realities of those involved in the project's implementation, making the process more reliable and valid.

Once the coding schema was trained and automated, the algorithms were applied to the full corpus of transit system comments, thereby providing the transit system with the capability to automate the production of structured descriptive and evaluative results in topic thematic counts, probabilities, sentiment scores, and other metrics. Using Microsoft's PowerBI data visualization software, a dynamic, interactive evidence-review application and a set of evaluation dashboards were developed as a preliminary findings deliverable, as well as a proof-of-concept prototype reporting and decision-support tool.

In conclusion, the NLU for evaluation approach, utilizing recent advancements in NLP and transformer models, has the potential to transform program evaluation by providing a more efficient and effective method of analyzing large amounts of unstructured qualitative text data. The ATM process, which involves a human-in-the-loop approach, allows for accurate and reliable qualitative data analysis, providing stakeholders with structured descriptive and evaluative results. The approach has been successfully demonstrated in a big data science for evaluation project, where it automated the extraction of valuable insights from a customer database to improve a transportation system's customer experience.

With the ability to build evidence-based causal evaluation models, decision support tools, and recommender engines, NLU for evaluation holds immense promise for advancing the field of program evaluation.

Evaluating Complex Programs and Policies[5]

There is widespread recognition in the development community that most development programs are *complex*, and that consequently the evaluation of development programs and policies is also complex. There are at least four main dimensions of complexity:

- the interactions among the multiple stakeholders involved in a program,
- the influence of multiple external factors (economic, political, social, demographic, environmental, etc.),
- the nature of the interventions themselves, and
- the non-linear processes of causality and change.

Development interventions vary in the level of complexity on each of these dimensions. Bamberger and Mabry (2020 Chapter 16) developed a checklist with a set of indicators to rate the level of complexity of each of these dimensions on a set of indicators (rated from 1 = very low complexity to 5 = very high complexity).[6]

The checklist was used in the evaluation of an OXFAM program to promote the access of women to justice in Lebanon (Lombardini, Garwood & Hassnain

(2018)[7]. Lewin et al. (2017) used a somewhat related approach to assess the complexity of medical interventions in systematic reviews.[8]

Despite the widespread recognition that many programs are complex, most development evaluations continue to use linear designs (such as most experimental and quasi-experimental designs) that are unable to capture complexity. Part of the reason for this is the conservatism of many evaluators who continue to use familiar linear evaluation designs, but a major factor is that many complexity-responsive evaluation designs require the ability to collect and analyze much larger amounts and more diverse kinds of data.

Large volumes of data are required to assess the influence of multiple external factors and to conduct longitudinal analysis to track trends. Large programs may also have as many as 100 different stakeholders (multiple government agencies, each with a number of different departments and levels involved, multiple donor agencies and implementing partners in addition to civil society organizations, community organizations, research institutions, consultants,and academia, as well private sector agencies). Many of the evaluation techniques, such as social network analysis, systems dynamics, and geospatial analysis, also require access to powerful analytical tools, and these are now becoming available.

The data science tools discussed in the first section of this chapter now make it affordable and much easier to collect and analyze the large volumes of data required for many kinds of complexity analysis, so that we can expect to see a steady increase in the use of complexity-responsive evaluations.[9]

There will also be an increase in the use of systems analysis tools to model and analyze the complex processes of interaction that drive many large development programs.

Discussion

We agree with Lazer (2021), and with Nielsen, Mazzeo Rinaldi, and Petersson (Chapter 1; 2025), that data science is transforming the social sciences. However, evaluators have been slower to adopt this new technology, and although, to the best of our knowledge, no recent statistics are available, it is our impression that most program evaluations are not using data science, and of those that do, many are still only using it as a source of data, with only a few using data analytics (see Nielsen, 2023, 2025). However, as both Chapter 1 and the present chapter show, there is a wide range of data science tools and techniques available to strengthen all stages of the evaluation process – and their use is increasing.

One of the developments which is providing an impetus to the adoption of data science by evaluators is the increasing ease with which administrative data can be harvested and transformed for use by evaluators. While less dramatic than the recent developments in generative AI, the creation of integrated databases and user-friendly text analytics software vastly increases the amount and kinds of data available for more sophisticated data analytics.

One of the major challenges for many evaluation studies is developing ways to model causality, and consequently we focus on AI-based causal modeling, complemented by data harvesting to illustrate how data science is advancing the frontiers of evaluation. Chapter 11 (York, 2025) also provides an example illustrating how causal modeling is used to address the key questions of concern to many evaluation clients.

Another important area, which currently receives relatively little attention from evaluators, concerns the evaluation of complex programs and policies. While evaluators and clients agree that most evaluations include dimensions of complexity, most evaluations continue to be based on conventional, linear models.

One of the reasons for this is that many approaches to complexity require the collection and processing of large volumes of data, often covering broader systems, and the longer time horizons within which programs are implemented and over which their impacts and sustainability must be assessed.

Most conventional evaluation data collection and analysis methods have difficulty addressing these analytical challenges, particularly as many techniques such as systems dynamics and social network analysis require the collection and analysis of continuous streams of real-time data. One of the exciting potentials of data science is the ability to collect and analyze these kinds of data (Bamberger & Zazueta, 2024).

Natural language understanding (NLU), using large language transformer models, now offers the evaluation community a cost-effective and efficient opportunity to structure and integrate qualitative data into mixed methods evaluation. NLU can be applied to large datasets of narrative data to analyze, structure, and understand large quantities of text, including an entire corpus of text that would have been too large for human researchers to analyze manually and comprehensively. These NLU algorithms and methods not only augment the quantitative data available but also expand the breadth of causal modeling opportunities for the evaluation of complex programs by adding qualitative measures of community contexts, program experiences, and outcomes. While the opportunity for the use of NLU to advance mixed methods evaluations is expanding, evaluators need to be mindful to address inherent biases in data and the resulting transformation and use of NLU for qualitative analysis.

However, it is also important to recognize and address many of the limitations of big data. These issues include data quality and data bias, algorithmic bias, and the dark side of big data, which includes cyber crime, hate speech, and the increasingly sophisticated social research on ways to manipulate attitudes and behavior.

Conclusions

Big data and data analytics are playing an increasingly important role in social and economic research evaluation. To date, the evaluation profession has been slower to adopt these new research technologies (see York & Bamberger, 2020), and there is currently a need to bridge the gap between data scientists and evaluators, including in training programs for the next generation of evaluators. There is also a need for the agencies funding and commissioning program and policy evaluations to adapt the requests for proposals (RFPs) to encourage, or at least permit, the use of big data for data collection and analysis in the evaluations they commission.

To accelerate this integration, strategic alliances must be formed across academia, industry, and government to share knowledge, develop competencies, and create collaborative opportunities (see also Chapter 13, Nielsen, 2025). By infusing traditional evaluation practices with innovative data science methodologies, we can greatly enhance the evaluative processes and outcomes.

As we progress, it is essential to cultivate an evaluative culture that not only harnesses the descriptive and predictive power of big data but also remains vigilant about maintaining the highest standards of ethical practice. This means being proactive in the identification, understanding, and resolution of any potential biases inherent in big data and the algorithms used to analyze it.

Moreover, the evolution of big data should not be seen as a replacement for traditional methods but as a complementary force that enriches the evaluation toolkit. The combination of traditional evaluation expertise and big data analytics promises a more nuanced understanding of program dynamics, enabling evaluators to provide more strategic and evidence-based recommendations.

In conclusion, while we stand at the cusp of a new horizon in evaluation science, it is our collective responsibility to ensure that the transition to big data-informed evaluation is both seamless and principled. Embracing the advancements in data analytics while adhering to our professional and ethical standards will allow us to illuminate the pathways to social progress and policy effectiveness with greater clarity and confidence.

Notes

1 A/B testing is widely used in marketing research to compare an outcome (e.g. number of online clicks, in-store purchases) for a group that received a treatment (placement of the product, changing the font style of an ad) and a group that did not.

2 The characteristics proposed by Ashfaque (2020) include: Volume, Velocity, Variety, Veracity, Validity, Volatility, Variability, Viability, and Visualization.

3 Examples of reflexivity include: people often communicate differently online than they do in person, certain groups may try to hide their identity (e.g., changing their photo on social media, using VPN networks so their communication cannot be tracked).

4 For a fuller discussion of values in evaluation, see Tashakkori, Johnson, and Teddlie (2021), Fundamentals of mixed-methods research, Chapters 1–3.

5 This section is based on Bamberger and Zazueta's "Evaluating complex development programs: Integrating complexity thinking and systems analysis" 2025 in Newcomer and Mumford (editors) *Research Handbook on Program Evaluation. 2025*

6 This 2-part blog published by 3ie provides an overview of complexity and introduces the complexity checklist and how to use it. Blog Part 1: https://www.3ieimpact.org/blogs/understanding-real-world-complexities-greater-uptake-evaluation-findings. Blog Part 2: https://3ieimpact.org/blogs/building-complexity-development-evaluations.
7 Women's empowerment in Lebanon www.oxfam.org.uk/effectiveness.
8 https://bmcmedresmethodol.biomedcentral.com/articles/10.1186/s12874-017-0349-x.
9 An example of the increasing affordability of data was the use of satellites and remote sensors to collect the data required for a rigorous quasi-experimental design to evaluate the effectiveness of a program to protect forest cover in Mexico. For previous evaluations, it had only been possible to collect data on a few local indicators. Satellite images now make it possible to collect information on a large number of indicators, covering longer periods of time and much larger areas, at a much lower cost (Global Environment Facility 2015).

References

Ashfaque, J. (2020). The 10 Vs of Big Data. Retrieved from Research Gate Feb 7 2020 https://www.researchgate.net/publications/339107749

Bamberger, M., & Mabry, L. (2020). *Real World evaluation: Working under budget, time, data, and political constraints*. Thousand Oaks, CA: Sage Publications.

Bamberger, M., & York, P. (2020). *Transforming evaluation in the 4th industrial revolution: Exciting opportunities and new challenges.* Evaluation Matters, African Development Bank 2nd Quarter https://idev.ifdb.org/sites/default/files/documents/files/EM%20Q2-2020-article1-

Bamberger, M., & Zazueta, A.E. (2024). Evaluating complex development programs: integrating complexity thinking and systems analysis. In K.E. Newcomer and S.W. Mumford (eds.). *Research Handbook on Program Evaluation* (pp. 348–370). Edward Elgar Publishing.

Bruce, K., Gandhi, V., & Vandelanotte, J. (2025). Emerging Technology and Evaluation in International Development. In S.B. Nielsen, F. Mazzeo Rinaldi and G.J. Petersson (eds.). *Artificial intelligence and evaluation. Emerging technologies and their implications for evaluation* (pp. 13–36). London: Routledge. https://doi.org/10.4324/9781003512493

D'ignazio, C., & Klein, L. F. (2020). *Data feminism*. Boston, MA: MIT Press.

Global Environment Facility [GEF] (2015). *Impact evaluation of GEF support to protected areas and protected area systems*. 49th GEF Council Meeting. October 20–22, 2015, Washington, DC.

Lazer, D., Hargittai, E., Freelon, D., Gonzalez-Bailon, S., Munger, K., Ognyanova, K., & Radford, J. (2021). Meaningful measures of human society in the twenty-first century. *Nature*, 595(7866), 189–196. https://doi.org/10.5167/uzh-207351

Leeuw, F. (2016). Understanding what is being evaluated: Theory-based evaluation. In M. Bamberger, M. Vaessen, and E. Raimondo (eds.). *Dealing with complexity in development evaluation.* Thousand Oaks, CA: Sage Publications.

Lewin, S., Hendry, M., Chandler, J., Oxman, A.D., Michie, S., Shepperd, S., Reeves, B.C., Tugwell, P., Hannes, K., Rehfuess, E.A., Welch, V., Mckenzie, J.E., Burford, B., Petkovic, J., Anderson, L.M., Harris, J., & Noyes, J. (2017). Assessing the complexity of interventions within systematic reviews: Development, content and use of a new tool (iCAT SR). *BMC Medical Research Methodology,* 17, Article number: 76.

Lombardini, S., Garwood, R. & Hassnain, H. (2019). *Women's Empowerment in Lebanon. Impact evaluation of the project 'Women's Access to Justice' in Lebanon.* Effectiveness Review Series 2017/18. Great Britain: OXFAM. DOI: 10.21201/2019.4764

Nielsen, S.B. (2023). Disrupting evaluation? Emerging technologies and their implications for the evaluation industry. *New Directions for Evaluation,* 178–179, 47–57. https://doi.org/10.1002/ev.20558

Nielsen, S.B. (2025). The Evaluation Industry and Emerging Technologies. In S.B. Nielsen, F. Mazzeo Rinaldi and G.J. Petersson (eds.). *Artificial intelligence and evaluation. Emerging technologies and their implications for evaluation* (pp. 266–286). London: Routledge.

Nielsen, S.B., Mazzeo Rinaldi, F., & Petersson, G.J. (2025). Evaluation in the Era of Artificial Intelligence. In S.B. Nielsen, F. Mazzeo Rinaldi and G.J. Petersson (eds.). *Artificial intelligence and evaluation. Emerging technologies and their implications for evaluation* (pp. 1–12). London: Routledge. https://doi.org/10.4324/9781003512493

Salganik, M.J. (2019). *Bit by bit: Social research in the digital age.* Princeton, NJ: Princeton University Press.

Sarker, I.H. (2021). Data science and analytics: An overview from data-driven smart computing, decision-making and applications perspective. *SN Computer Science,* 2(377). https://doi.org/10.1007/s42979-021-00765-8

York, P. (2021). Unlocking real-time evidence for practitioners: How evaluation and data analytics are generating on-demand, actionable evidence for front-line practitioners at First-Place for Youth and Gemma Services. Retrieved from Project Evident: https://projectevident.org/wp-content/uploads/2022/09/FPFY-GemmaActionableEvidenceCaseStudyJul21.pdf

York, P. (2025). The Future of Evaluation Analytics: Case Studies of Structural Causal Modeling in Action. In S.B. Nielsen, F. Mazzeo Rinaldi and G.J. Petersson (eds.). *Artificial intelligence and evaluation. Emerging technologies and their implications for evaluation* (pp. 219–241). London: Routledge. https://doi.org/10.4324/9781003512493

York, P., & Bamberger, M. (2021). *Measuring results and impact in the age of big data: The nexus of evaluation, analytics, and digital technology.* New York, NY: The Rockefeller Foundation.

4 Ethics and Equity in Data Science for Evaluators

Nathan Greenstein and Sung-Woo Cho

Introduction

Data science unlocks powerful tools for practitioners across disciplines. Just as these tools can help good-faith actors do valuable work that would be impossible or impractical without them, they can help inattentive or bad-faith actors do new forms of harm. Such harm can range from invasion of privacy to overt discrimination, and its effects are often intertwined with downstream outcomes of interest to evaluators. In one alarming example, a Black man in the state of Michigan was wrongfully arrested due to faulty facial recognition and held for thirty hours before release. As we will explore, intervention from evaluators might have averted this troubling outcome.

Several factors make ethics and equity important in the context of data science. First, many data science tools operate with broader reach and narrower human oversight than traditional alternatives. Increasing reach can heighten the consequences of problematic practices, and reducing oversight can limit experts' ability to notice and correct problems. Second, some practitioners assume that data science tools will attend to ethics and equity automatically, or that employing machine intelligence renders ethics and equity obsolete. As we explore, neither assumption is reliably true; ample literature documents that data science can amplify human failings and introduce problems of its own (Mehrabi et al., 2021). Finally, data science tools complicate questions of accountability: when data and algorithms contribute to a breach of ethics or equity, they cannot make reparations or participate in the justice system as individuals and organizations can. Together, these factors create an urgent need to invest in the evaluation and monitoring of data science tools, with a focus on ethics and equity.

This need has not escaped notice. Since 2018, the European Commission has published guidelines for the ethical use of artificial intelligence (High-Level Expert Group on Artificial Intelligence, 2018; Proposal, 2021). UNESCO and the OECD have also issued recommendations of their own (UNESCO, 2021; OECD Legal Instruments, 2019). Further, when detected, ethics and equity violations can provoke media attention. Although data science has matured more quickly than governing bodies have responded, trends suggest a societal shift

DOI: 10.4324/9781003512493-4

toward stricter regulation – and keener public consciousness – of ethics and equity.

This leaves conscientious evaluators a twofold responsibility: (a) to wield data science tools responsibly **in their own work**, and (b) to study the ethics and equity of any data science tools involved in their **objects of evaluation**. This chapter aims to help evaluators rise to this challenge.

This chapter's three-part structure approaches ethics and equity in a manner that aligns with evaluators' skills and responsibilities. The first section focuses on theory. It explores ways to think about ethics and equity, and ways that violations can occur. Subsequent sections apply this foundation to the practice of evaluation. The second section discusses how evaluators might adopt data science tools responsibly in their own work, and the third section considers how evaluators might study and monitor existing applications of data science through an ethics and equity lens. The three sections are best approached in order.

Defining and Violating Ethics and Equity in Data Science

This section inventories several ways to conceive of ethics and equity in the context of data science. The section contains three parts, each representing one facet of data science. The first focuses on *big data*, the raw material of many data science tools. The subsequent parts each address a broad use case of such tools: the second covers *interpreting and generating information*, and the third covers *making real-world decisions*. We recommend approaching the parts in order.

No framework for ethics and equity is universal. This section cannot be exhaustive, and it is not the sole "correct" framework. Here, we hope to help evaluators navigate this uncertainty, and we encourage independent reflection with other perspectives, including the literature cited throughout the chapter.

Big Data

As detailed in this volume by Bruce et al. (2025), "big data" is produced in great volume, at high velocity, and in a variety of structures. Examples include social media posts, credit card transactions, and satellite images. Several ways that evaluators harness big data are reviewed in the preceding chapters of this volume, then explored further in the case studies.

In the context of ethics and equity, two traits of big data stand out. First, big data is often collected in relatively passive ways: drawing on satellite imagery, for example, is less invasive than conducting field observations. Second, big data often represents more people than traditionally sampled sources. Many more people post on social media, for example, than will respond to a survey. These differences can lead practitioners to disregard ethics and equity inappropriately. By one line of thinking, if data is collected non-invasively, there is no opportunity for ethics violations to take place. By a second line of thinking, big data's scale renders it inherently democratic and equitable. Unfortunately, neither notion is necessarily true, leaving evaluators with an important role to play. The following paragraphs highlight four key ideas: *informed consent*, *representation*, *accuracy*, and *scope*.

Informed consent: It is not controversial that invading others' privacy without consent can be unethical, and that doing so with disproportionate impact on vulnerable groups can be both unethical and inequitable. This extends to digital privacy, an internationally recognized right (Nyst & Falchetta, 2017). However, these concepts can be improperly dismissed in the context of big data. For instance, practitioners may believe that if information is publicly available or easy to access, it has no bearing on privacy. Arguably, though, if data is truly of value to the person seeking to harvest it, then it is of value to the people it describes (Ioannidis, 2013, p. 40). It may also be tempting to believe that, if data is anonymized and collected at sufficient scale, no individual will stand out enough for their privacy to be compromised. Unfortunately, anonymized data can sometimes be reidentified (Emam et al., 2011), and violating a person's privacy is not excused by also violating their neighbor's.[1]

Social media offers several examples. In 2008, Harvard researchers released data from students' Facebook profiles (Zimmer, 2010). Despite efforts to anonymize the data, it was swiftly reidentified, sparking backlash (Parry, 2011). Students were neither notified of data collection nor asked for consent. Subsequently, Facebook studied 700,000 of its users, claiming that they consented when creating their accounts (Kramer et al., 2014).

When considering big data, evaluators should assess whether the collection process meets an adequate standard of informed consent. This should be explored generally and comparatively: if a vulnerable group is given less opportunity to consent than its peers, or if loss of privacy could harm its members disproportionately, then a breach of ethics may also be a breach of equity. Some informed consent requirements are imposed by laws like the General Data Protection Regulation (European Commission, n.d.). Such regulatory standards may or may not be stringent enough for a given situation.

Representation: Often, big data most heavily represents those who produce the most of it. Credit card data, for example, likely underrepresents groups who are more likely to use cash. Censorship, propaganda, and disinformation can also affect representation. For example, certain voices may be lost to the restriction of social media sites (Sundara Raman et al., 2020), erased by biased moderation (Haimson et al., 2021), or drowned out by propaganda and disinformation campaigns (Salaverría & León, 2022). In contrast, in traditional research, investigators often seek out harder-to-reach groups to achieve a representative sample. Consequently, when relying on big data, researchers risk systematically excluding people who are underrepresented in the data in question (Lerman, 2013).

For example, the city of Boston collected data on roads needing maintenance from a smartphone app available to citizens. The results likely overrepresented neighborhoods frequented by younger and higher-earning individuals more likely to own smartphones. Absent evaluation and monitoring, such a program risks deepening inequity by funneling infrastructure investment away from the vulnerable populations underrepresented in the data (Barocas & Selbst, 2016, p. 685). When considering big data, evaluators should explore who it describes and who may have been denied a voice.

Accuracy: Inaccurate data, when acted on, can harm those it (mis)represents. This risk is heightened in big data contexts where data is collected with limited human oversight, without consent, or without a mechanism to submit corrections (Zimmer, 2010, p. 322). An ethics issue becomes an equity issue when data represents vulnerable groups less accurately than their peers (Barocas & Selbst, 2016, p. 684). This can arise from human discrimination – for example, technicians exercising less care toward some groups – or passively, parallel to mechanisms discussed under "representation." Regardless, the resulting data can create or amplify inequity and compound the risk of unethical treatment due to inaccuracy.

For example, many facial recognition systems perform best for White, male faces (Buolamwini & Gebru, 2018). Therefore, when relying on facial recognition data, conclusions may be disproportionately error-prone for vulnerable groups. Without intervention, these errors can lead to harm (Buolamwini & Gebru, 2018, p. 3). Therefore, when considering the use of big data, evaluators should consider risks posed by inaccuracies, especially those disproportionately impacting vulnerable groups. Accuracy is difficult to regulate, but evaluators can develop standards and monitoring strategies appropriate to a given situation.

Scope: Guidelines for ethical research emphasize minimizing potential harm (NCPHSBBR, 1979), which can involve accessing "the minimum amount of information necessary to complete the study" (VPRI, n.d., sec. F.1). This is uncontroversial in conventional research, but it can be improperly ignored in big data, especially when drawn from smart devices (Li et al., 2016). When any risk is posed by collecting or using big data, including risk to privacy, evaluators should explore whether fewer participants, fewer data points per participant, or data aggregated at a coarser level might suffice.

Interpreting and Generating Information

Data science tools can make meaning out of data whose structure, scale, or scope renders it infeasible to interpret by hand. Tools may also leverage patterns in data to generate new material, as in text-to-speech models and large language models like ChatGPT. This section focuses on uses of such tools where data is reshaped into structured information that becomes an object of study or deliverable. Other use cases, where practitioners make decisions about how people, places, or things are treated, are covered separately below. In both cases, evaluators have an important role to play in ensuring that data science tools are used ethically and equitably.

It is tempting to view interpreting or generating information with data science as an objective operation without ethics or equity implications. Further, it may seem that because data science tools are less prone to prejudice and fatigue than humans, their output is more ethical and equitable. Unfortunately, this is not always the case. It falls to evaluators to assess and monitor such use cases. The following paragraphs highlight four conceptions of ethics and equity to consider: *problematic data*, *human bias*, *pre-trained tools*, and *misinformation.*

Problematic data easily produces problematic results, even when no new data is collected. For example, reinterpreting data collected without consent risks further harm, perhaps by deepening invasion of privacy or compounding downstream effects (Zimmer, 2010, p. 315). Similarly, drawing new conclusions from data previously collected with deficient representation or accuracy might amplify the risk of harm introduced during collection (Shankar et al., 2017, pp. 4–5). Therefore, absent evaluation and monitoring, interpreting or generating information can aggravate breaches of ethics and equity, even when no new data is collected. When considering machine learning to interpret or generate information, evaluators should interrogate the source data with respect to the concerns raised in the first part of this section.

Human bias constitutes a specific case of problematic data. Abundant evidence indicates that humans make biased judgments at vulnerable groups' expense, including when well intentioned and unaware of their biases. This is true in domains ranging from healthcare (FitzGerald & Hurst, 2017) to banking (Korver-Glenn, 2018) to criminal justice (Kovera, 2019). Such bias matters in the context of interpreting and generating information because some data science tools are trained to replicate human judgments. For example, evaluators might use data science tools to expedite screening for an evidence review (Rathbone et al., 2015), or to prioritize relevant texts for qualitative coding (Mills De La Rosa et al., 2021, p. 5). Unfortunately, when reviewing abstracts, humans are likely biased against low-income countries (Skopec et al., 2020). Data science tools trained on human decisions often reproduce such biases (Mehrabi et al., 2021), perhaps resulting in an inequitable evidence review.

Of course, bias is harmful in human actions, but data science tools often involve increased reach, reduced oversight, and uncertain accountability. Therefore, when considering data science to interpret and generate information, evaluators should investigate whether the tools have been trained on biased human judgments, and if adequate countermeasures have been taken.

Pre-trained tools are commonly used to analyze text, tag photos, process faces, and more. Even if all project-specific data meets ethics and equity standards, problematic pre-trained models can undermine results, and practitioners can unwittingly propagate ethics and equity violations. This has been observed in multiple contexts: language models show gender bias (Bolukbasi et al., 2016), image recognition models show location bias (Shankar et al., 2017), and facial analysis models show race and gender bias (Buolamwini & Gebru, 2018).

Therefore, evaluators should consider two instances each of problematic data and human bias: one as applied to project-specific data, and one as applied to the data baked into pre-trained tools.[2]

Misinformation: Conclusions drawn from inaccurate information can lead to harm. For example, a scan of fabricated blog posts might give policymakers an inaccurate understanding of public sentiment, or an evidence review based on faulty literature might recommend a flawed intervention. Further, because downstream consequences often affect vulnerable groups disproportionately, inaccuracies can create or entrench inequities. Regrettably, data science tools can generate inaccurate information, either unintentionally or as devices of disinformation campaigns. Examples range from believable citations of imaginary research articles (Walters & Wilder, 2023) to compelling but fabricated videos of public figures (Hancock & Bailenson, 2021).

Misinformation from data science tools is especially problematic because it is generated very efficiently and because it may be more convincing than misinformation generated by humans (Spitale et al., 2023). Further, risks are heightened by the familiar dynamics of broad reach, narrow oversight, and unclear accountability associated with data science tools. More generally, such tools can behave unpredictably, in ways that change without warning, and in ways that humans do not fully understand (Bowman, 2023). Therefore, when data science tools are used to generate information, evaluators should investigate the risk of harm from misinformation and whether sufficient protections against it are in place.

Making Real-World Decisions

In many applications, data science tools assist or replace humans in deciding how people, animals, places, or things will be treated. Examples include determining which job applicants should advance based on their resumes (Dastin, 2022), prioritizing which incoming university students receive early outreach from advisors (Greenstein & Crider-Phillips, 2023), and targeting resource-constrained public health interventions based on drone imagery (Liu et al., 2022). Some data science tools can draw on more information than any human decision-maker could synthesize, or exploit patterns too complex or unruly for humans to detect. With this in mind, some practitioners assume that machine-made decisions are free from problems caused by human bias, fatigue, and neglect. Unfortunately, this assumption is not reliable. Evaluators, therefore, should scrutinize applications of these tools for threats to ethics and equity.

Evaluators have not widely adopted data science tools in this way in their own work. However, potential use cases do exist: an evaluation agency might use data science to screen job applicants, target costly in-person visits to specific sites judged least likely to comply with a program model, or sort through recordings of calls to a complaint hotline to determine which are most likely to require follow-up. Moreover, evaluators can and should be called upon to evaluate such data science applications in other fields, making it important to understand relevant conceptions of ethics and equity.

This section builds upon the two before it, as most conceptions of ethics and equity found above remain relevant here. However, this decision-making context's heightened potential for immediate harm merits additional development. Here, we extend the conceptions of ethics and equity introduced above, and introduce new conceptions specific to real-world decision-making. To that end, we cover *problematic data*, *human bias*, *transparency*, and *deployment strategy*.

Problematic data can lead to problematic decisions. Evaluators should explore any relevant data for ethics and equity issues surrounding consent, representation, accuracy, and scope, as described in the first part of this section. In addition, if data being harnessed has itself been interpreted or generated by data science tools, that process should be evaluated for problematic data, human bias, and misinformation, as discussed above. Thoughtful analysis is critical here, as even minor flaws in underlying data can lead to significant harm when data science tools are used for real-world decision-making.

As noted, many facial recognition tools perform less accurately for vulnerable groups (Buolamwini & Gebru, 2018). Research may link this inequity to the unfair representation of vulnerable groups in training data (Klare et al., 2012, pp. 1798–1799). When facial recognition tools are used to make real-world decisions, such as whether someone should be charged with a crime, the consequences can become severe. A Black man in Michigan, for example, was wrongfully imprisoned on the basis of faulty facial recognition software (Barrett, 2020, p. 246; Hill, 2020). Had the police department's use of facial recognition been evaluated, representation issues in the tool's training data could have been identified and addressed, preventing this alarming outcome.

However, problems can arise from data even when a decision-making tool's training is ethical and equitable. Typically, after a tool is deployed, it makes decisions about people or things it has never encountered before. If these new entities are described problematically, the resulting decisions can be flawed. Consider, for example, a model designed to make medical recommendations based on patient history. Research shows that individuals belonging to vulnerable groups can be less likely to trust medical professionals enough to disclose sensitive information (Bernstein et al., 2008), leading to accuracy disparities in health data. One can imagine a hypothetical data science tool trained on data where this issue has been overcome, such as histories from select culturally responsive clinics with high levels of patient trust. Even if the tool is unbiased, when reapplied in a different context, such as a more typical clinic that vulnerable patients trust less, new data fed into the tool could be less accurate for these vulnerable patients. The recommendations issued to them might thus be systematically less appropriate, deepening health disparities. Evaluators, therefore, should extend their search for problematic data to include new data passed to established tools during day-to-day operations.

Human bias: As noted previously in this section, human decision-makers are often biased, and their biases are often expressed without their knowledge or intent. Mirroring the dynamic described above, when a data science tool is trained with decisions made by humans, human biases are easily reproduced: without intervention, the tool cannot differentiate between the biased human behavior it observes and the equitable behavior practitioners would like it to adopt. Further, the familiar factors of increased reach, reduced oversight, and uncertain accountability mean that data science tools can amplify human biases far beyond simply replicating them. In a decision-making context, this can lead to incalculable harm.[3]

In a seminal example, investigators found racial bias in COMPAS, a tool used to predict recidivism and inform criminal justice decisions across the United States. Racial bias in past human decisions, such as where to focus policing and surveillance, is believed to be at fault (Angwin et al., 2016; Mehrabi et al., 2021, p. 5). Unchecked, the result is unfair treatment of non-White Americans in the justice system, with profound and lasting impacts on the people involved and those who depend on them. With thoughtful evaluation of ethics and equity, authorities could have envisioned a decision-making tool for this purpose that begins to counteract entrenched human biases, instead of emulating them and doling them out with mechanical efficiency.

Transparency: Many data science tools are considered "black boxes," where a decision is rendered with little available insight into why or how it was reached. This is acceptable in some contexts, but it can pose obstacles to ethics and equity in others. For example, pathologists have begun to use data science to inform diagnostic decisions. However, some argue that medical professionals have an ethical duty to justify why and how a diagnosis was reached,[4] and to be accountable for the results (Tosun et al., 2020). Not all data science tools meet this standard. Selecting more explainable tools can also support equity, because when a tool's reasoning and level of confidence are exposed, human actors have more opportunity to note and correct bias. Evaluators, therefore, should consider whether data science tools used for decision-making offer sufficient transparency to meet ethics and equity standards. In the common scenario where increased transparency comes hand in hand with decreased accuracy, evaluators should consider both benefits and costs.

Deployment strategy: In some cases, a given tool can be used for ethical and equitable decision-making in one context but implemented problematically in another. ShotSpotter, for example, is a gunshot detection product deployed on the streets of numerous US cities. The product records audio, and when a noise is detected, it uses data science to help decide whether to notify police that a gunshot has occurred. Regrettably, many deployments concentrate the product in neighborhoods of color, amplifying the existing burden of racially disproportionate policing and surveillance (Stanley, 2021; MacArthur Justice Center, n.d.). The tool's adoption thus cements existing inequities, despite the fact that it detects gunshots fired by people of all races with equal accuracy. Monitoring and evaluation could inform an alternative deployment strategy designed to overcome this flaw and avert further harm to potentially vulnerable communities.

Using Data Science Ethically and Equitably for Evaluation

As illustrated throughout this book, data science has much to contribute to evaluation. This section aims to help evaluators apply data science tools to their work in a manner that prioritizes ethics and equity. To that end, it puts the theoretical framework offered in the previous section to concrete use. Below, we detail the following process by which evaluators can assess the ethics and equity of a proposed use of data science: (a) identify conceptions of ethics and equity that are relevant, (b) assess the risks of the proposal by way of each conception, (c) identify potential solutions and/or changes to the proposal, and (d) decide whether or not to proceed. Because evaluators currently use data science tools mainly to collect and process data, this section focuses on big data and using data science tools to interpret and generate information (both introduced in this chapter's first section).

Because the framework provided in the previous section is most useful when situated within a particular context, we encourage evaluators to begin by thoroughly developing their proposed use of data science tools. By concretely specifying goals, technical details, and implementation strategies, evaluators will be able to approach ethics and equity more seriously.

In addition, to undertake this exercise meaningfully, evaluators should pause to answer a series of challenging questions that frame the remaining work. The questions, as follows, serve to establish specific standards for ethics and equity as situated alongside the proposal being considered.

1. If breaches of ethics or equity occur, who is at greatest risk of harm or neglect? Who stands to benefit?
2. What baseline level of risk to ethics or equity is presented by the status quo or traditional alternative to this proposal?
3. What level of risk to ethics or equity is justified by the potential benefits of this proposal?
4. Which additional voices should be sought out to faithfully explore the preceding questions? How can consensus best be built?

Addressing these questions with rigor can demand research and deliberation, especially when done for the first time. Nevertheless, we urge evaluators to invest the necessary patience and resources, approaching the exercise as they would any other important methodological decision point. Additionally, we ask evaluators to address these questions *before* proceeding further, to prevent observed levels of risk from retroactively influencing the standards they are judged against.

From this point, evaluators can proceed to apply the framework given in this chapter's first section. Typically, this process begins with identifying conceptions of ethics and equity that might be relevant. The proposal can then be explored by way of each relevant conception. We suggest (a) determining whether any risk to ethics or equity is posed; then, if so, (b) choosing how the abstract risk ought to be quantified,[5] and (c) conducting research and/or reflection to estimate the risk's magnitude.

Where risks are identified, evaluators can consider technical solutions and/or amendments to the original proposal. Fortunately, technical measures can mitigate certain risks, including those posed by various biases in source data (Mehrabi et al., 2021, p. 15), gender bias in language models (Bolukbasi et al., 2016, pp. 11–14), location bias in image recognition models (Yang et al., 2020), and racial bias in facial analysis models (Klare et al., 2012, pp. 1798–1799). Other risks can often be addressed by improving the original proposal, perhaps to strengthen informed consent requirements or provide safeguards against human bias.

Ultimately, after risks have been identified and solutions have been explored, evaluators must make a final judgment on whether or not to implement the (amended) proposal. This choice should be made with respect to the ethics and equity standards that were established in advance through the four framing questions given above. To that end, the group of stakeholders involved in establishing the standards should typically be re-engaged to guide the final decision.

This process can demand both research and thoughtful (perhaps philosophical) deliberation, especially when single definitive answers are not apparent. For this reason, we stress the importance of patience, introspection, and consensus-building. This work is difficult, but we hope evaluators will come to see it as a worthwhile measure against creating or reproducing harm and inequity.[6] The exercise at the end of this chapter, inspired by the data science use case described in Chapter 9 of this volume (Mazzeo Rinaldi et al., 2025), offers a point of entry into this process.

Evaluating Applications of Data Science for Ethics and Equity

Just as data science can serve evaluators, evaluation can make important contributions to data science. Recent technological progress has created abundant evaluation and monitoring opportunities. Prominent among them, and the focus of this section, is the ethics and equity impact of data science tools. Evaluators are uniquely positioned to work in this often-ignored space. After using the framework laid out here to set appropriate standards for ethics and equity, core evaluation skills transfer elegantly to the remaining work. In essence, the task is to observe an intervention, measure specific ethics and equity impacts against said standards, communicate findings, and recommend (then monitor) changes and/or follow-up interventions. Further, because this work benefits immensely from diverse perspectives, evaluators will be well-served by their ability to work alongside multiple disciplines and engage multiple forms of lived and subject matter expertise.

This section's structure reflects the two broad use cases of data science introduced previously: the first part covers interpreting and generating information, and the second covers making real-world decisions.

Interpreting and Generating Information

In the previous section, we suggested a process for evaluators considering data science to interpret or generate information in their own work. The recommended process is similar when evaluating an existing intervention that involves interpreting or generating information with data science tools: (a) identify conceptions of ethics and equity that are relevant, (b) assess the ethics and equity impact of the intervention by way of each conception, (c) identify potential improvements, and (d) monitor the success of any solutions implemented, as well as the ongoing ethics and equity impacts of the intervention as a whole.

We refer readers to the previous section for an outline of how this process is best carried out. However, we note two key ways that the process differs in this context. First, when an intervention using data science has already been implemented, evaluators cannot always anticipate and prevent problems before they arise. Rather, they should seek to identify real and potential problems, measure or estimate their (potential) impact, and use their findings to improve the intervention. Because this difference aligns well with evaluators' core competencies – defining, measuring, suggesting follow-up, and monitoring – we do not explore it further here.

The second key difference is that evaluating data science used for non-evaluation purposes often requires additional help from individuals with subject matter expertise or lived experience in the relevant domain(s). For example, when evaluating the ethics and equity of an intervention that uses language modeling to shape disaster response (Ragini et al., 2018), evaluators might consult emergency management professionals and people who have been affected by disasters. Once again, given traditional evaluation's familiarity with seeking out such perspectives, we do not address this in depth.

Making Real-World Decisions

When evaluating the ethics and equity of an intervention that uses data science tools to make real-world decisions, the recommended process is similar to that described above. The key difference is one additional question that evaluators must explore in the first stage of the process, while defining the standards for ethics and equity that they will apply. For the sake of completeness, all questions are included below, but only Question 2, emphasized, is new.

1. If breaches of ethics or equity occur, who is at greatest risk of harm or neglect? Who stands to benefit?
2. *What kind of behavior constitutes biased, harmful, or unjust decision-making in this context?*
3. What baseline level of risk to ethics or equity is presented by the status quo or traditional alternative to this intervention?
4. What level of risk to ethics or equity is justified by the potential benefits of this intervention?
5. Which additional voices should be sought out to faithfully explore the preceding questions? How can consensus best be built?

Given the similarities between the process we recommend in this context and the processes we have described above, our focus here is on how this additional question can be explored.

Often, as a prerequisite to being considered ethical and equitable, a decision-making tool should yield reliable results for both privileged and vulnerable people. This is important in the context of data science because virtually all decision-making tools involve some level of error compared to the observed outcomes or human decisions they are trained to replicate. If this level of error is higher when a tool is applied to vulnerable people, the tool can be said to have a *"level of service" disparity*. This kind of inequity can be measured by comparing a tool's level of error[7] for the vulnerable groups identified in Question 1 against its error for a specific comparison group or its overall performance. Sample level of service standards include, "The tool's precision for people of color must be within five percentage points of its overall precision for everyone," or "The tool's R^2 value for people with disabilities may not be lower than its R^2 value for people without disabilities." As always, to avoid improper influence, a concrete standard should be set before proceeding with the process.

Other forms of biased, harmful, or unjust decision-making depend on the context of a given intervention. Most importantly, evaluators must determine whether producing systematically *different decisions for different groups* is problematic or desirable. For example, an intervention intending to help farmers make decisions about agricultural practices might involve a data science tool designed to predict crop yields (van Klompenburg et al., 2020). This tool might naturally, on average, predict lower yields for farmers without access to modern fertilization or irrigation technology. In this case, requiring the tool to suggest comparable decisions for comparably situated farms, ignoring their level of access to technology, would likely undermine its accuracy for everyone, hindering its ability to help farmers make informed use of the resources they have. Therefore, different decisions for different groups could be seen as desirable, and evaluators could focus on whether the tool provides a fair level of service to farmers with limited access to modern technologies.

Conversely, consider a tool used to decide the price of a new life insurance policy (Jain et al., 2019). Such a tool would be considered inequitable if it systematically issued higher prices to Black applicants than to comparable White applicants, even if race were correlated with a policy's true cost to the insurance company (Gaulding, 1994). Therefore, evaluators might seek to ensure that this tool does not issue systematically different decisions for different racial groups.

In cases like this, where different decisions for different groups are problematic, evaluators must determine how to measure disparity and what degree of disparity can be tolerated. For example, when considering a tool used to expedite a company's hiring process (Dastin, 2022), evaluators might require it to recommend women for interviews at least 0.95 times as often as men,[8] regardless of any population-level differences between male and female applicants. Alternatively, evaluators could require the tool to offer interviews at comparable rates for applicants of all genders *within* any given set of qualifications. Because the former standard aims to equalize outcomes with respect only to gender, it might be seen as taking a stronger position on equity by seeking to ease the effects of systemic inequities, such as access to education or glass ceilings at former places of employment. Conversely, by working within population-level differences, the latter standard might be seen as taking a weaker position, setting systemic factors aside but seeking to ensure that the specific employer in question does not engage in inequitable hiring.

The underlying question being considered here – "What kind of behavior constitutes biased, harmful, or unjust decision-making in this context?" – is a complex one. Ultimately, selecting the most appropriate standard falls to evaluators, subject matter experts, people with lived experience, and other stakeholders they engage. However, a range of tools and philosophies exist to guide this process of formalizing fairness and equity, and a helpful review is given in Mitchell et al. (2021). We recognize that defining equity is challenging, and we again stress the importance of patience, introspection, and consensus-building.

Once this question and the four others that accompany it have been answered as concretely as possible, evaluators are clear to proceed to the next stage of the process. At this point, we recommend applying the framework given in this chapter's first section with respect to the standards for ethics and equity that have been set. Then, much as described above, where (potential) problems are identified, evaluators should document them, seek to measure or estimate their downstream impacts, and propose follow-up interventions or improvements to the original, plus relevant monitoring measures and future re-evaluation.

Specific improvements could consist of alternative data sources, procedural changes, or bias mitigation strategies. The mitigation strategies referenced in the previous section can be helpful, as can a separate branch of techniques specific to decision-making. These techniques can help data scientists re-tune existing tools to comply (or approach compliance) with equity standards. A review of such strategies and situations where they have been applied is given in Mehrabi et al. (2021, pp. 13–25). Finally, such mitigation and other improvements or follow-up interventions can themselves be monitored and evaluated to assess whether they achieve the desired effects, using the standard tools of evaluation and repetition of the process described in this section.

Conclusion

Properly wielded, the tools of data science can do immense good in the world. With their help, an elected official might rapidly understand feedback from a more inclusive range of citizens. A university might predict negative academic outcomes and intervene with support, helping vulnerable students graduate. And a human services agency might direct scarce resources with unprecedented efficiency, improving more lives. Put differently, data science can be harnessed to combat the structural inequities and unethical conditions that persist in the world.

However, without conscientious planning, evaluation, and monitoring, data science tools can cement these same inequities and introduce ethics violations of their own. With no malicious intent, the elected official might exclude the voices of constituents without internet access. The university might nudge minority students away from challenging courses of study. And the human services agency might reproduce the biases held by decision-makers of the past. In many cases, these outcomes are the default. In data science, like elsewhere, it is easier to propagate the world's imperfections than it is to push back against them.

But evaluation, we suggest, is concerned with pushing back against imperfection. The aim of exploring how an intervention meets a need is often, at heart, to discover how a flaw in the world can best be overcome. This is not, we argue, rightly separable from attending to issues of ethics and equity. By taking on this difficult work, evaluators can play a transformative role in putting data science to work for good. We hope that this chapter helps evaluators scrutinize data science more confidently and adopt it responsibly in their own work.

Although this chapter cannot be exhaustive, it aims to equip any evaluator who interacts with data science with a framework to consider ethics and equity. Some readers may go on to pursue deeper data science expertise, and others may go on to serve as informed and conscientious facilitators of partnerships with data scientists. Regardless, we thank you for reading.

Exercise: Data Science and Public Opinion on the Russia–Ukraine War

This exercise aims to practice applying the concepts discussed throughout this chapter. We focus on using data science ethically and equitably for evaluation, as discussed in this chapter's second section, but engaging with this exercise will also help prepare readers to apply the processes discussed in its third section.

Here, we envision a hypothetical proposal from evaluators considering data science to interpret information, inspired by the work of Mazzeo Rinaldi et al. (2025) in Chapter 9. We describe the proposal, offer brief orientation toward how a review of ethics and equity might begin, and suggest next steps.

Our proposal comes from a hypothetical evaluation group studying peace and conflict. This group is beginning a project to understand public opinion surrounding the Russia–Ukraine war in affected areas. You find yourself on the project team. Findings will be compiled in a report to be referenced by scholars and practitioners in the future. To measure public opinion, your team plans to use several traditional methods, such as polls, surveys, and analysis of discourse from elected officials. However, the team recognizes that much communication and self-expression occurs digitally. Further, it is challenging to conduct polls and surveys in conflict zones. Therefore, your team proposes supplementing traditional methods with big data, which you will interpret using data science tools. Specifically, you plan to capture all tweets that contain any of several relevant hashtags posted over a period of several years. Then, using a data science tool called *emotion detection*, you will interpret the text of the tweets by quantifying which emotions they appear to express. The results will be analyzed and incorporated into the project's final report.

Before proceeding with this data science proposal, your team is evaluating its impact on the project's ethics and equity. You begin by posing the four questions given near the top of this chapter's second section, in order to establish ethics and equity standards against which the proposal can be judged. Here, we offer some brief orientation on how these questions might be approached and suggest next steps toward answering them.

1. If breaches of ethics or equity occur, who is at greatest risk of harm or neglect? Who stands to benefit?
    (a) Because the project aims to deepen understanding of public opinion, at-risk groups might include those who have not traditionally been given a strong voice to express their opinions. Such groups are often already vulnerable, such as (but not limited to) gender, racial, ethnic, linguistic, or religious minorities. People with limited access to information, education, or connectivity may also be at risk.
    (b) Similarly, groups who stand to benefit might include powerful people whose interests conflict with those of vulnerable groups – in other words, actors who would benefit from the opinions of vulnerable people remaining unheard.
    (c) *Next Steps: Working with other stakeholders, refine these suggestions, adding any other relevant groups. Translate these descriptions into specific populations or actors found in the areas you intend to study. As thoroughly as possible, sketch out how this proposal might impact them.*

2. What baseline level of risk to ethics or equity is presented by the status quo or traditional alternative to this proposal?
   (a) The at-risk groups identified above are likely underserved by traditional public opinion research, so parallel risks may exist in alternatives to this proposal.
   (b) These risks can be mitigated in poll and survey research because practitioners can target specific hard-to-reach groups and use demographic data to weight responses, bringing them closer to a balanced sample. However, the same risks may be aggravated in cases where not all groups are reachable, as can happen under conflict. Further research is needed to determine whether vulnerable groups are more or less likely to be neglected if this proposal is carried out.
   (c) *Next Steps: Working with other stakeholders, search the literature for work that explores – and ideally quantifies – the magnitude and impacts of those risks in traditional public opinion research. Acknowledge the level of certainty you are able to achieve, whatever it may be.*
3. What level of risk to ethics or equity is justified by the potential benefits of this proposal?
   (a) You believe that analyzing digital spaces is important to the project and worry that it will be challenging to conduct the desired amount of poll and survey research while the conflict persists. Therefore, some amount of risk is likely justified.
   (b) Because findings will be communicated in a long-form report, you have some ability to describe the methods used and guide readers on how results can safely be interpreted, including warnings of potential limitations. This may increase your risk tolerance.
   (c) *Next Steps: Working with other stakeholders, further explore how much risk can be tolerated in exchange for the proposal's expected benefits. Quantify the answer as much as possible, be it by choosing concrete metrics and setting tolerable thresholds, or by envisioning a variety of scenarios and labeling them as acceptable or unacceptable. Acknowledge the level of specificity you are able to achieve, whatever it may be.*
4. Which additional voices should be sought out to faithfully explore the preceding questions? How can consensus best be built?
   (a) Valuable input could likely be given by representatives of the groups whose opinions you seek to measure, especially individuals who belong to the vulnerable groups identified in Question 1. Representatives of the intended audience of the project's report may also be helpful.
   (b) *Next Steps: Develop a strategy to recruit, compensate, and acknowledge other stakeholders. Consider how to integrate their input into this process without burdening them unduly or inviting the possibility that they could be blamed for the ill effects of the project, as that*

responsibility remains with the project team. Consider how consensus can be reached or approached in situations where different stakeholders have opposing views.

Next Steps: After tackling the preceding questions, evaluators are clear to move on through the framework given in this chapter's second section. In brief, this involves (a) identifying conceptions of ethics and equity that are relevant, (b) assessing the risks of the proposal by way of each conception, (c) identifying potential solutions and/or changes to the proposal, and (d) with other stakeholders' help, deciding whether or not to proceed. The remainder of this process is left as an exercise to the reader, although nudges toward specific conceptions of ethics and equity are given in Table 4.1.

Table 4.1 Suggested conceptions of ethics and equity relevant to the exercise

Category	*Conception*	*Relevance*
Big data	Informed consent	It is debatable whether users of social media have consented to their posts being used for research. See discussion in the first section of this chapter
	Representation	Not all groups have equal access to technology and connectivity. Some governments block access to social media. Social media data lacks the necessary demographic information to assess representation
	Accuracy	Censorship and misinformation may affect accuracy in ways that are difficult to measure. This may impact vulnerable groups disproportionately
	Scope	The research could perhaps be conducted with fewer tweets, such as through a smaller set of hashtags, a shorter period of time, or by sampling only one day of each week
Interpreting and generating information	Problematic data	(See previous rows in this table)
	Pre-trained tools	Emotion detection tools are typically pre-trained. They may not perform equally well for all groups, perhaps varying by language, dialect, or writing level. Not all tools offer transparency into how they are trained

Notes

1 Questions of ownership and rightful access can be explored similar to those of informed consent. If a dataset includes creative works, for example, have creators consented to their work being used for this purpose? Have practitioners accessed and processed the work ethically? Are there inequities in how creators are acknowledged or compensated for their work?

2 Complex tools like ChatGPT are pre-trained on vast and secret data. Some reports even suggest that developers of such tools decline to keep training records in order to avoid confronting ethical issues (Schaul et al., 2023). In such cases, evaluators must weigh a tool's benefits against the risks incurred by its uncertain impact on ethics and equity.

3 Although less prevalent, bias can sometimes emerge purely from the mathematical inner workings of data science algorithms (Mehrabi et al., 2021, p. 7). Therefore, even if all data is deemed to be free of bias, it is still necessary to evaluate decision tools' output. More on this in the third section.

4 There are limits to humans' ability to rigorously account for our own decision-making. By some definitions, certain data science tools could thus be said to offer more transparency than humans. When considering a data science tool's transparency, evaluators may wish to also identify the level of transparency provided by the status quo or alternatives to the tool.

5 Quantifying risk depends on both context and the conception of ethics and equity being considered. Some methods may be abstract, such as ranking population groups in order of how much they might be harmed by a particular violation of privacy. Others may be concrete, such as directly comparing groups' share of representation in data to their true share of the target population. We ask evaluators to push themselves toward accountability by being as specific as possible during this step.

6 It is reasonable to observe that this section calls for a deeper assessment of ethics and equity than evaluators typically apply to traditional methods. This may be true, but we question the assumption that traditional levels of scrutiny are or were adequate. Moreover, we emphasize that the potential harm caused by violations of ethics and equity can be magnified in a data science context, as addressed in this chapter's introduction.

7 There are multiple ways to measure error (Botchkarev, 2019; Naidu et al., 2023), and the metrics used during a tool's development will not always be appropriate for measuring the level of service parity. Evaluators, likely in collaboration with data scientists, should consider which outcomes have the greatest bearing on ethics and equity, and then select metrics that capture these outcomes.

8 Historically, similar standards were set at 0.8 in the United States (Feldman et al., 2015, p. 2).

References

Angwin, J., Larson, J., Mattu, S., & Kirchner, L. (2016, May 23). Machine Bias. *ProPublica*. https://www.propublica.org/article/machine-bias-risk-assessments-in-criminal-sentencing?token=p-v0T1xjfOJ8jrJHzc08UxDKSQrKgWJk

Barocas, S., & Selbst, A. D. (2016). Big Data's Disparate Impact. *California Law Review*, *104*(3), 671–732.

Barrett, L. (2020). Ban Facial Recognition Technologies for Children—And for Everyone Else. *Boston University Journal of Science and Technology Law*, *26*(2), 223–285.

Bernstein, K. T., Liu, K.-L., Begier, E. M., Koblin, B., Karpati, A., & Murrill, C. (2008). Same-Sex Attraction Disclosure to Health Care Providers Among New York City Men Who Have Sex with Men: Implications for HIV Testing Approaches. *Archives of Internal Medicine*, *168*(13), 1458–1464. https://doi.org/10.1001/archinte.168.13.1458

Bolukbasi, T., Chang, K.-W., Zou, J. Y., Saligrama, V., & Kalai, A. T. (2016). Man Is to Computer Programmer as Woman Is to Homemaker? Debiasing Word Embeddings. *Advances in Neural Information Processing Systems*, *29*, 1–25.

Botchkarev, A. (2019). A New Typology Design of Performance Metrics to Measure Errors in Machine Learning Regression Algorithms. *Interdisciplinary Journal of Information, Knowledge, and Management*, *14*, 45–76.

Bowman, S. R. (2023). *Eight Things to Know about Large Language Models* (arXiv:2304.00612). arXiv. https://doi.org/10.48550/arXiv.2304.00612

Bruce, K., Vandelanotte, J., & Gandhi, V. (2025). Emerging Technology and Evaluation in International Development. In S. B. Nielsen, F. Mazzeo Rinaldi, & G. J. Petersson (Eds.), *Artificial Intelligence and Evaluation. Emerging Technologies and Their Implications for Evaluation* (pp. 13–36). London: Routledge. https://doi.org/10.4324/9781003512493.

Buolamwini, J., & Gebru, T. (2018). Gender Shades: Intersectional Accuracy Disparities in Commercial Gender Classification. *Proceedings of the 1st Conference on Fairness, Accountability and Transparency*, 77–91. https://proceedings.mlr.press/v81/buolamwini18a.html

Dastin, J. (2022). Amazon Scraps Secret AI Recruiting Tool That Showed Bias against Women. In *Ethics of Data and Analytics* (pp. 296–299). Auerbach Publications. https://doi.org/10.1201/9781003278290-44

Emam, K. E., Jonker, E., Arbuckle, L., & Malin, B. (2011). A Systematic Review of Re-Identification Attacks on Health Data. *PLoS ONE*, *6*(12), e28071. https://doi.org/10.1371/journal.pone.0028071

European Commission. (n.d.). *How Should My Consent Be Requested?* Retrieved October 23, 2023, from https://commission.europa.eu/law/law-topic/data-protection/reform/rights-citizens/how-my-personal-data-protected/how-should-my-consent-be-requested_en

Feldman, M., Friedler, S., Moeller, J., Scheidegger, C., & Venkatasubramanian, S. (2015). Certifying and Removing Disparate Impact. *arXiv:1412.3756 [Cs, Stat]*. http://arxiv.org/abs/1412.3756

FitzGerald, C., & Hurst, S. (2017). Implicit Bias in Healthcare Professionals: A Systematic Review. *BMC Medical Ethics*, *18*(1), 19. https://doi.org/10.1186/s12910-017-0179-8

Gaulding, J. (1994). Race Sex and Genetic Discrimination in Insurance: What's Fair. *Cornell Law Review*, *80*(6), 1646–1694.

Greenstein, N., & Crider-Phillips, G. (2023, April 13). *Predicting Risk Earlier: Machine Learning to Support Success and Combat Inequity. AERA Online Paper Repository*. AERA 2023 Annual Meeting, Chicago, IL. https://doi.org/10.3102/2005024

Haimson, O. L., Delmonaco, D., Nie, P., & Wegner, A. (2021). Disproportionate Removals and Differing Content Moderation Experiences for Conservative, Transgender, and Black Social Media Users: Marginalization and Moderation Gray Areas. *Proceedings of the ACM on Human-Computer Interaction*, *5*(CSCW2), 466, 35. https://doi.org/10.1145/3479610

Hancock, J. T., & Bailenson, J. N. (2021). The Social Impact of Deepfakes. *Cyberpsychology, Behavior, and Social Networking*, *24*(3), 149–152. https://doi.org/10.1089/cyber.2021.29208.jth

High-Level Expert Group on Artificial Intelligence. (2018). *Ethics Guidelines for Trustworthy AI* [Text]. European Commission. https://ec.europa.eu/futurium/en/ai-alliance-consultation

Hill, K. (2020, June 24). Wrongfully Accused by an Algorithm. *The New York Times*. https://www.nytimes.com/2020/06/24/technology/facial-recognition-arrest.html

Ioannidis, J. P. A. (2013). Informed Consent, Big Data, and the Oxymoron of Research That Is Not Research. *The American Journal of Bioethics*, *13*(4), 40–42. https://doi.org/10.1080/15265161.2013.768864

Jain, R., Alzubi, J. A., Jain, N., & Joshi, P. (2019). Assessing Risk in Life Insurance Using Ensemble Learning. *Journal of Intelligent & Fuzzy Systems*, *37*(2), 2969–2980. https://doi.org/10.3233/JIFS-190078

Klare, B. F., Burge, M. J., Klontz, J. C., Vorder Bruegge, R. W., & Jain, A. K. (2012). Face Recognition Performance: Role of Demographic Information. *IEEE Transactions on Information Forensics and Security*, *7*(6), 1789–1801. https://doi.org/10.1109/TIFS.2012.2214212

Korver-Glenn, E. (2018). Compounding Inequalities: How Racial Stereotypes and Discrimination Accumulate Across the Stages of Housing Exchange. *American Sociological Review*, *83*(4), 627–656.

Kovera, M. B. (2019). Racial Disparities in the Criminal Justice System: Prevalence, Causes, and a Search for Solutions. *Journal of Social Issues*, *75*(4), 1139–1164. https://doi.org/10.1111/josi.12355

Kramer, A. D. I., Guillory, J. E., & Hancock, J. T. (2014). Experimental Evidence of Massive-scale Emotional Contagion through Social Networks. *Proceedings of the National Academy of Sciences*, *111*(24), 8788–8790. https://doi.org/10.1073/pnas.1320040111

Lerman, J. (2013). Big Data and Its Exclusions. *Stanford Law Review*. https://www.stanfordlawreview.org/online/privacy-and-big-data-big-data-and-its-exclusions/

Li, Y., Dai, W., Ming, Z., & Qiu, M. (2016). Privacy Protection for Preventing Data Over-Collection in Smart City. *IEEE Transactions on Computers*, *65*(5), 1339–1350. https://doi.org/10.1109/TC.2015.2470247

Liu, Z. Y.-C., Chamberlin, A. J., Tallam, K., Jones, I. J., Lamore, L. L., Bauer, J., Bresciani, M., Wolfe, C. M., Casagrandi, R., Mari, L., Gatto, M., Diongue, A. K., Toure, L., Rohr, J. R., Riveau, G., Jouanard, N., Wood, C. L., Sokolow, S. H., Mandle, L., … De Leo, G. A. (2022). Deep Learning Segmentation of Satellite Imagery Identifies Aquatic Vegetation Associated with Snail Intermediate Hosts of Schistosomiasis in Senegal, Africa. *Remote Sensing*, *14*(6), Article 6. https://doi.org/10.3390/rs14061345

MacArthur Justice Center. (n.d.). *ShotSpotter Is Deployed Overwhelmingly in Black and Latinx Neighborhoods in Chicago*. End Police Surveillance. Retrieved January 23, 2023, from https://endpolicesurveillance.com/burden-on-communities-of-color/

Mazzeo Rinaldi, F., Celardi, E., Miracula, V., & Picone, A. (2025). Artificial Intelligence and Text Analysis in Evaluating Complex Social Phenomena. The Russia-Ukraine Conflict. In S. B. Nielsen, F. Mazzeo Rinaldi, & G. J. Petersson (Eds.), *Artificial Intelligence and Evaluation. Emerging Technologies and Their Implications for Evaluation* (pp. 168–195). London: Routledge. https://doi.org/10.4324/9781003512493

Mehrabi, N., Morstatter, F., Saxena, N., Lerman, K., & Galstyan, A. (2021). A Survey on Bias and Fairness in Machine Learning. *ACM Computing Surveys*, *54*(6), 115, 35. https://doi.org/10.1145/3457607

Mills De La Rosa, S., Greenstein, N., Schwartz, D, & Lloyd, C. (2021). *Machine Learning in Workforce Development Research: Lessons and Opportunities*. U.S. Department of Labor, Chief Evaluation Office. https://www.dol.gov/sites/dolgov/files/OASP/evaluation/pdf/Machine%20Learning%20in%20Workforce%20Development%20Research_Lessons%20and%20Opportunities_issue%20brief.pdf

Mitchell, S., Potash, E., Barocas, S., D'Amour, A., & Lum, K. (2021). Algorithmic Fairness: Choices, Assumptions, and Definitions. *Annual Review of Statistics and Its Application*, *8*(1), 141–163. https://doi.org/10.1146/annurev-statistics-042720-125902

Naidu, G., Zuva, T., & Sibanda, E. M. (2023). A Review of Evaluation Metrics in Machine Learning Algorithms. In R. Silhavy & P. Silhavy (Eds.), *Artificial Intelligence Application in Networks and Systems* (pp. 15–25). Springer International Publishing. https://doi.org/10.1007/978-3-031-35314-7_2

NCPHSBBR. (1979). *The Belmont Report*. New York: Department of Health, Education, and Welfare.

Nyst, C., & Falchetta, T. (2017). The Right to Privacy in the Digital Age. *Journal of Human Rights Practice*, *9*(1), 104–118. https://doi.org/10.1093/jhuman/huw026

OECD Legal Instruments. (2019). *Recommendation of the Council on Artificial Intelligence*. OECD. https://legalinstruments.oecd.org/en/instruments/OECD-LEGAL-0449

Parry, M. (2011, July 10). *Harvard Researchers Accused of Breaching Students' Privacy*. The Chronicle of Higher Education. https://www.chronicle.com/article/harvard-researchers-accused-of-breaching-students-privacy/

Proposal for a Regulation of the European Parliament and of the Council Laying Down Harmonised Rules on Artificial Intelligence (Artificial Intelligence Act) and Amending Certain Union Legislative Acts, COM/2021/206 final, CNECT. (2021). https://eur-lex.europa.eu/legal-content/EN/TXT/?uri=celex%3A52021PC0206

Ragini, J. R., Anand, P. M. R., & Bhaskar, V. (2018). Big Data Analytics for Disaster Response and Recovery Through Sentiment Analysis. *International Journal of Information Management*, *42*, 13–24. https://doi.org/10.1016/j.ijinfomgt.2018.05.004

Rathbone, J., Hoffmann, T., & Glasziou, P. (2015). Faster Title and Abstract Screening? Evaluating Abstrackr, a Semi-automated Online Screening Program for Systematic Reviewers. *Systematic Reviews*, *4*(1), 80. https://doi.org/10.1186/s13643-015-0067-6

Salaverría, R., & León, B. (2022). Misinformation Beyond the Media: 'Fake News' in the Big Data Ecosystem. In J. Vázquez-Herrero, A. Silva-Rodríguez, M.-C. Negreira-Rey, C. Toural-Bran, & X. López-García (Eds.), *Total Journalism: Models, Techniques and Challenges* (pp. 109–121). Springer International Publishing. https://doi.org/10.1007/978-3-030-88028-6_9

Schaul, K., Chen, S. Y., & Tiku, N. (2023, April 19). Inside the Secret List of Websites that Make AI Like ChatGPT Sound Smart. *Washington Post*. https://www.washingtonpost.com/technology/interactive/2023/ai-chatbot-learning/

Shankar, S., Halpern, Y., Breck, E., Atwood, J., Wilson, J., & Sculley, D. (2017). No Classification without Representation: Assessing Geodiversity Issues in Open Data Sets for the Developing World. *NIPS 2017 Workshop: Machine Learning for the Developing World*. https://doi.org/10.48550/arXiv.1711.08536

Skopec, M., Issa, H., Reed, J., & Harris, M. (2020). The role of geographic bias in knowledge diffusion: A systematic review and narrative synthesis. *Research Integrity and Peer Review*, *5*(1), 2. https://doi.org/10.1186/s41073-019-0088-0

Spitale, G., Biller-Andorno, N., & Germani, F. (2023). AI Model GPT-3 (Dis)informs us Better Than Humans. *Science Advances*, *9*(26), eadh1850. https://doi.org/10.1126/sciadv.adh1850

Stanley, J. (2021, August 24). Four Problems with the ShotSpotter Gunshot Detection System. *American Civil Liberties Union News & Commentary*. https://www.aclu.org/news/privacy-technology/four-problems-with-the-shotspotter-gunshot-detection-system

Sundara Raman, R., Shenoy, P., Kohls, K., & Ensafi, R. (2020). Censored Planet: An Internet-wide, Longitudinal Censorship Observatory. *Proceedings of the 2020 ACM SIGSAC Conference on Computer and Communications Security*, 49–66. https://doi.org/10.1145/3372297.3417883

Tosun, A. B., Pullara, F., Becich, M. J., Taylor, D. L., Fine, J. L., & Chennubhotla, S. C. (2020). Explainable AI (xAI) for Anatomic Pathology. *Advances in Anatomic Pathology*, *27*(4), 241. https://doi.org/10.1097/PAP.0000000000000264

UNESCO. (2021). *Recommendation on the Ethics of Artificial Intelligence*. United Nations. https://unesdoc.unesco.org/ark:/48223/pf0000381137

van Klompenburg, T., Kassahun, A., & Catal, C. (2020). Crop Yield Prediction Using Machine Learning: A Systematic Literature Review. *Computers and Electronics in Agriculture*, *177*, 105709. https://doi.org/10.1016/j.compag.2020.105709

VPRI. (n.d.). *Participant Privacy, Data Disposition, and Data Confidentiality*. Human Subjects Research. Retrieved January 27, 2023, from https://research.uoregon.edu/manage/research-integrity-compliance/human-subjects-research/f-participant-privacy-data-disposition-and-data-confidentiality

Walters, W. H., & Wilder, E. I. (2023). Fabrication and Errors in the Bibliographic Citations Generated by ChatGPT. *Scientific Reports*, *13*(1). Scopus. https://doi.org/10.1038/s41598-023-41032-5

Yang, K., Qinami, K., Fei-Fei, L., Deng, J., & Russakovsky, O. (2020). Towards Fairer Datasets: Filtering and Balancing the Distribution of the People Subtree in the Imagenet Hierarchy. *Proceedings of the 2020 Conference on Fairness, Accountability, and Transparency*, 547–558. https://doi.org/10.1145/3351095.3375709

Zimmer, M. (2010). "But the Data Is Already Public": On the Ethics of Research in Facebook. *Ethics and Information Technology*, *12*(4), 313–325. https://doi.org/10.1007/s10676-010-9227-5

5 Extracting Meaning from Textual Data for Evaluation

Lessons from Recent Practice at the Independent Evaluation Group of the World Bank

Virginia Ziulu, Harsh Anuj, Ariya Hagh, Estelle Raimondo, and Jos Vaessen

Introduction

Assessing the relevance and effectiveness of development interventions tends to be a challenging task due to the complexity of the social, cultural, institutional, economic, and political contexts in which they are embedded. In this chapter, we attempt to showcase how textual data can be a promising tool to aid evaluators in the evaluation of development interventions.

Textual data refers to any form of unstructured textual information, including web pages, documents, open-ended survey questions, social media posts, feedback forms, news, and reviews. Unstructured data lack a clear structure that can be easily read and understood by a computer, unlike structured data which can be more easily tabulated, stored in (relational) databases, or used for further analysis (Manning, 2009).

Text analytics is becoming increasingly ubiquitous due to the staggering amount of unstructured textual data that is generated continuously, as well as advances in computational resources to process and analyze these data. Within this context, automation is vital to fully leverage text data efficiently and effectively, as computers can analyze natural language[1] data without fatigue and in a consistent manner. This allows us to unearth facts, relationships, and assertions (i.e., knowledge) that would otherwise remain buried in the mass of textual big data.

One of the main advantages of textual data in the field of evaluation is that it provides rich and detailed information that might not be captured by traditional data sources such as census and survey data. Evaluators and researchers in fact have at their disposal a large and rich repository of textual information about economic and social activity from which they can extract and encode data that can be analyzed both descriptively and causally using modern analytical tools

DOI: 10.4324/9781003512493-5

(Gentzkow et al., 2019; Bronjecki et al., 2017). For example, textual data can capture the complexity of social and cultural contexts, as well as the perceptions and attitudes of different groups of beneficiaries toward a policy intervention. Additionally, textual data can help identify unexpected outcomes or unintended consequences of a policy intervention, which may not be evident from other indicators.

The inherently complex nature of natural language poses, however, some very specific challenges for working with textual data. This includes many linguistic phenomena such as vagueness, metaphors, sarcasm, idioms, and ambiguities in the use of language. Furthermore, accommodating the context-specific nuances of text proves challenging for text-based applications, as meanings and interpretations heavily depend on the surrounding linguistic context. Consequently, as stated by Russell (2016), when working with text data it is not possible "*to speak of a single meaning for a sentence, but rather of a probability distribution over possible meanings.*" Another difficulty specific to textual data is that the whole corpus of available text is spread across a large variety of languages and dialects. It may be the case that the collection of textual data needed to assess a particular policy intervention's results may be embedded in documents that are written in different languages. Consequently, manual or automatic translation is often needed to harmonize different sources of textual data. Furthermore, text data might not always fully capture the diversity of beneficiaries' views, a situation that needs to be ascertained before commencing a specific analysis. Lastly, text data is typically high-dimensional, as it tends to consist of a large number of features (or dimensions). Each feature may correspond to a different word or sequence of words in the text, and the number of possible features can be extremely large, even for relatively short texts. The high dimensionality of textual data is an important consideration that can influence the selection of appropriate methods and computing resources.

In this chapter, we discuss the increasing use of text as data in the framework of evaluations conducted by the Independent Evaluation Group (IEG) of the World Bank Group (WBG). These examples focus on using text-based techniques that either indirectly (portfolio identification, portfolio analysis) or directly (evaluative synthesis) contribute to responding to relevance and effectiveness questions. The multiple documents regularly produced at the WBG throughout the project cycle (such as Project Appraisal Documents, Implementation Status and Results Reports, and Implementation Completion and Results Reports for lending projects; and publications and working papers for non-lending projects) as well as evaluation documents produced at IEG (such as past evaluation reports, Implementation Completion Report Reviews, and Project Performance Assessment Reports), and external text-based data (such as social media posts, research publications, and project documents from other donors) present an opportunity for IEG to learn about project design and performance through (semi-)automated and systematic mining and analysis.

In the next section, we provide a brief overview of different text analytics techniques. Subsequently, we illustrate applications of some of these techniques for different types of evaluative analysis: identification of the evaluand, portfolio analysis of projects, and evaluative synthesis. In a final section, we reflect on some of the ongoing challenges and opportunities for using text analytics for evaluation.

Text Analytics: A Brief Overview

As noted earlier, textual data are highly unstructured and not easily understandable by computers. In fact, most text analytic applications are designed to work with numerical data and cannot directly process text in its raw form. Therefore, most applications require that text data be converted into a numerical representation. Converting text to numerical form involves several processing steps, including tokenization (splitting text into individual words) and encoding (assigning a numerical value to each word) (Gatto & Bundi, 2025, Mazzeo et al., 2025).

There are different techniques that can be used for extracting meaning from textual data. Among these, the most relevant techniques are text mining and natural language processing (NLP), although there is some overlap between the two.

Text mining focuses on the discovery and extraction of non-trivial knowledge from text (Kao et al., 2007). NLP, on the other hand, is a branch of artificial intelligence which combines machine learning[2] and statistical models with computational linguistics and focuses on developing algorithms that can understand and generate natural language. NLP typically takes into consideration the grammatical and semantic structure of text, as well as the lexical relationships between different parts of a text. Consequently, NLP can help answer questions that go beyond frequency tables, such as identifying the main topics in a collection of documents or identifying the main sentiment (i.e., positive, negative) in a document. In contrast with text mining, NLP aims to extract a fuller meaning representation from textual data (Kao et al., 2007).

NLP techniques, such as the machine learning techniques that they rely upon, can be broadly classified as either unsupervised or supervised (Barber, 2012). Unsupervised learning is applied to unlabeled or untagged text data and aims to detect patterns in text. Unsupervised NLP techniques include topic modeling and text clustering. On the other hand, in supervised learning, the starting point is labeled or tagged data (i.e., the output class(es) for each document are known in advance). Supervised techniques aim to model the relationship between the input and the output so that the model can be applied to new unlabeled data to predict the output. That is to say, the overall goal of supervised techniques is to achieve an accurate prediction. Supervised techniques include text classification and text summarization (Næss et al., 2025).

The delineation between supervised and unsupervised algorithms is, however, not always clear. Some techniques, such as sentiment analysis and machine translation, can be applied using both a supervised and an unsupervised approach. It is also possible to encounter semi-supervised approaches, which combine a small amount of labeled data with a large amount of unlabeled data during the model training phase.

Topic modeling is an unsupervised technique used to uncover hidden patterns or topics in a large collection of text data. It automatically analyzes and categorizes textual data into different groups or topics based on the frequency and distribution of the words used. This technique has been applied for example to understand success and growth factors in global renewable energy projects (Kumar et al., 2022) to model the nexus between poverty, ecology, and the environment (Cheng et al., 2018), and to understand equity through the mining of social media data (Cintron et al., 2022).

Text clustering is an unsupervised algorithm that groups similar texts together based on their content or features. It involves the automatic discovery of clusters of texts that share common characteristics, such as topics or sentiments. This technique has been used for example to identify topics in nuclear waste treatment patents (Suh et al., 2020), and to cluster short text responses for mobile educational activities to enhance student engagement (Tseng et al., 2018).

Text classification, a supervised technique, involves automatically assigning one or more predefined categories or labels to a given document or text. Examples of applications include the use of text classification algorithms to classify news articles on hazards for disaster management in India (Gopal et al., 2020) and the classification of flood tweets with contextual hydrological information to improve flood detection and monitoring (de Bruijn et al., 2020).

Text summarization is a supervised technique that involves generating a shorter version of a given document while preserving its most important information and meaning. This technique has been used for example to summarize in a clear and easy-to-understand way results stories included in over 120,000 non-standardized grant reports (Ahlsén et al., 2019).

Sentiment analysis uses supervised or unsupervised learning to automatically identify and extract the emotional or subjective tone from text. The goal of sentiment analysis is to determine whether a given text expresses a positive, negative, or neutral sentiment toward a particular topic. Sentiment analysis could be useful, for example, to help identify the perceptions and attitudes of beneficiaries toward a specific policy intervention. For example, this technique has been applied to analyze the sentiment toward the Syrian conflict using tweets (Lucić et al., 2020), to conduct an emergency response and early recovery assessment on the aftermath of the 2019 Albanian earthquake (Contreras et al., 2022), and to understand sentiment polarity regarding COVID-19 vaccines (Christensen et al., 2021).

Machine translation is a technique that involves using computers to automatically translate text from one natural language to another. This technique can be implemented using a supervised or unsupervised approach. In the context of development policy interventions, machine translation can help overcome language barriers and ensure that policymakers and practitioners are capturing the perspectives and experiences of all beneficiaries. It has been applied, for example, to develop fast and affordable translation systems for resource-poor[3] languages such as Mapuche in Chile and Quechua in Peru (Llitjós et al., 2005).

In the context of IEG's evaluative work, practical applications of text analytics have primarily focused on three areas: identification of the evaluand, portfolio analysis, and evaluative synthesis. Though different text analytics methods can offer a variety of efficiencies related to the practice of evaluation, arguably the most pertinent one has involved the classification of large quantities of text using supervised and/or unsupervised approaches. This is also the focus of the examples presented in this chapter. Traditional text categorization has heavily relied on desk review and manual coding, demanding extensive effort and time from subject experts. This is often time-consuming and tends to be unscalable. Automatic text classification, in contrast, offers a scalable solution (Bravo et al., 2023).

Using Text Analytics for Evaluation: Illustrations from Recent IEG Evaluations

Example 1: Identifying a Complex Evaluand Using Text Mining and Supervised Machine Learning

Context of Use

IEG's recent thematic[4] evaluation titled *"The Development Effectiveness of the Use of Doing Business Indicators, Fiscal Years 2010–20"* sought to assess *"[...] the relevance of Doing Business (DB) (doing the right things) and its effectiveness (doing things right) in motivating countries to reform their legal and regulatory environment for business and identifying areas for reform"* (World Bank, 2022, p. xii). The relevance question sought to *"... [examine] the relevance of [DB] indicators to country contexts and priorities, substantive dimensions of the areas they cover, and [WBG] strategic and operations priorities"* (World Bank, 2022, p. 78), while the effectiveness question sought to assess the extent to which *"... reforms measured by the DB indicators [are] linked to development outcomes such as job creation and economic growth in WBG client countries"* (World Bank, 2022, p. 39).

One key channel through which DB was expected to have affected development outcomes in World Bank (WB) client countries was its role in informing and influencing WB lending projects (World Bank, 2022, p. 6). Therefore, to understand the relevance and effectiveness of DB as operationalized in WB projects in client countries, it was first necessary to identify all WB projects that belong to this portfolio. Such projects generally referred to DB indicators in either the project objectives, components, or results frameworks (World Bank, 2022, p. 85).

This was not a straightforward exercise, since 5,710 lending projects were approved by the WB during the evaluation period, and a manual coding of all these projects would have been prohibitively expensive for the IEG. In this case, manual coding mainly refers to the activity wherein evaluators read the various documents available for each project and decide whether it was DB-informed or not.

The main project design document type for WB lending projects is the Project Appraisal Document (PAD), which would have to be reviewed for each project to understand its rationale, objectives, and results framework. This document type is, on average, 92 pages long, implying a total of over half a million pages of text to be reviewed by the evaluation team. That would amount to around 100,000 pages of text per evaluation team member (assuming an evaluation team size of five people).

In addition to the time-consuming task of reading documents, evaluators must spend time on various steps such as identifying, downloading, and organizing relevant documents. Beyond the human resource cost, there's also the issue of bias in human coding. The challenge is not just the manual review of text quantity but ensuring quality within resource constraints. The expansive evaluation topic, driven by the broad DB program scope, amplifies the time and error risks of a purely human coding process, demanding rigorous intercoder reliability testing.

How the Application Contributes to Answering the Evaluation Questions/Doing the Task

IEG decided to leverage text mining and NLP to conduct the exercise of portfolio identification efficiently and accurately. The process allowed the evaluation team to concentrate on their strength – applying judgment to classify a small number of projects – while the algorithms focused on their strength of processing a large volume of documents and accurately predicting project classifications.

How It Was Done

IEG deployed a stepwise approach, which combined keyword searches and supervised NLP on the one hand and manual review (on subsets of the population of documents) on the other. The process is highly iterative in nature. Figure 5.1 schematically describes this process.

Initial portfolio identification using a search taxonomy. A first portfolio identification exercise was conducted, which involved the development of a search taxonomy that was used to search the text of all project titles, development

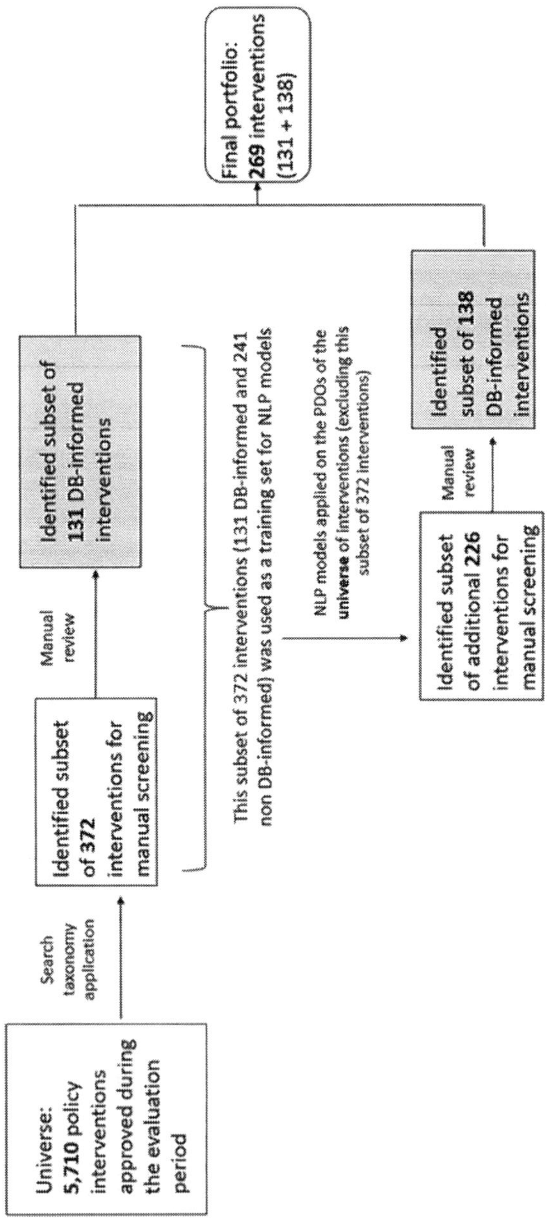

Figure 5.1 Portfolio identification scheme.

Source: Independent Evaluation Group.

objectives, component titles, and results framework indicator titles (World Bank, 2022, p. 87). This search taxonomy was constructed around the DB's twelve regulation areas or topics (World Bank, 2020, p. 78). This initial identification resulted in a long list of 372 projects, which were then manually screened to arrive at 131 projects coded as being DB-informed. This approach identified half of the final DB-informed portfolio of a total of 269 projects.

This initial exercise highlighted the challenge of capturing relevant projects by using just a search taxonomy, given the complexity of the evaluand. To illustrate what we mean here, consider just one of the twelve areas (World Bank, 2020, p. 3) of regulation covered by DB: "getting credit." There are multiple pathways through which a WB project can seek to affect changes in government regulations that improve the private sector's access to credit. For each of these pathways, there are probably a few different ways of describing it in words. Thus, the task of first identifying all the different pathways related to "getting credit," then all the different ways in which these pathways can be phrased and are being phrased can be quite cumbersome yet at the same time does not provide comprehensive results. In addition, there are many instances in which words such as "getting credit" (or similar words) are used in a document without this being related to the context of the DB regulation areas.

Second portfolio identification using supervised NLP. The list of 372 projects (of which 131 were identified as being DB-informed) that had previously been manually reviewed was used as a training sample for an ensemble of supervised text classification models.[5] No additional training dataset was developed for this task. Specifically, the models were trained using as input the text corresponding to project development objectives (PDOs), which tend to be short paragraphs of one to five sentences. The text was pre-processed using a relatively standard text preprocessing pipeline comprised of stopword[6] removal and lemmatization.[7] After preprocessing, the text was converted to sparse numerical representations using a Term Frequency–Inverse Document Frequency (TF-IDF) weighting scheme. The sparse numerical representations were input into the models, enabling them to learn and predict the probability of a project being DB-informed. These models were then applied to the PDOs from over 5,000 lending projects that had been approved during the evaluation period (and were not in the list of 372 projects that had been reviewed previously) to predict the probability of each project being DB-informed. Since three unique models were used for inference, three probability values were assigned to each project.

Supplementary keyword searches. Additionally, IEG developed a more general search taxonomy[8] for the theme of DB, without attempting to capture all the possible words or phrases associated with the DB's twelve reform areas and forty-one indicators. This search taxonomy was applied to: (a) PDOs from over 5,000 lending projects that had been approved during the evaluation period; and (b) the full text of 3,727 disclosed PADs and program documents. This excluded those projects which were in the list of 372 projects that had been reviewed previously. An automated bulk download protocol was used to download the documents, thereby reducing the transaction costs related to gathering documents. The frequencies of terms from the search taxonomy were used to assign relative "relevance scores" to each project based on the PDO and full document searches respectively.

Final portfolio identification. Cut-off values for the model-generated probabilities and the "relevance scores" were used to generate a long list of projects which were to be manually reviewed by team members to determine which ones were DB-informed. That is, a project could be included in this long list because of a high probability assigned by the models or because of a high number of results from the keyword searches in the PDOs and full documents. The cut-off values for the model probabilities and term frequencies were determined by the team while keeping in mind the trade-off between the completeness of the portfolio identification and the time required for manual review. As a result of this step, an additional 226 projects were identified for manual screening, from which another 138 DB-informed projects were identified and added to the final portfolio.

What We Can Learn

First, the approach played a pivotal role in the team's precise identification of the evaluand, laying the groundwork for the relevance and effectiveness assessments in the evaluation's portfolio analysis. By leveraging text mining and machine learning techniques, the team was able to double the size of the evidence base (of relevant projects and their documentation), which would have likely been impossible with traditional portfolio identification approaches. Doubling the evidence base enhanced the breadth and validity of findings from subsequent analyses.

Second, IEG experienced that developing a sizeable and high-quality training dataset can be challenging. Resource limitations impose constraints on both the size and quality of the training data that can be produced. Consequently, this impacts the precision and accuracy of outputs from supervised NLP models, which typically demand ample high-quality training data.

Third, close communication between the data scientist and evaluators is essential to develop a shared understanding around the application. Over time, as evaluators and data scientists develop a better understanding of each other's work, jointly working on data science applications becomes easier.

Fourth, investment of time and resources in innovative data science applications eventually pays off. The classification models developed for this task were reused in another IEG evaluation (World Bank, 2022, p. 93). Furthermore, the same model development could be applied to other tasks in other evaluations.

Fifth, this particular exercise constituted an early pilot in IEG in the use of supervised text classification for portfolio identification, which helped to demonstrate the utility of the approach.

Example 2: Portfolio Monitoring and Analysis Using NLP: Human Capital Project Just-in-time Note

Context of Use

This analysis was conducted as part of an evaluative exercise titled "Monitoring a multifaceted agenda: A Just in Time Note on the footprint of the Human Capital Project on the Human Development portfolio." This note aimed to trace the footprint of the Human Capital Project (HCP)[9] on the Human Development (HD) portfolio, develop a rigorous assessment methodology/tool that could be used to monitor progress for the remaining years of the HCP and inform its future evaluation, and test the use of supervised NLP to improve upon traditional portfolio review and analysis (PRA) approaches.

More specifically, the analysis sought to test two main hypotheses. If the HCP has provided an impetus to pursue specific human capital outcomes and embed multisectoral activities in the HD portfolio to advance the human capital agenda, then the following patterns might be observed: a quantitative increase in the number of operations that pursue specific human capital outcomes that are at the core of the HCP, and a qualitative shift in the design of HD operations with further emphasis on key human capital outcomes and cross-cutting themes that are promoted by the HCP (e.g., gender equality, digital solutions, human capital measurement, and institutional strengthening). Within this conceptual framework, the analysis aimed to answer the following questions: (a) did the size of the HD portfolio increase with the introduction of HCP, (b) did the design of the HD portfolio become more focused on pursuing specific human capital outcomes core to the HCP agenda, and (c) did the HD portfolio become more focused on cross-cutting themes core to the HCP agenda?

A potential approach to gather data to answer these questions is traditional PRA. This approach relies on WB sector and theme codes (World Bank, 2016) to identify projects with the desired focus and then proceeds with a manual review of each project. This conventional approach can be effective for portfolio monitoring tasks on topics that are well aligned with the WB's sector or theme taxonomies.

However, for many PRA tasks there is no crisp alignment with the existing sector or theme taxonomies, making conventional approaches either particularly inefficient and costly (due to the need for extensive manual review) or inaccurate (due to inconsistency in coding). This approach tends to yield both errors of inclusion and errors of exclusion. Besides this issue of alignment, there are also other issues with the tagging of WB projects to the sector and theme taxonomies that lead to inaccurate data. For example, one major issue is the lack of incentives for teams working in one domain to tag their projects to other domains that are outside of their purview. Over the years, IEG evaluations have documented various instances where this incompleteness in the tagging of projects to the sector and theme codes has been addressed using search taxonomies to identify projects by string searches.

How the Application Contributes to Answering the Questions/Doing the Task

Given the increasing cross-sectoral and multidimensional nature of the HCP portfolio, the reliance on sector and thematic codes using a traditional PRA approach would not have been sufficient to accurately and efficiently identify the HCP footprint on the HD portfolio. To circumvent the limitations of traditional PRA, IEG developed a novel approach which builds on a rigorous application of supervised NLP to identify trends in the HCP portfolio.

How It Was Done

The HD portfolio under consideration consisted of 984 lending operations. To be able to answer the questions outlined above, the portfolio was segmented into two time periods, a before-HCP period (FY15-FY18) and an after-HCP period (FY18-FY20). It was also segmented into two country groupings: countries that had joined the HCP by October 2018 (HCP countries) and countries that had still not joined by then (non-HCP countries). Figure 5.2 illustrates the composition of the HD portfolio selected for this application.

The sequential steps used to complete this task are outlined below. Figure 5.3 illustrates this process schematically.

Initial portfolio split. This portfolio was split into two sets: (a) a set of 183 projects, which was manually coded and used to build training and testing sets for the NLP model, and (b) a set of 801 uncoded projects for which the team aimed to predict codes by applying an NLP classification model.

Codebook. The team developed a codebook to precisely and systematically capture and categorize the main themes of the HCP. The design of the codebook was based on several sources of information, including the use of unsupervised NLP on the training set to help identify the main topics prevalent in the overall portfolio, an in-depth review of the HCP global and regional plans, and interviews with task team leaders. This resulted in 30 codes across human capital outcomes and cross-cutting areas. The codebook was tested, calibrated, and then used in the preparation of the training set.

Training set. The team used the content analysis software NVivo to prepare the training set for the subsequent supervised machine learning tasks. The team reviewed PDOs and component sections of the PADs of the 183 projects that were part of the sample and coded their content by following the prepared codebook. This process led to the identification of 1,125 segments of text, each of which was mapped to one or several codes (labels). To ensure that the text extracted for each of the codes was distinct enough, the cosine similarity[10] was calculated between each pair of codes. Codes with a high cosine similarity (not distinct enough) were merged into one category. The 1,125 segments of text included in the training set were subsequently randomly split into the following two subsets using an 80:20 ratio: (a) a training set, which is used to train different classification models and observe their performance, and (b) a testing

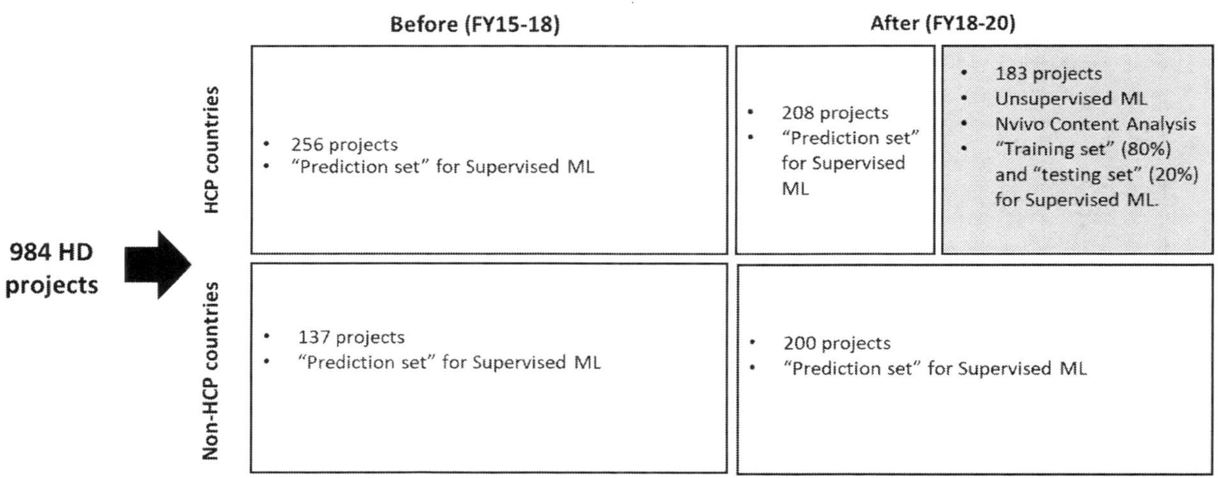

Figure 5.2 Initial portfclio composition.

Source: Independent Evaluation Group.

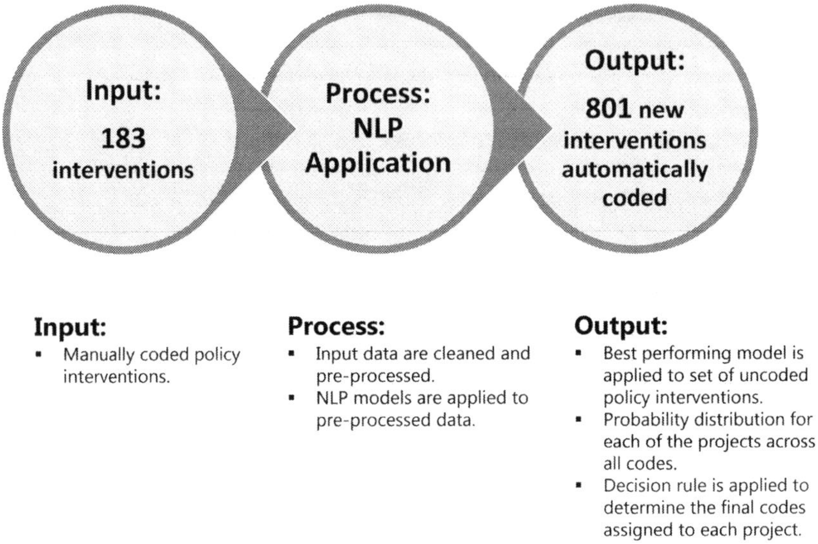

Figure 5.3 Text classification process.

Source: Independent Evaluation Group.

set, which is used to determine how well the chosen model performs outside the model sample.

Classification models. The team applied multiple classification models on the training set (logistic regression[11], K-nearest neighbors[12], support vector machine[13], decision tree[14], random forest[15], naïve Bayes[16], and stochastic gradient descent classifier[17]), iterating through different hyperparameters[18] for each model. Subsequently, the accuracy score was calculated for each model and the classifier with the highest accuracy score on the training data was selected (in this case, logistic regression with a 76.1 percent accuracy on the training set). The best-performing model was applied to the testing set (which until now had remained unseen by the models), resulting in an overall accuracy on the testing set of 72 percent.

Model application to unknown data. The best performing model was then applied to a separate portfolio of 801 projects, the prediction set, for which the team had the text from the PDOs and components but not their labels or codes. The output of applying the classification model to the prediction set is a probability distribution for each project across all 30 codes.

Final assignment of codes to each project. Several decision rules were considered to identify the cut-off point to select the codes to be mapped to each project in the prediction set. The objective was to select a cut-off point that allowed to minimize type I and type II errors, while accurately reflecting projects' multi-sectoriality. Keeping these considerations in mind, it was decided to keep those codes with a cumulative prediction greater than 80 percent.

Comparative analysis. Once the final codes were ready for both the training and the prediction sets, a comparative analysis was performed using frequency counts of the codes and similarity scores (cosine similarity) with a non-parametric permutation test. Four types of comparisons were performed: (a) before vs. after comparisons to assess the extent to which the design of HD projects differs "before" and "after" the HCP launch, (b) HCP countries vs. non-HCP countries comparisons to investigate whether the design of HD projects in HCP countries differs from those in non-HCP countries, (c) before vs. after comparisons within HCP countries, and (d) before vs. after comparisons within non-HCP countries.

What We Can Learn

The approach piloted and tested by IEG demonstrated a strong potential for several portfolio analysis and monitoring tasks with significant efficiency gains over a more traditional approach which would solely rely on manual coding. The main advantages of this approach for the HCP included: (a) the combination of manual coding (generation of a training set in NVivo) and supervised NLP optimizes coding accuracy while providing efficiency in analyzing a large portfolio with multiple categories for classification, and (b) the trained model can be run on new projects or projects that go further back in time at almost no cost (apart from processing and cleaning of data), allowing this methodology to be of practical use for the HCP team to continue tracking and monitoring progress in operationalizing HCP priorities. The latter is an important consideration, as the accuracy of supervised learning models generally tends to improve as additional data is fed into them. Beyond the HCP analysis, this approach can also potentially be replicated for other cross-sectoral topics that do not relate to human development but for which WB sector and theme taxonomies do not offer a robust enough identification framework.

This experiment also allowed IEG to distill some caveats and limitations for the application of this approach for portfolio analysis and monitoring tasks. First, the accuracy and validity of model outputs are largely a function of the quality of input data – especially the codebook and training set – used. To acquire high-quality data for model development, it is essential to ensure distinct categories are identified and a carefully calibrated training set is developed. This is a relatively manual process that tends to require multiple interactions among different members of the team, including both evaluators and data scientists. Second, data processing (e.g., identifying stopwords, and removing acronyms) is a critical step and should be done carefully and in collaboration with domain experts. Finally, as the conditions for applying a more rigorous causal design (e.g., a differences-in-differences model) were not in place in this case, the depth of the analysis is somewhat limited and more of a descriptive nature. There is potential, however, to explore, in a different setting, the use of a similar approach in combination with quasi-experimental techniques to assess causality.

Example 3: Evaluative Synthesis Using Supervised and Unsupervised ML: Project Insights[19]

Context of Use

IEG aimed to review a large corpus of project evaluation documents with the objective of identifying the critical types of challenges and impediments that influence private sector projects (co-)financed by the International Finance Corporation.[20] Traditionally, the analysis of implementation challenges involved manual identification, review, and categorization of text from project documents by evaluation officers. This process includes a qualitative review of a broad swath of performance indicators, taking advantage of evaluators' established experience in diagnosing critical challenges and impediments to project performance. However, what this process offers in nuance comes at a significant cost in terms of time and effort expended. To address these challenges, the WB has been exploring the use of text analytics for the development of taxonomies of delivery challenges and for tagging projects to this taxonomy (Ortega et al., 2022).

How the Application Contributes to Answering the Questions/Doing the Task

IEG piloted an approach leveraging recent advances in NLP applications to overcome some of the above challenges for IFC projects. This approach is based on the use of NLP to efficiently parse through evaluative evidence from evaluation documents, grouping text fragments according to a curated taxonomy of implementation challenges commonly faced by private sector engagement projects. This semi-automated analysis of private sector project evaluation documents served two major goals. First, to build an automatic classifier to efficiently categorize the vast quantity of existing evaluative evidence into distinct clusters, and second, to help minimize issues related to inter-coder reliability and evaluator subjectivity in classification by carefully training and calibrating an NLP model.

How It Was Done

The main steps implemented by IEG to identify performance challenges are described below. Figure 5.4 illustrates this process schematically.

Initial taxonomy. An initial taxonomy of topics and issues usually faced by private sector projects was generated based on prior expert knowledge. First, human experts discussed and shared what were the main issues faced in their

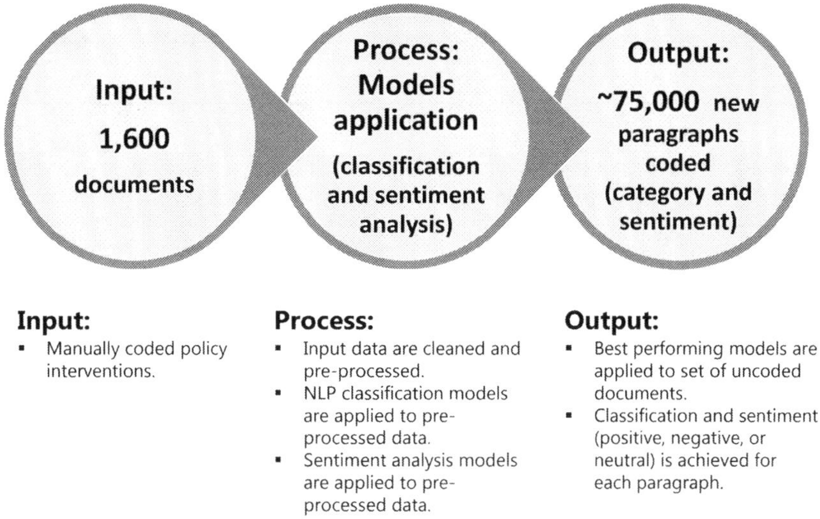

Figure 5.4 Text classification and sentiment analysis process.

Source: Independent Evaluation Group.

respective sectors and generated a draft taxonomy of the most common factors faced by projects in their sectors, including categories and sub-categories.

Taxonomy refinement. Relevant keywords and concepts were extracted from evaluation documents (IEG Evaluative Notes), creating a document-term matrix. The resulting matrix was then used to take stock of the frequency and saliency of various topics typically observed in private sector projects, generating conceptual categories that served to further disaggregate the initial taxonomy into more granular dimensions. This process culminated in a taxonomy comprising 5 categories (country, market, sponsor, project-specific, and IFC-controllable) and 51 subcategories.

Pre-processing. Input data was sourced from approximately 1,600 evaluation documents produced between 2008 and 2022. Text data were subjected to several standard processing steps such as stemming,[21] lemmatization, and stopword removal to prepare the corpus for classification. The goal of this step is to reduce inflectional forms and to remove words from the vocabulary that do not have explanatory power. Bigrams (a sequence of two adjacent words from a sentence) were subsequently selected to tokenize the documents, using frequency-based methods to extract the 1,000 most frequent words according to the TF-IDF information retrieval statistic. The resulting tokens comprised the document-term matrix used for training and classification.

Train/test split. From the extracted document-term matrix, 80 percent of the observations (accounting for some 4,000-odd issues) were selected to serve as a training sample. The remaining 20 percent of the observations were kept separately to be used for model validation.

Model training. IEG provided each of the models with a list of examples for each of the categories and subcategories in the taxonomy that the algorithm took as input to categorize the new paragraphs with the help of Naïve Bayes, random forest, support vector machine, and multi-layer neural networks. Naïve Bayes was the best-performing algorithm on this dataset and therefore this model was selected to categorize unlabeled paragraphs, assigning a probability of the paragraph being part of a category/subcategory. The average accuracy of the model was approximately 70 percent, with some categories having more than 90 percent accuracy.

Sentiment analysis. Furthermore, a pre-trained sentiment analysis algorithm was applied to each paragraph to assign sentiments between −1 and +1 (−1 if totally negative and +1 if totally positive). Sentiment analysis was performed to determine whether each factor affected the project positively or negatively.

Model validation. Two methods were used to validate the results of the model: Latent Dirichlet Allocation (LDA)[22] and Word2Vec. LDA aims to identify latent topics in a corpus of text, while Word2Vec aims to learn word associations from a large corpus of text. The rationale for using these approaches was to compare the topics/clusters identified by these models to the existing categories in the taxonomy.

Final output. The final model was used to classify approximately 75,000 paragraphs overall. This resulted in a table that included the key factors (in terms of the categories and subcategories included in the taxonomy) that affected each project, the specific paragraphs that support the inclusion of each factor, and the sentiment associated with each paragraph.

What We Can Learn

This experiment showed that automated evaluative synthesis can streamline private sector project evaluations by minimizing manual efforts in categorizing and assessing the impact of implementation challenges, thereby saving evaluators time.

Furthermore, the developed taxonomy allows evaluators to access the entire universe of insights from projects, generating actionable identification of main factors and lessons to help improve future project design and implementation. Taken together, the efficiency gains and data accessibility benefits generated by NLP serve to generate a virtuous circle whereby evaluators and practitioners can better integrate past issues into future practice.

As with any other form of analysis, and as noted in the previous examples, the accuracy of results is contingent on the quantity and quality of input data, as well as the presence of adequate model training and validation. As expected, well-defined categories and subcategories in the taxonomy (such as legal obstacles, political risk, and market pricing) tend to generate fewer false positives than broader ones. In contrast, where categories were imprecisely specified *a priori*, the model faced greater difficulties in converging on the correct categories and had to be further refined following an iterative process. The team also noted that more rigorous data cleaning at the initial stage would be able to improve the accuracy of the classification model.

Another important lesson from this experiment is the dynamic nature of these models. To guarantee the continued relevance and adaptability of the NLP model in the face of evolving implementation challenges, periodic evaluations of model performance are crucial, prompting adjustments to categories and subcategories in the taxonomy as needed.

Challenges and Opportunities for Using Text Analytics for Evaluation

The above examples illustrated the potential for text-based techniques to contribute to the field of evaluation. Furthermore, we are already seeing some impressive developments in the form of more complex large language models (LLM)[23] which are showing superior performance in many areas and could potentially lead to useful applications in evaluation. There is also a growing amount of research focusing on multi-modal learning models (e.g., combining text and imagery data), which could lead to new and more nuanced insights to answer evaluation questions on effectiveness or relevance.

These techniques are, however, not devoid of limitations and present multiple challenges to their implementation. A failure to consider these limitations could result in inaccurate or inappropriate results, especially when these results are the basis for decision-making.

A first set of challenges is around data requirements. Machine learning algorithms require a large volume of good quality data in order to extract meaningful insights (in comparison, text analytics is better suited for extracting insights from smaller text datasets). There are also more specific data requirements which depend, to a great extent, on the specific technique to be applied. Supervised learning requires, as a starting point, data that are accurately labeled. Mapping segments of text to labels is, however, a laborious and somewhat manual exercise. Some approaches that have been proposed to deal with this issue include crowdsourcing or outsourcing the generation of labels (Paul et al., 2018; Zhao et al., 2014). However, these alternatives also have their limitations due to the specialized domain knowledge required to accurately label data. Unsupervised learning, on the other hand, does not require labels as input data and can therefore be applied more straightforwardly on raw data. However, the absence of labels does not allow for an automated assessment of model accuracy, which introduces a different type of challenge (Awwad et al., 2020).

A second set of challenges is around potential biases in the data, many of which can be difficult to detect (Leslie, 2019; Hovy et al., 2021). This issue can be very insidious as the application of models on biased data has a substantial risk of perpetuating these biases. In the case of classification algorithms – such as in the examples previously described – biases could be introduced in the data at the labeling stage due to human error, misunderstanding of the labels, or disagreement with the codebook. Furthermore, the codebook itself might be biased. Biases can also appear at the data collection stage. LLMs – which can be used for tasks such as classification and sentiment analysis – are also susceptible to biases. These models are pre-trained on large corpora of text[24]; however, the underlying dataset can be biased in terms of demographic representativeness and/or usage of language. Another example of potential biases is in the use social media data.[25] Research has shown that data collected from social media tend not to be representative of the overall population as different socio-economic groups gravitate more to specific platforms or may not be actively present on social media platforms (Olteanu et al., 2019) (for instance, a study showed that Twitter users are skewed toward male and urban demographics (Mislove et al., 2011)). A closely related issue is that of uneven access to both social media and traditional media, which can vary greatly across countries, especially in those cases where access to the press is controlled or restrictive.

A third set of challenges arises from the lack of interpretability of complex models. From an interpretability angle, NLP models can be classified as opaque or transparent. An opaque or "black box" model refers to those models that are not easily understandable just by looking at their parameters. A typical example of opaque models is LLMs. On the other hand, model transparency allows for a human-level understanding of the inner workings of the model (Molnar, 2020; Belle et al., 2020). There is generally a tradeoff between accuracy and interpretability, with opaque models generally achieving higher accuracy at the expense of interpretability. The use of opaque models can be acceptable for some applications but might not be advisable in other cases. Therefore, the decision of whether an opaque or a transparent model is best suited for the task needs to be carefully considered at the early stages of the process.[26]

A fourth set of challenges concerns the institutional context in which the work is conducted. Through implementation, IEG has also learned that in order to successfully implement complex data science applications, data science experts and evaluators (including domain experts) need to closely work together. More specific input and guidance from the evaluation team have led to innovative new questions, new lines of inquiry, and an improved ability to overcome technical issues and in the end better outcomes from these applications.

A final set of challenges concerns the use of LLMs. IEG has rigorously experimented with these models to assess their objective performance in completing different pre-analysis, analysis, and post-analysis evaluation tasks.[27,28,29] While several of these experiments yielded satisfactory results, it is important to acknowledge that several risks are still present. In addition to their black-box nature and the existence of potential biases in LLMs, a specific issue with these models is their tendency to "*hallucinate*" (e.g., Alkaissi et al., 2023; Curran et al., 2023). As generative models do not have any knowledge of "ground truth," their responses might sound plausible but be factually incorrect. The inadvertent use of incorrect information could very well lead to inaccurate findings.

Conclusions

As more and more relevant textual data become available, computing resources continue to develop, and new algorithms to access and process these data become more sophisticated and powerful (including LLMs and other generative algorithms), it is expected that text analytics will continue to play a large role in the context of evaluation.

The experience of IEG has shown that the use of textual data in evaluations can generate important benefits in terms of enhancing the efficiency, quality, and breadth of evaluative work under the right circumstances.

Improvements in the *efficiency* of the evaluation process are related to using automated and semi-automated data extraction and analysis techniques and other machine-assisted techniques. For example, semi-automated portfolio identification and analysis can streamline the labor-intensive manual classification of project documents, helping draw insights from a larger corpus of prior knowledge in a more efficient manner.[30]

The use of textual data can potentially enhance the *quality* of findings through the application of innovative techniques that offer higher rates of accuracy than traditional methods. For example, NLP-based data classification and extraction can outperform manual coding when using a well-defined and structured coding scheme (and when training data sets are sufficiently large and representative of the broader universe of textual data).[31]

The use of textual data can also increase the *breadth* of evaluative inquiry. Examples include the use of new forms of text analytics in the realm of the internet (e.g., using social media data to conduct network or sentiment analyses).

However, the application of these techniques requires a systematic and rigorous approach to data collection, processing, and analysis, as well as careful consideration of the cultural and social context in which the policy interventions are implemented (Holm et al., 2025). As the use of textual data continues to grow, evaluators must continue to adapt their methods and approaches to ensure that the data is used both effectively and ethically.

Finally, reflecting on the applications of textual data for evaluation practice within the context of this book's three main questions, we believe that text-based techniques will have a transformational effect in the evaluation practice by enhancing evaluation work, allowing for some functions to be performed differently and more efficiently, and enhancing synthesis capabilities as larger and better text databases become available. Furthermore, as data science techniques become more complex and specialized, we believe that close collaboration between evaluators and data scientists throughout the application process is critical to ensure obtaining useful and robust findings. Furthermore, increasing evaluators' data science literacy is instrumental toward achieving a more fruitful collaboration (Question 1[32]). Finally, as the evaluation community continues to experiment with and test specific data science applications, we believe that evaluation has the potential to become a testing site to operationalize new knowledge (Question 3[33]).

Notes

1 A language that has evolved naturally as a means of communication among people, as opposed to programming languages, which are designed to be understood by machines (Wolfram, 2010).

2 "[Machine learning is the] field of study that gives computers the ability to learn without being explicitly programmed" (Samuel, 1959, as cited in Géron, 2022).

3 With respect to machine translation, "resource poor" refers to the lack of a large corpus of text in electronic form or a lack of native speakers trained in computational linguistics for a specific language.

4 Thematic evaluations are multi-level, multi-project evaluations that rely on elaborate mixed-methods designs that usually combine synthetic analyses at the overall portfolio level with in-depth analyses at the country, project, or other levels of analysis. (https://ieg.worldbankgroup.org/evaluations).

5 The models applied were logistic regression, support vector machine (SVM), and a multi-layer perceptron.

6 Stopwords are commonly used words in a given language (such as "a," "the," and "and").

7 Lemmatization considers the morphological context in which each word is used and converts the word to its meaningful base form, which is called a lemma. For example, after lemmatization, the word "was" is transformed to "be."

8 It included the following phrases: "business climate," "business development," "business environment," "business regulation," "develop private sector," "development of private sector," "doing business," "enhance competitiveness," "enterprise growth," "improve competitiveness," "investment climate," "private sector competitiveness," "private sector development," "private sector growth," "private sector led growth," "private sector-led growth," "reforming regulation," "registering property," "regulation reform," "regulatory environment," and "regulatory reform."

9 The Human Capital Project is a World Bank project designed to support countries through a customized package of data, policies, and interventions to accelerate human development outcomes.

10 Cosine similarity is a metric that quantifies the similarity between documents by measuring the inner angle between the vector representation corresponding to each document.

11 Logistic regression is a supervised machine learning algorithm primarily used for classification tasks where the objective is to estimate the probability of an instance belonging to a given class.

12 K-nearest neighbors is a non-parametric supervised learning classifier that relies on proximity metrics to predict the categorization of each individual data point.
13 Support vector machine is a supervised learning algorithm that relies on finding the optimal hyperplanes that separate data points into different classes.
14 Decision tree is a non-parametric supervised learning algorithm consisting of a hierarchical tree structure. A decision tree makes decisions by splitting the data into branches that provide the best separation between the classes.
15 Random forest is an ensemble machine learning algorithm that builds multiple decision trees and combines their outputs to make more robust and accurate predictions.
16 Naïve Bayes is a probabilistic supervised machine learning algorithm. To classify a new data point, the algorithm combines the conditional probabilities of the observed features for each class using Bayes' theorem and the class with the highest calculated probability is chosen as the predicted class.
17 Stochastic gradient descent implements a stochastic gradient descent learning routine (optimization technique).
18 Hyperparameters refer to configurations that are used to control the learning process.
19 Bravo et al., 2023.
20 The International Finance Corporation (IFC) is a member of the World Bank Group. It focuses on advancing economic development and improving the lives of people by encouraging the growth of the private sector in developing countries.
21 Stemming usually refers to a crude heuristic process that simply removes or stems the last few characters of a word. Stemming operates on a single word without knowledge of the context.
22 Latent Dirichlet Allocation (LDA) is a probabilistic model used for topic modeling in NLP. The goal of LDA is to identify topics within a large corpus of text documents by analyzing the frequency of words that appear together in the documents.
23 Large Language Models refer to complex language models, most commonly based on complex neural network architectures comprising a very large number of parameters (millions or even billions). These models are pre-trained on large amounts of data. Some examples of LLMs include BERT, ChatGPT, LaMBDA, and LLaMA.
24 For example, BERT is trained on a book corpus and English Wikipedia, and ChatGPT is trained on text databases from the internet (such as books, webtexts, Wikipedia, articles, and other pieces of writing).
25 Social media data is a very common type of textual data and one which has been used in evaluations. See, for example, World Bank. 2020. The World's Bank: An Evaluation of the World Bank Group's Global Convening. Independent Evaluation Group. Washington, DC: World Bank.
26 The field of XAI (Explainable Artificial Intelligence) is developing post-hoc interpretability models which would essentially allow to transform opaque models into transparent models without sacrificing accuracy. This could be an alternative for those cases when state-of-the-art performance is needed without losing explanatory power.
27 https://ieg.worldbankgroup.org/blog/setting-experiments-test-gpt-evaluation.
28 https://ieg.worldbankgroup.org/blog/fulfilled-promises-using-gpt-analytical-tasks.
29 https://ieg.worldbankgroup.org/blog/unfulfilled-promises-using-gpt-synthetic-tasks.
30 See, for example, Aggarwal et al., 2015; Burscher et al., 2015; Grimmer et al., 2013.
31 See, for example, Hillard et al., 2008; Okori et al., 2011.
32 Question 1: What requisite skills do evaluators need?
33 Question 3: What contribution can evaluation make to AI and vice versa?

References

Aggarwal, C. C., & Aggarwal, C. C. (2015). *Mining Text Data* (pp. 429–455). Springer International Publishing. https://doi.org/10.1007/978-1-4614-3223-4.

Alkaissi, H., & McFarlane, S. I. (2023). Artificial hallucinations in ChatGPT: Implications in scientific writing. *Cureus*, *15*(2). https://doi.org/10.7759/cureus.35179.

Ahlsén, M., Edberg, C., & Larsson, M. (2019). *Automating Text Summarization with Machine Learning: Extraction of End Results From Result Reports*. Uppsala University.

Awwad, Y., Fletcher, R., Frey, D., Gandhi, A., Najafian, M., & Teodorescu, M. (2020). *Exploring Fairness in Machine Learning for International Development*. CITE MIT D-Lab.

Barber, D. (2012). *Bayesian Reasoning and Machine Learning*. Cambridge University Press. https://doi.org/10.1017/cbo9780511804779.

Belle, V., & Papantonis, I. (2021). Principles and practice of explainable machine learning. *Frontiers in big Data*, 39. https://doi.org/10.3389/fdata.2021.688969.

Bravo, L., Hagh, A., Joseph, R., Kambe, H., Xiang, Y., & Vaessen, J. (2023). *Machine Learning in Evaluative Synthesis: Lessons from Private Sector Evaluation in the World Bank Group*. https://doi.org/10.1596/40054.

Broniecki, P., & Hanchar, A. (2017, October). Data Innovation for International Development: An Overview of Natural Language Processing for Qualitative Data Analysis. In *2017 International Conference on the Frontiers and Advances in Data Science (FADS)* (pp. 92–97). IEEE. https://doi.org/10.1109/fads.2017.8253201.

Burscher, B., Vliegenthart, R., & De Vreese, C. H. (2015). Using supervised machine learning to code policy issues: Can classifiers generalize across contexts?. *The ANNALS of the American Academy of Political and Social Science*, *659*(1), 122-131. https://doi.org/10.1177/0002716215569441.

Cheng, X., Shuai, C., Liu, J., Wang, J., Liu, Y., Li, W., & Shuai, J. (2018). Topic modelling of ecology, environment and poverty nexus: An integrated framework. *Agriculture, Ecosystems & Environment*, *267*, 1–14. https://doi.org/10.1016/j.agee.2018.07.022.

Christensen, B., Laydon, D. J., Chelkowski, T., Jemielniak, D., Vollmer, M., Bhatt, S., & Krawczyk, K. (2021). Quantifying changes in vaccine coverage in mainstream media as a result of COVID-19 outbreak. *medRxiv*, 2021-11. https://doi.org/10.1101/2021.11.07.21266018.

Cintron, D. W., & Montrosse-Moorhead, B. (2022). Integrating big data into evaluation: R code for topic identification and modeling. *American Journal of Evaluation*, *43*(3), 412–436.

Contreras, D., Wilkinson, S., Alterman, E., & Hervás, J. (2022). Accuracy of a pre-trained sentiment analysis (SA) classification model on tweets related to emergency response and early recovery assessment: The case of 2019 Albanian earthquake. *Natural Hazards*, *113*(1), 403–421. https://doi.org/10.1007/s11069-022-05307-w.

Curran, S., Lansley, S., & Bethell, O. (2023). Hallucination is the last thing you need. *arXiv preprint arXiv:2306.11520*.

de Bruijn, J. A., de Moel, H., Weerts, A. H., de Ruiter, M. C., Basar, E., Eilander, D., & Aerts, J. C. (2020). Improving the classification of flood tweets with contextual hydrological information in a multimodal neural network. *Computers & Geosciences*, *140*, 104485. https://doi.org/10.1016/j.cageo.2020.104485.

Gatto, L., & Bundi, P. (2025). The Use of Quantitative Text Analysis in Evaluations. In S.B. Nielsen, F. Mazzeo Rinaldi and G.J. Petersson (Eds.), *Artificial Intelligence and Evaluation. Emerging Technologies and Their Implications for Evaluation* (pp. 144–167). London: Routledge. https://doi.org/10.4324/9781003512493

Gentzkow, M., Kelly, B., & Taddy, M. (2019). Text as data. *Journal of Economic Literature*, *57*(3), 535–574. https://doi.org/10.1257/jel.20181020.

Géron, A. (2022). *Hands-on Machine Learning with Scikit-Learn, Keras, and TensorFlow*. O'Reilly Media, Inc.

Gopal, L. S., Prabha, R., Pullarkatt, D., & Ramesh, M. V. (2020, October). Machine learning based classification of online news data for disaster management. In *2020 IEEE Global Humanitarian Technology Conference (GHTC)* (pp. 1–8). IEEE. https://doi.org/10.1109/ghtc46280.2020.9342921.

Grimmer, J., & Stewart, B. M. (2013). Text as data: The promise and pitfalls of automatic content analysis methods for political texts. *Political Analysis*, *21*(3), 267–297. https://doi.org/10.1093/pan/mps028.

Hillard, D., Purpura, S., & Wilkerson, J. (2008). Computer-assisted topic classification for mixed-methods social science research. *Journal of Information Technology & Politics*, *4*(4), 31–46. https://doi.org/10.1080/19331680801975367.

Holm, J., Newman-Griffis, D., & Petersson, G. J. (2025). Big Data for Big Investments: Making Responsible and Effective Use of Data Science and AI in Research Councils. In S.B. Nielsen, F. Mazzeo Rinaldi and G.J. Petersson (Eds.), *Artificial Intelligence and Evaluation. Emerging Technologies and Their Implications for Evaluation* (pp. 120–143). London: Routledge. https://doi.org/10.4324/9781003512493

Hovy, D., & Prabhumoye, S. (2021). Five sources of bias in natural language processing. *Language and Linguistics Compass*, *15*(8), e12432. https://doi.org/10.1111/lnc3.12432.

Kao, A., & Poteet, S. R. (Eds.). (2007). *Natural Language Processing and Text Mining*. Springer Science & Business Media. https://doi.org/10.1007/978-1-84628-754-1.

Kumar, M., & Ng, J. (2022). Using text mining and topic modelling to understand success and growth factors in global renewable energy projects. *Renewable Energy Focus*, *42*, 211–220. https://doi.org/10.1016/j.ref.2022.06.010.

Leslie, D. (2019). Understanding artificial intelligence ethics and safety. *arXiv preprint arXiv:1906.05684*.

Llitjós, A. F., Aranovich, R., & Levin, L. (2005). Building Machine translation systems for indigenous languages. In *Second Conference on the Indigenous Languages of Latin America (CILLA II), Texas, USA*.

Lucić, D., Katalinić, J., & Dokman, T. (2020). Sentiment analysis of the Syrian Conflict on twitter. *Media Studies*, *11*(22), 46–61. https://doi.org/10.20901/ms.11.22.3.

Manning, C. D. (2009). *An introduction to information retrieval*. Cambridge University Press. https://doi.org/10.1007/s10791-009-9096-x.

Mazzeo Rinaldi, F., Celardi, E., Miracula, V., & Picone, A. (2025). Artificial Intelligence and Text Analysis in Evaluating Complex Social Phenomena. The Russia-Ukraine Conflict. In S.B. Nielsen, F. Mazzeo Rinaldi and G.J. Petersson (Eds.), *Artificial Intelligence and Evaluation. Emerging Technologies and Their Implications for Evaluation* (pp. 168–195). London: Routledge. https://doi.org/10.4324/9781003512493

Mislove, A., Lehmann, S., Ahn, Y. Y., Onnela, J. P., & Rosenquist, J. (2011). Understanding the demographics of Twitter users. In *Proceedings of the International*

AAAI Conference on Web and Social Media (Vol. 5, No. 1, pp. 554–557). https://doi.org/10.1609/icwsm.v5i1.14168.

Molnar, C. (2020). *Interpretable Machine Learning*. Lulu.com.

Næss, T., Prabhu, C., Mjaaland, M., Holtermann, H., & Skage Engebretsen, L. (2025). Text Mining and Machine Learning in an Evaluation of Police Handling of Cybercrime in Norway. In S.B. Nielsen, F. Mazzeo Rinaldi and G.J. Petersson (Eds.), *Artificial Intelligence and Evaluation. Emerging Technologies and Their Implications for Evaluation* (pp. 103–119). London: Routledge. https://doi.org/10.4324/9781003512493

Okori, W., & Obua, J. (2011, July). Machine Learning Classification Technique for Famine Prediction. In *Proceedings of the World Congress on Engineering* (Vol. 2, No. 1, pp. 4–9). https://doi.org/10.1080/08839514.2011.611930.

Olteanu, A., Castillo, C., Diaz, F., & Kıcıman, E. (2019). Social data: Biases, methodological pitfalls, and ethical boundaries. *Frontiers in Big Data, 2*, 13. https://doi.org/10.3389/fdata.2019.00013.

Ortega Nieto, D., Hagh, A., & Agarwal, V. (2022). *Delivery Challenges and Development Effectiveness*. https://doi.org/10.1596/1813-9450-10144.

Paul, A., Jolley, C., & Anthony, A. (2018). *Reflecting the Past, Shaping the Future: Making AI Work for International Development*. Center for Digital Development, USAID.

Pedregosa, F. (2011). Scikit-learn: Machine learning in python. *Journal of Machine Learning Research, 12*, 2825.

Russell, S. & Norvig, P. (2016). Artificial Intelligence. A Modern Approach. Upper Saddle River, NJ: Pearson Education Inc.

Samuel, A. L. (1959). Some studies in machine learning using the game of checkers. *IBM Journal of Research and Development, 3*(3), 210–229. https://doi.org/10.1147/rd.33.0210.

Suh, J. W., Sohn, S. Y., & Lee, B. K. (2020). Patent clustering and network analyses to explore nuclear waste management technologies. *Energy Policy, 146*, 111794. https://doi.org/10.1016/j.enpol.2020.111794.

Tseng, Y. H., Lee, L. H., Chien, Y. T., Chang, C. Y., & Li, T. Y. (2018, July). Multilingual short text responses clustering for mobile educational activities: A preliminary exploration. In *Proceedings of the 5th Workshop on Natural Language Processing Techniques for Educational Applications* (pp. 157–164). https://doi.org/10.18653/v1/w18-3723.

Zhao, Y., & Zhu, Q. (2014). Evaluation on crowdsourcing research: Current status and future direction. *Information Systems Frontiers, 16*, 417–434. https://doi.org/10.1007/s10796-012-9350-4.

Wolfram, S. (2010). *Programming with Natural Language Is Actually Going to Work*. Wolfram Blog.

World Bank. (2016). *Doing Business 2020*. The World Bank. https://doi.org/10.1596/978-1-4648-1440-2.

World Bank. (2020). *An Evaluation of the World Bank Group's Global Convening*. The World Bank. 10.1596/IEG147246.

World Bank. (2022). *The Development Effectiveness of the Use of Doing Business Indicators, Fiscal Years 2010–20*. The World Bank. https://doi.org/10.1596/IEG1661660.1596/IEG166166.

6 Text Mining and Machine Learning in a Performance Audit of Police Handling of Cybercrime in Norway

Tom Næss, Helge Holtermann, Carolin Prabhu, Lars Skage Engebretsen, and Mari Mjaaland

Introduction

Technological development has opened new opportunities for criminal activity. While traditional theft has declined, crimes such as online fraud and identity theft have risen sharply in Norway. The shift in crime poses a challenge to national police authorities, as it requires new investigation methods. In 2021, we (the Office of the Auditor General of Norway) therefore conducted a performance audit[1] of the national police's efforts to combat cybercrime.[2,3] The aim was to assess whether the police had an overview of, investigated, and solved cases of cybercrime in accordance with the Police Act adopted by the Norwegian parliament.

Cybercrime is an important area of policing as society is undergoing a digital transformation where a large share of our lives is spent online. For cybercrime scholars, the lack of statistics has been a challenge (Bossler & Berenbum, 2019). In Norway, official statistics have not included cybercrime. Since 2018, the police have tried to identify such crimes by registering the *modus operandi* of reported crimes. However, registration practices have varied across police districts, and the statistics have not been reliable according to the police themselves. In this performance audit, one important task was therefore to identify and get an overview of cybercrime.

New technologies give opportunities not only for criminals but also for evaluators and auditors. In a performance audit of the Norwegian police's efforts to combat cybercrime, we made use of two rather novel tools to identify cybercrime: text mining and machine learning. Text mining is a specific application within natural language processing (NLP) focused on extracting information from text, while natural language understanding (NLU) is a more advanced aspect of NLP focused on the deeper understanding and interpretation of language.

DOI: 10.4324/9781003512493-6

Text mining includes various techniques for automatically extracting information from text to enable quantitative analysis of patterns, while machine learning comprises statistical tools that allow models to automatically improve by finding patterns in data and deriving rules to interpret new data (see also Bruce, Gandhi & Vandelanotte, 2025, Gatto & Bundi, 2025, Mazzeo et al., 2025, & Ziulu et al., 2025). We used these tools to identify cybercrime among all criminal cases reported in a year, which enabled us to assess police performance in mapping and handling cybercrime. To our knowledge, such analyses had not previously been done in Norway before.

In this chapter, we use our example to illuminate what value text mining and machine learning may add to evaluations. We also discuss what it requires to make use of such tools. What competence do evaluators and performance auditors need, and how may evaluators and performance auditors collaborate with data scientists to benefit from these tools?[4]

In our case, the main benefit of these tools was to allow the use of population rather than sample data, which provided the opportunity for a more fine-grained assessment of how the police performed in investigating and solving cybercrimes. However, employing these tools was time-demanding, and we spent somewhat more resources than normal on this performance audit. Scant previous experience with such tools in our organization might partly explain this. This was a risk we were willing to take to learn how these methods could be applied. The case also suggests that collaboration between (in-house) data scientists and auditors can be a fruitful model for utilizing such new methods, as properly using them requires considerable expertise. At the same time, we found it useful for auditors to have competence in quantitative methods more broadly.

The chapter is structured in the following way: first, we define cybercrime and describe the type of competence involved in the performance audit. In the second part, we describe how manual coding, text mining, and machine learning were deployed in the audit. In the final part, we discuss the challenges involved, success criteria, and lessons learned related to the main research questions of this book.

What Is Cybercrime?

The lack of a clear definition of cybercrime has been a problem for the police and for researchers. Cybercrime can also be referred to as computer crime, data crime, and ICT-related crime. Literature reviews by Akdemir, Sungur, and Basaranel (2020) and Phillips et al. (2022) posit that the two most cited academic definitions of cybercrime are:

- "computer-mediated activities which are either illegal or considered illicit by certain parties and which can be conducted through global electronic networks" (Thomas & Loader, 2000);

- "any crime that is facilitated or committed using a computer, network, or hardware device" (Gordon & Ford, 2006).

Phillips et al. (2022) conclude that "the lack of clarity surrounding the term cybercrime has a significant impact on society, cybercrime policy, legal intervention and academic research." In Norway, ambiguity surrounding the concept of cybercrime also creates challenges. The term "cybercrime" was interpreted in different ways by police districts, special agencies, and national authorities. This ambiguity has probably impeded efforts to get an overview of the scope of cybercrime and may also have made it difficult to create effective strategies and measures to combat such crimes.

The term "cybercrime" is consistently used in this chapter. Cybercrime is criminalized under Norwegian law and generally covers two types of crime.[5] This definition of cybercrime was applied in the audit:

- Crime that targets computer systems and technology includes hacker attacks, data breaches, computer attacks, sabotage, industrial espionage, and blocking of internet services. Such crimes are often referred to as "cyber-dependent crime."
- Crime in which key elements of the criminal act are carried out using computer systems, equipment, or networks. Such crimes were previously committed in the physical space but now take place online. Examples include buying and selling drugs, sharing material related to abuse, ID theft, fraud, and violation of privacy.

Various terms are often used to refer to specific types of cybercrime. *Online sexual abuse*, for example, may be referred to as online abuse, online sexual offenses, and grooming. The terms may vary depending on the type of sexual offense that is committed. *Financial cybercrime* denotes cybercrime committed with the intention of achieving financial gain, such as fraud, ID theft, and ransomware.

Supreme Audit Institutions, Performance Audit, and Evaluation

Supreme Audit Institutions (SAIs) are independent public bodies whose main role is to audit a government's use of public funds. SAIs are normally mandated by the constitution and laws of the country. The means at their disposal to oversee the use of public funds and secure democratic accountability are financial, compliance, and performance audits (INTOSAI, 1977). Of these means, performance audits share the most traits with evaluations. Performance audits are independent and objective examinations that aim to contribute to improved economy, efficiency, and effectiveness in the public sector (INTOSAI, 2019).

The Innovation Lab and the Audit Team

In 2018, the Office of the Auditor General of Norway established an Innovation Lab with the purpose of exploring and utilizing data science in audit (Beckstrom, 2020; Otia et al., 2022). The Innovation Lab consists of data scientists with various backgrounds, including experience in performance auditing/evaluation, coding, and machine learning. Besides the automation of standard tasks to enable more analytic work for auditors, this unit assists with tasks related to data acquisition (e.g., web scraping) and data analysis (e.g., text mining). It also offers internal training in coding and analysis tools to gradually transform the audit office to utilize available information more efficiently. The Innovation Lab is interconnected with the audit units through collaboration on common projects and tools. As an in-house unit, it also allows for the exchange of innovative ideas where success is not guaranteed.

The cybercrime audit was led by one of the performance audit divisions, and the auditors did not have expertise in machine learning or text mining. The audit team had competence in quantitative methods, though, and was able to understand what benefits the new technologies could yield.

The use of machine learning to analyze and classify police records started as an experiment. It had the potential to enhance the audit but was not essential for its completion. We ended up using these tools, and three data scientists from the innovation lab contributed significantly to the audit by carrying out the text mining and machine learning parts of the audit in collaboration with auditors.

Manual Coding, Text Mining, and Machine Learning

The purpose of the audit was to assess the extent to which the Norwegian police had an overview of, investigated, and solved cybercrime. The main audit questions were: 1. What overview do the police have of cybercrime? 2. Do the police investigate and solve cybercrime? 3. What factors prevent cybercrime from being solved? 4. How well do the Ministry of Justice and the Police Directorate steer the police's handling of cybercrime? Text mining and machine learning were applied as methods for answering the first two audit questions.

At the outset of the audit, we did not envisage using text mining and machine learning.[6] However, over the course of the planning (pre-analysis) phase, we found that these tools would be useful for answering audit questions 1 and 2.

Since 2018, the police had registered whether ICT was a relevant mode for every reported crime case.[7] We were informed, however, that the registration could not necessarily be trusted. To further explore this and find a reliable estimate of the amount of cybercrime, we considered using reference to *modus operandi* in the criminal case records.[8] However, these records often did not contain sufficient information, registration practices differed from one case to another, and more details about the event could often be found in the case reports. We therefore decided to retrieve and use the full case reports by the victims of cybercrime as a source of information.

From this point, several approaches were possible. By taking a sizeable sample of case reports and manually coding whether each event is cybercrime (according to our definition), we could gain an overview of the share of cybercrime among reported crimes in one year. However, we also aimed to do more detailed analyses and look at the extent to which cybercrime was investigated and solved within subcategories, such as financial cybercrime and online sexual abuse.[9] To do this, we would need a very large sample – probably several thousand cases – to get sufficiently precise estimates. Since manually reading and coding case reports is time-consuming, we found that using text mining and machine learning tools on the whole population of case reports might be a more efficient approach. To use these tools in a sensible way, however, we needed some labeled data – that is, we needed a sample of case reports that were coded according to a clear definition of cybercrime. We therefore chose to first manually code cybercrime in a random sample of about 1,000 cases, and then use these coded data to train a machine learning model to classify all reported cases as cybercrime or not.

The learning aspect was also an important basis for the decision to apply text mining and machine learning in this audit. This is in accordance with the international standards for performance audit issued by the INTOSAI. According to the *Performance Audit Standard*, ISSAI 3000, "the auditor shall be willing to innovate throughout the audit process."[10] It further says that performance auditors can identify opportunities to develop innovative audit approaches for collecting, interpreting, and analyzing information. An important reason for our decision to apply these techniques, and use the necessary resources, was to gain experience with these methods.

Manual Coding of Reported Crimes

The sample of manually coded cases was deployed to train and test a machine learning model. To provide an acceptable level of certainty, there must be a certain number of cases. The more cases, the better the model, but at the same time, this is time consuming work. It was decided between auditors and data scientists to classify 1,000 cases. Before we saw the end product, we were not sure if the machine learning model would be good enough to be used. Fortunately, we would in any case be able to use the result of the manually coded cases to estimate the proportion of cybercrime within the different case categories applied by the police.

The reports registered (reporting documents) with the police were not easily available. Reports were stored in a Documentum[11] database as objects in a file system. Retrieving cases for one year took a long time and came with several difficulties. All documents from the year 2018 were retrieved by a script developed by the Police IT Services. From a total of 296,345 cases, a selection of 1,072 cases was randomly sampled for manual review and classification. The sample of cases for manual review was selected with two objectives in mind:

1. To estimate the proportion of cybercrime within the different case categories the police use.
2. To enable the development and testing of a machine learning model for classifying all cases in 2018 as either cybercrime or *not* cybercrime.

To achieve the first objective, each case category had to be adequately represented in the sample. This was solved by stratifying the sample according to case categories. To achieve the second objective, it was necessary to have enough cases of cybercrime in the sample. Some case categories contained very few instances of cybercrime. We therefore chose to select a *disproportionately stratified sample* of 1,072 cases. The sample contained fewer cases from the traffic and physical vandalism categories because there was reason to assume that there were fewer instances of cybercrime in these categories[12] than the other six case categories.[13]

Before we started, we agreed upon the definitions of several important concepts, such as "cybercrime," "computer systems," "computer networks" and "key elements of the course of action," and how to operationalize them. We also made a code form. This form included the categories "case id," "crime category," our categorization of whether the case was cybercrime or not (*modus operandi*), and a column for comments or justification for coding practice.

To ensure that coding rules were equally understood, we conducted a pilot session in which four people individually coded the same 30 randomly selected cases. This was used as a starting point for obtaining a uniform understanding of the criteria. These cases were not included in the sample. Then, 100 cases were reviewed by the same four people (individually) to assess correlation between the coders. These cases were included in the final sample. The correlation coefficients for the measures were between either a "significant" degree of correlation (between 0.61 and 0.80) or "near-perfect" correlation (between 0.81 and 1). The discrepancies primarily concerned the coding of cases as "do not know." Of the 21 cases with discrepancies between one or more of the coders, there were only four cases that one or more of the coders coded as "Cybercrime," while others coded "not Cybercrime." This indicates that the discrepancies primarily concern how much information is required for being able to classify a case. In cases with sufficient information available, there is very little discrepancy in terms of the classification.

The rest of the sample of 898 cases (two casefiles ended up being empty and were taken out of the sample) was divided into two, with two groups of two people coding the same cases individually before comparing. Deviations in coding were jointly reviewed by the four coders to determine the final coding of the case. In addition to the 998 cases coded, we added 74 to be used for testing the machine learning model. Since these 74 cases were selected using the same selection criteria as the other cases, these cases were included in the manual classification used for estimating the proportion of cybercrime.

Modus operandi is a code in the criminal case register that is used to describe the method used by the perpetrator to commit the offense. Examples of this are the ICT *modus operandi* codes "By using a computer system" and "By exploiting errors/weaknesses in digital authentication solution." We did not consider the police coding of the *modus operandi*, as we did not want this to affect our categorization. We used the text of the reporting document as a starting point and coded the case as *cybercrime* when there were clear indications of cybercrime based on the text. Cases were coded as *not cybercrime* when there were clear indications of a modus operandi other than cybercrime. Finally, cases were coded as *do not know* if an ICT modus operandi was conceivable, but the information available did not allow for a certain conclusion to be drawn.

Table 6.1 shows how the cases are divided into case categories in the final sample and in the population of reported cases.

For six types of cases, three of which occurred relatively frequently, it was still difficult to definitively decide on the coding. In order to determine the final status of these cases, the opinions of a reference group consisting of five computer forensic investigators from the police and five police prosecutors from police districts and special agencies were obtained. It was decided to apply a conservative interpretation of the feedback so as not to overestimate the proportion of cybercrime.

Extracting Text Data and Preparation of Text for Quantitative Analysis

Text mining is a method for extracting text from large collections of documents. In this audit, text mining was used to extract the text from a total of 334,544[14]

Table 6.1 Population and sample of reported cases in 2018 by case category

Case category	*Reported cases 2018*		*Sample of cases*	
	Number	*Percentage*	*Number*	*Percentage*
Other	42,626	13.4	150	14.0
Drugs	35,309	11.1	148	13.8
Sexual offence	8,406	2.6	150	14.0
Vandalism	16,912	5.3	88	8.2
Traffic	54,107	17.0	86	8.0
Theft/illicit gain	99,447	31.2	150	14.0
Violence	32,716	10.3	150	14.0
Financial	29,425	9.2	150	14.0
Total	318,948	100.0	1,072	100.0

Source: National Police Directorate, BL/Strasak.

cases reported to the police in 2018, where 286,726 documents were retrieved from these cases. For 47,814 cases, the text only consisted of case descriptions and modus operandi summaries, and four only had *modus operandi* summaries.

The starting point for the classification was the text in the report to the police, as well as short textual descriptions of the case (summary of modus operandi and case description). The work with the model involved identifying and extracting relevant text from the reporting document (text mining). The text was reviewed and divided into words (tokens).[15] The frequency of each word in a text is then calculated relative to the frequency in all texts, i.e., of all cases in the entire population, including non-labeled cases of unknown class. This approach, called TF-IDF (*term frequency-inverse document frequency*), means that words that are more specific to a document (e.g., a report of cybercrime) and that are also unusual in other documents, are weighted higher.

The text was then broken up into individual words and converted to numerical variables. These were used as a basis for training a model based on known classes (*cybercrime* or *not cybercrime*).

Development of a Machine Learning Model

The manual text review described in the section on manual coding of reported crimes above resulted in the classification of about 1,000 cases. This manual coding was used as labels for the supervised training of a machine learning model.

Because different words can be used to describe similar things, grouping words into synonyms can make a significant difference in classification. In the context of cybercrime, it was for example meaningful to define synonyms for websites, social media, or verbs that describe actions related to cybercrime. Synonyms were used to merge words that occur infrequently. For example, social media platforms like Facebook, Snapchat, and TikTok were grouped as "social media." Meaningful synonyms strongly depend on the context of the application: For example, a synonym included a large variety of drugs, the difference of which would have been significant in a medical context but not regarding the likelihood of an essential cyber component in the crime. Involvement of domain experts in this part of the *feature engineering* is therefore important to ensure a good understanding of the subject matter. The cooperation between auditors and data scientists in feature engineering was essential to developing a machine learning algorithm that was aligned with the subject matter understanding of the definition of cybercrime. This definition was developed in cooperation with police experts and facilitated the development of synonyms that could be applied in the machine learning model. This collaboration in finding a meaningful definition maximized the chances of getting useful output from the statistical model.

Synonyms were also used to merge words that occurred too infrequently for a model to pick up on but which are a clear indicator of whether or not a case concerns cybercrime (e.g., *hack*, *computer system*) or not (e.g., *accounting violation*, *bookkeeping*). A special "synonym" was developed to identify sentences from what we call standard forms, i.e., forms that can be filled to report a crime. All words that are part of the standard text on the form could thus be removed, such that only the sentences written on the form itself were retained. Standard forms were sometimes filled out by hand, and handwriting was not possible to read into machine-readable font with optical character recognition (OCR). Where it was possible to identify from the form that a physical object had been stolen or lost, this information was treated as an artificial word and added to the synonym for infrequent words.

The model was trained on 75 input variables, most of which reflected the relative frequency of specific words in the text (TF-IDF). The choice of variables for the model, known as features, had a major influence on the outcome of the final model. In order to find words with large distinctive power, words were separately ranked by their relative frequency in the two classes of texts: 150 words with the highest TF-IDF weighting were first extracted from cybercrime cases, and 150 words with the highest TF-IDF weighting were extracted from cases classified as not cybercrime. Words found in both lists were then removed, and 70 words with the greatest difference were then selected from the two lists. We combined this list of feature words with a manually compiled list from domain experts that included words to be included or excluded. All features were weighted before being used in the model. The weighting was carried out by taking the natural logarithm of the frequency of a word in a document, relative to the total number of words in the document.

Machine Learning: Model Development and Performance Testing

Four different kinds of model architectures were tested:

- Naïve Bayes – a comparatively simple model that has the advantage that it is possible to understand how a certain input leads to a certain model output (also called "white box" type of model).
- Random forest, XGBoost, and Support Vector Machine – three more complex model types that are generally deemed more powerful, but more difficult to explain (the so-called "black box" type of models).

A "white box" model is generally preferred if it performs sufficiently well due to the possibility of explaining the model's prediction. This is particularly important if the model is to be used for decisions on single cases with consequences for individuals. The "black box" models need more effort to reconstruct[16] how particular input values lead to the model's output, information that can be important or even legally required if the model's prediction guides, e.g., certain decisions in public administration. For our application of the model to produce statistics on the total number of cybercrime cases, accurate prediction of the classes is more important than the potential to explain single case predictions.

Because the data contained many more examples of non-cyber than cybercrime, model performance was assessed by Mathew's correlation coefficient (MCC), which is symmetric between the two classes and independent of which class is regarded as the "positive" outcome:

$$MCC = \frac{TP \cdot TN - FP \cdot FN}{\sqrt{(TP + FP)(TP + FN)(TN + FP)(TN + FN)}}$$

TP, TN, FP, and FN refer to true positives, true negatives, false positives, and false negatives, respectively. MCC values can vary between −1 (perfect misclassification) and 1 (perfect classification).

The simple model (Naïve Bayes) was discarded due to a significantly worse MCC than all three other models and because explanation of single predictions was not crucial in the audit.[17] The three remaining models showed similar initial performance during training, so the model type that generalized best to unseen cases (Support Vector Machine) was deemed most promising and optimized further. The final model reached an MCC of 0.79 on unseen data (validation data).

In addition to MCC, we calculated the proportions of true cybercrime and non-cybercrime that are correctly classified, as well as the proportion of true cybercrime in the predicted cybercrime cases.[18] To get a conservative estimate of whether police *modus operandi* classification underestimates the amount of cybercrime, we deemed it more acceptable to misclassify a true cybercrime case as non-cybercrime (false negative) than the other way around. This way, our prediction gives a lower bound for the true amount of cybercrime. If the working hypothesis had been that the police overestimate cyber-related crime, the conservative approach would have been to minimize false negatives over false positives to get an upper bound on the true amount of cybercrime.

In all steps of the machine learning model development, seemingly technical decisions about meaningful input features, suitable model architecture, and sensible performance metrics depend on domain knowledge as well as the intended usage and final goal of the application. A team that combines qualifications in evaluation and technical development (data scientists) is essential to find the most appropriate solution for the task.

Table 6.2 shows the results from the machine learning model in terms of the number of cybercrimes identified.

Sensitive Data and Information Security

Our approach required the use of sensitive data, including detailed information on reported crimes, which involves some challenges. As a supreme audit institution, we have access to governmental data and a secure infrastructure in place

Table 6.2 Number of reports and percentage of cybercrime estimated per crime category

Type of crime	*Number of reports*	*Percent cybercrime**
Financial	29,578	55.0
Sexual offence	8,438	16.6
Other	40,003	5.8
Violence	33,171	1.6
Environment	1985	1.1
Illicit gain/theft	100,854	0.8
Investigation cases	12,017	0.5
Drugs	36,292	0.2
Working environment	863	0.1
Vandalism	17,068	0.1
Traffic	54,275	0.0
Total	334,544	-

Source: Office of the Auditor General of Norway.
*Percentage of cybercrime in the entire population estimated from machine learning prediction.

for handling sensitive data. We stored and analyzed the crime data on servers specifically set up for sensitive data. Only designated personnel were allowed to access these servers, and analysis tools had to be security assessed. For programming languages with open-source packages, as used here, this means separate security assessment of single libraries and restrictions on the novelty of the analysis approach.

Challenges, Success Criteria, and Lessons Learned

Challenges with Data Quality and Time Consumption

Machine learning was considered a suitable approach in this performance audit due to the large number of police reports that would otherwise have to be read and manually classified. Another aspect is the well-defined task of binary classification based on a clear definition of cybercrime, which makes machine learning a suitable tool.

An essential precondition for using machine learning is the availability of enough data with sufficient quality. Data was not easily available in this audit, however. The case management system used in the Norwegian police dated back to the early 1990s, and police reports were stored as PDFs in a database. We therefore spent considerable time simply obtaining the documents (mostly PDFs) and extracting text from them. Only after these steps, we could begin with the time-demanding analytical tasks of reviewing the text and dividing it into tokens, labeling and training a model, and testing machine learning models. As a result, we spent more resources on this audit than we normally do. However, this cost was accepted to gain experience and knowledge of how these methods could be applied in future audits.

Data quality is essential for any type of analysis, including the use of text mining and machine learning (Gudivada et al., 2017). Data quality depends on practices in the organization responsible for gathering, sorting, and storing data. The quality of and access to data must be properly investigated before deciding on the research design. If possible, how data is stored, as well as the quality and richness of the data, should be assessed early on. The original idea was to use the *modus operandi* registered by police officers when entering a report in the police's case management system. It turned out, however, that the quality of the data was not satisfactory, so we opted for the more time-consuming approach of retrieving and analyzing police reports. While the challenges of one type of data may not apply to another, we believe this audit has made our organization better prepared to assess whether the questions and data available allow for the gainful use of text mining and machine learning tools.

When we opted for retrieving police reports, other challenges materialized. What seemed like a trivial task was quite complex. There were data quality issues, and checking the data for faults and making it ready for analysis were resource intensive. Retrieving data from legacy IT solutions is a difficult task. However, we agreed that learning by doing is a good principle. Text mining and machine learning are new tools available for use in audits, and it is only by adopting these methods in our work that we learn how to use them. Technological development also creates new tools that make it easier to retrieve data from old applications. This is of course important to consider regarding our own capacity and resources, but one should also keep in mind the work it entails for the audited entity. If we had a tight deadline that could not be pushed, we would not have been able to include these kinds of methods, as the police would probably not have been able to deliver the data soon enough.

Another important aspect of data quality is label quality. Manual coding was a time-consuming effort due to the necessity of obtaining objective labels that all auditors could agree on and that reflected a common understanding of cybercrime. In addition, we needed to code a considerable number of cases within different types of crimes to be able to identify cybercrime within these crime types. However, we found that, in our case, about 1,000 coded cases were sufficient to train a machine learning model with solid predictive performance.

Success Criteria

To establish a robust foundation and gain a comprehensive grasp of the information at hand, it was imperative to *conduct early assessments of the data*. These initial assessments revealed quality issues with the police data. To ensure the integrity of our analyses, we decided to rely on the full case reports provided by victims or entities reporting crimes rather than the criminal case register. While the criminal case register was more easily available, it proved to be of low quality. Opting for the case reports required more resources, but it was necessary to ensure sufficiently high validity.

Several prerequisites were essential for employing text mining and machine learning. Perhaps the most critical was the presence of *a well-defined problem or question that lends itself to text mining and machine learning techniques*. In our case, the task was to classify criminal cases into cybercrime and non-cybercrime, which in turn was essential for evaluating the police's overview of and handling of this type of crime. This is a suitable task for algorithmic models and guided every aspect from data collection and preparation to algorithm selection and interpretation of results.

Another prerequisite was the availability of the necessary skills and resources. Assembling an *interdisciplinary team with competence on the subject matter, evaluation methods, statistics, and data science* was instrumental to succeed in the use of machine learning and text mining in this evaluation. The application of such methods in a performance audit has provided valuable lessons for everyone working on the audit, data scientists and auditors alike. Together, we gained experience with these methods, and the methods gave us information of importance to the evaluation and to the Norwegian police. Before the evaluation, the Norwegian police did not have reliable data on cybercrime. Neither did they know to what extent such cases were being investigated and solved.

Furthermore, we learned that *quality assurance is pivotal*. This encompasses for example measuring the performance of several machine learning models to decide which model to use. Quality assurance was in many cases based on domain knowledge from evaluators as well as technical knowledge from data scientists. The combination of knowledge was applied to make decisions on input features, model architecture, and performance metrics. A team that combines qualifications in evaluation and data science was essential to find the most appropriate solutions. Another important aspect that contributed to the quality of the analysis was spending time on building common understanding of how to classify cybercrime, how to agree upon definitions, and how to operationalize key concepts.

To mitigate challenges related to uncertainty around the concept of cybercrime, we employed a *reference group* from the evaluated entity (the police) to advise us on the classification of cases we were uncertain of. By using a reference group that consisted of people working in the police, even if this was an expert group, we can make the results more acceptable to others who do not fully understand the method.

Collaboration Between Auditors/Evaluators and Data Scientists

The collaboration between auditors and data scientists was essential to make good use of the new methods in this audit. For a discussion on the emerging collaboration practices in general between evaluators and data scientists, see Nielsen (2025) in this volume. Since we are all employees at the Office of the Auditor General of Norway, we could collaborate closely and have frequent discussions. Our roles and tasks were also not very clearly separated. For instance, the auditors provided important input to the text mining and machine learning analyses, and the data scientists helped discuss how to use the machine learning results to answer the audit questions. Nonetheless, the auditors and data scientists contributed in somewhat different ways with their respective competencies.

The main contribution from data scientists was the use of text mining and machine learning to classify criminal cases. Mastering such tools takes considerable time, and without the data scientists' existing competence, we would not have been able to use them in this audit. The data scientists also provided important input in the initial discussions about how to approach the classification problem. Without their input, the idea of using these tools may never have materialized.

The auditors handled many other tasks in the audit, including research design, data collection, analysis, and writing. These tasks required both subject matter knowledge as well as competence in performance audit and social science methods more broadly, which the auditors possessed. However, the auditors also contributed to the task of identifying cybercrime. They assisted in identifying data quality issues early on, which was helpful in guiding the selection of relevant data sources. Further, they used their subject matter knowledge to form a systematic definition of cybercrime and a clear coding scheme. They also conducted the manual coding of cases, which provided a basis for developing the machine learning model. And in the evaluation of the algorithmic model, the auditors contributed to the interpretation and analysis of results to be certain that these were in line with the objectives of the audit.

The iterative process of cooperation between data scientists and auditors was in our experience overall important and contributed in a major way to the result. To ensure that machine learning and text mining were appropriately used and integrated with the rest of the audit, we needed not only competence in the text mining and machine learning tools but also subject matter knowledge and competence in audit methods. In other words, the data scientists were vital to employing these advanced methods, while the auditors were vital to maximizing their value to the broader evaluation.

Notes

1 Performance audits are, broadly speaking, evaluations of government activities focusing on economy, efficiency, and/or effectiveness.
2 The Office of the Auditor General of Norway is a Supreme Audit Institution (SAI).
3 Cybercrime includes crimes aimed at computer systems and/or networks and crimes where key elements of the action are carried out using computer systems and/or networks.

4 These questions are also dealt with in York and Bamberger (2025) and Nielsen, S.B. (2025).

5 Ministry of Justice and Public Security (2015) *The Ministry of Justice and Public Security's strategy for combating cybercrime,* issued by the Ministry of Justice and Public Security on 26 June 2015.

6 We conducted a number of interviews and a survey of Digital Policing Units, and reviewed relevant documentation. These approaches were mainly aimed at exploring what factors impede the investigation of crime and the effectiveness of the efforts of the Ministry of Justice and the Police Directorates to combat cybercrime.

7 Since 2018, the Norwegian police have introduced their own *modus operandi* codes in the BL case processing system in order to identify cases of cybercrime. Examples are "When using a computer system" and "When exploiting errors/weaknesses in digital authentication solution."

8 The electronic criminal case processing system in Norway is called BasisLøsning (Basic Solution) or BL.

9 We also had an idea of breaking down these numbers by police districts to compare performance among the districts. This idea was abandoned partly because there were problematically few districts.

10 INTOSAI, 2019 ISSAI 3000 *Performance audit standard.* INTOSAI Standards are issued by the International Organisation of Supreme Audit Institutions (INTOSAI) as part of the INTOSAI Framework of Professional Pronouncements. For more information, visit www.issai.org.

11 Documentum is an enterprise content management platform owned by OpenText. The core platform manages content in a repository consisting of three parts: a content server, a relational database, and a place to store files. Items in the repository are stored as objects. The file associated with an object is usually stored in a file system; the object's associated metadata (file name, storage location, creation date, etc.) are stored as a record in a relational database. In our case, the reports we wanted were stored as files (pdf, gifs, etc.).

12 Interviews in the preliminary study and review of the statistics indicated that there was a low prevalence.

13 We divided the cases into eight case categories (Table 1). The categorization is the same as that used by the police, albeit with two adjustments: the categories "environment" and "working environment" were placed under the category of "other" because there were few cases in those categories.

14 The difference between the number 318,948 used in Table 1 and 334,544, which is being used here, is that sub-cases are included. Investigation cases and partial cases are not part of the official statistics and are not included in the number 318,948.

15 Tokens can consist of several words or phrases; for example, «social media» was one token. We will refer to words in the following for simplicity and readability.

16 One method to explain predictions of a "black box" model is with the help of the so-called surrogate models, which approximate parts of the black box model. Those explanations are, however, approximate and the validity of the surrogate model has to be justified.

17 The goal of our application was to compare statistics of cybercrime vs. non-cybercrime, based on the whole population rather than just a sample, with the police's own categorization. For example, if the police wanted to use a similar algorithm to classify cases for the purpose of deciding on further processing of individual cases, transparency of the decision-making process and thereby the working of the algorithm becomes important.

18 With cybercrime as the «positive» class, specificity measures the proportion of non-cybercrime cases that are correctly classified, sensitivity measures the proportion of

cybercrime cases that are correctly classified, and precision measures the proportion of ICT classifications that are correct. Our model reached specificity 0.99, sensitivity 0.79, and precision 0.79. All these measures are interesting; however, there is major uncertainty associated with both sensitivity and precision because they are heavily dependent on cybercrime classes that have few observations. Moreover, we deemed false negatives more acceptable than false positives. Therefore, more emphasis was placed on high specificity, i.e., fewer false positives, where cybercrime corresponds to the positive class.

References

Akdemir, N., Sungur, B., & Basaranel, B.U. (2020). Examining the challenges of policing economic cybercrime in the UK. *Güvenlik Bilimleri Dergisi* (*International Security Congress)*, Special Issue, 113–134. https://doi.org/10.28956/gbd.695956

Arthur A., Rydland L.T., & Amundsen K. (2012). The user perspective in performance auditing–a case study of Norway. *American Journal of Evaluation, 33*(1), 44–59. https://doi.org/10.1177/1098214011408283

Beckstrom, J. R. (2020). Auditing machine learning algorithms: A white paper for public auditors. *International Journal of Government Auditing, 48*(1), 40–41.

Bruce, K., Vandelanotte, J., & Gandhi, V. (2025). Emerging Technology and Evaluation in International Development. In S.B. Nielsen, F. Mazzeo Rinaldi and G.J. Petersson (eds.). *Artificial Intelligence and Evaluation. Emerging Technologies and Their Implications for Evaluation* (pp. 13–36). London: Routledge. https://doi.org/10.4324/9781003512493

Bossler, A. M., & Berenblum, T. (2019). Introduction: New directions in cybercrime research. *Journal of Crime and Justice, 42*(5), 495–499. https://doi.org/10.1080/0735648X.2019.1692426

Gatto, L., & Bundi, P. (2025). The Use of Quantitative Text Analysis in Evaluations. In S.B. Nielsen, F. Mazzeo Rinaldi and G.J. Petersson (eds.). *Artificial Intelligence and Evaluation. Emerging Technologies and Their Implications for Evaluation* (pp. 144–167). London: Routledge. https://doi.org/10.4324/9781003512493

Gordon, S., & Ford, R. (2006). On the definition and classification of cybercrime. *Journal in Computer Virology, 2*, 13–20, August 2006. https://doi.org/10.1007/s11416-006-0015-z

Gudivada, V., Apon, A., & Ding, J. (2017). Data quality considerations for big data and machine learning: Going beyond data cleaning and transformations. *International Journal on Advances in Software, 10*(1), 1–20. https://www.issai.org/pronouncements/intosai-p-1-the-lima-declaration/

INTOSAI. (1977). *INTOSAI-P1: The Lima Declaration.* The International Organisation of Supreme Audit Institutions (INTOSAI). https://www.issai.org/pronouncements/intosai-p-1-the-lima-declaration/

INTOSAI. (2019). *ISSAI 3000 Performance Audit Standard.* The International Organisation of Supreme Audit Institutions (INTOSAI). https://www.issai.org/wp-content/uploads/2019/08/ISSAI-3000-Performance-Audit-Standard.pdf

Mazzeo Rinaldi, F., Celardi, E., Miracula, V., & Picone, A. (2025). Artificial Intelligence and Text Analysis in Evaluating Complex Social Phenomena. The Russia-Ukraine Conflict. In S.B. Nielsen, F. Mazzeo Rinaldi and G.J. Petersson (eds.). *Artificial Intelligence and Evaluation. Emerging Technologies and Their Implications for Evaluation* (pp. 168–195). London: Routledge. https://doi.org/10.4324/9781003512493

Nielsen, S.B. (2025). The Evaluation Industry and Emerging Technologies. In S.B. Nielsen, F. Mazzeo Rinaldi and G.J. Petersson (eds.). *Artificial Intelligence and Evaluation. Emerging Technologies and Their Implications for Evaluation* (pp. 266–286). London: Routledge. https://doi.org/10.4324/9781003512493

Office of the Auditor General of Norway. (2021). *Undersøkelse av politiets innsats mot kriminalitet ved bruk av IKT [Study of Police efforts against cybercrime].* Dokument 3:5 (2020-2021). https://www.riksrevisjonen.no/rapporter-mappe/no-2020-2021/undersokelse-av-politiets-innsats-mot-kriminalitet-ved-bruk-av-ikt/

Otia, J. E., & Bracci, E. (2022). Digital transformation and the public sector auditing: The SAI's perspective. *Financial Accountability and Management*, *38*(2), 252–280. https://doi.org/10.1111/faam.12317Cit

Phillips, K., Davidson, J.C., Farr, R.R., Burkhardt, C., Caneppele, S., & Aiken, M.P. (2022). Conceptualizing cybercrime: Definitions, typologies and taxonomies. *Forensic Sciences, 2*, 379–398. https://doi.org/10.3390/forensicsci2020028

Thomas, D., & Loader, B. (2000). Introduction – Cybercrime: Law enforcement, security and surveillance in the information age. In D. Thomas & B. Loader (eds.). *Cybercrime: Law Enforcement, Security and Surveillance in the Information Age.* London: Routledge.

York, P. & Bamberger, M. (2025). The Applications of Big Data to Strengthen Evaluation. In S.B. Nielsen, F. Mazzeo Rinaldi and G.J. Petersson (eds.). *Artificial Intelligence and Evaluation. Emerging Technologies and Their Implications for Evaluation* (pp. 37–55). London: Routledge. https://doi.org/10.4324/9781003512493

Ziulu, V., Anuj, H., Hagh, A., Raimondo, E., & Vaessen, J. (2025). Extracting Meaning from Textual Data for Evaluation. Lessons from Recent Practice at the Independent Evaluation Group of the World Bank. In S.B. Nielsen, F. Mazzeo Rinaldi and G.J. Petersson (eds.). *Artificial Intelligence and Evaluation. Emerging Technologies and Their Implications for Evaluation* (pp. 78–102). London: Routledge. https://doi.org/10.4324/9781003512493

7 Big Data for Big Investments

Making Responsible and Effective Use of Data Science and AI in Research Councils

Jon Holm, Denis Newman-Griffis, and Gustav Jakob Petersson

Introduction

Only a few years ago, the revolution of Big Data and artificial intelligence (AI) was almost exclusively a phenomenon of the private sector (Petersson & Breul, 2018). Today, the public sector at large is showing increasing interest in new data sources and new processing techniques; new ways of creating, collecting, analyzing, and presenting data are being explored for new purposes. To enhance developments in public use of data, agencies, government investigations, and digital infrastructures are being launched, nationally as well as internationally (Misuraca & van Noordt, 2020).[1] Alongside these infrastructures, new policies and methods of scrutiny for data use are being developed to ensure that data are being used responsibly in service of the public (Janssen et al., 2020).

While researchers show growing interest in this shift toward data-driven policy, the research system itself is increasingly being analyzed in novel ways. In addition to sector-internal goals of using data to enhance productivity and quality of decision-making, increased political pressure to document that investments in research are relevant to society at large is driving a need for new ways to analyze research, as well as research policies and funding instruments (Reale et al., 2018).

The shift toward evaluation in terms of societal goals – and toward the use of data to drive this evaluation – is partly associated with a shift in how research policies motivate public spending on research. Such motivations can be formulated at the level of the individual (quest for knowledge), organizations (competitive universities and industries), or society at large (democracy, green transition, the United Nations' Sustainable Development Goals [SDGs]). Different levels of motivation may of course coincide. Still, the general tendency in the development of post-World War II research policy has been to lift the motivation for public spending on research to include societal goals. Since the turn of the millennium, investments in basic research are increasingly also

DOI: 10.4324/9781003512493-7

framed as a means to better understand and tackle societal challenges (Schot & Steinmueller, 2018).

This shift has particular implications for government research funders as instruments of public policy. Traditionally, targets for public research funding were mainly defined through research internal parameters such as recruitment and scientific capacity, as well as quality – parameters for which measurement and data are fairly well established and understood. Lifting the motivation for research policy to the societal level introduces new challenges and needs for richer information about research. On the one hand, it creates incentives to allocate funding by funding programs targeted at societal needs, and to focus measures on societal alignment. On the other hand, it is well established that some of the most important discoveries from research have come out of curiosity-driven foundational research, which often suffers in contemporary evaluation assessing clear near-term impact (Stokes, 1997). The standard reference for this type of documentation is the 1968 United States National Science Foundation's report on *Technology in Retrospect and Critical Events in Science* (TRACES) (Franck, 1969). Recent examples include research on mRNA, previously dismissed as frivolous, that enabled a rapid development of COVID-19 vaccines and was recognized with a 2023 Nobel Prize (Smith, 2022; Nobel Prize Outreach AB, 2023).

Funders are therefore faced with dual pressures. There is, first, a need to balance the policy drive for steering research activities toward predefined societal goals with the openness for unintended results and great societal benefits that may come from curiosity-driven research. On the other hand, the movement toward data-driven policy creates needs for data and insights aligned with each of these goals. Importantly, funders must ensure that this increasing use of data to support policy remains firmly aligned with their public mission and the benefit of society.

All this calls for methodological innovation among research evaluators. In this chapter, we examine opportunities for *data-driven learning* in research funding and evaluation, drawing on advances in data science approaches that can help research funders learn from the rich data on research applications and outcomes increasingly available to them. We illustrate how new affordances in machine learning (ML), which helps identify and leverage patterns in data, and artificial intelligence (AI), which helps apply observed patterns and expert knowledge in analyzing new data,[2] can add valuable tools to the funding and evaluation toolbox when effectively adapted. We further highlight key considerations in the process of AI/ML adoption and adaptation, and how approaching this process in a careful and context-sensitive way is key to achieving responsible use of data and AI that are aligned with ethical standards as well as societal benefit.

We first provide an overview of the changing landscape of research funding and evaluation and how this aligns with new data needs, including examples of how AI and ML approaches are already being explored in this space. We take two applications of ML in the Research Council of Norway as case studies of the process of integrating AI into funding and evaluation practices. Drawing on these case studies, we describe key lessons learned and questions raised for informing similar efforts. Our aim is for our discussion to serve as a starting point for exploring the process and practice of integrating the responsible use of AI and machine learning into the work of research funding and evaluation, with broader lessons learned that speak across contexts of AI use.

Challenges in Aligning Research to Societal Needs

The role of research policy is broadly speaking to motivate public spending on research. Since World War II, the general tendency in research policy has been to lift the motivation for public spending on research to increasingly higher levels (Schot & Steinmueller, 2018).

However, the models we have for understanding how research contributes to societal development are at best incomplete. The pathways from a research finding to its implications for society are complex, with many factors playing a role – both inside and outside of the research system. Expectations for research to help address societal challenges are followed by increased interest among research funders in new, growing, interdisciplinary fields of research as particularly impactful, both academically and societally (Gooch et al., 2017). New interdisciplinary fields of research are not easily mapped with traditional classification systems, and interdisciplinary research and impact are notoriously difficult to measure (Mansilla et al., 2006). At the same time, the peer review processes frequently adopted in research evaluation are under critique for being biased, conservative, and time-consuming, and thus motivating the need for additional measures to supplement peer review.

Since the turn of the millennium, investments in basic research are also increasingly framed as a means to better understand and tackle societal challenges (Shneiderman, 2018). One example of this framing is found in the Norwegian *Long-term plan for research and higher education 2019–2028* (Ministry of Education and Research, 2018): Research and higher education play a key role in the development of a society that is environmentally, socially, culturally, economically, and politically sustainable. Adequate knowledge is a prerequisite for making decisions that make it possible to sustain prosperity and welfare, preserve a planet at risk of overload, and protect fundamental values such as freedom and democracy.

The task of evaluation in a research council has thus in some parts come to resemble that of any agent of public policy: to document outputs, outcomes, and impacts of a portfolio of projects and assess to what extent the observed results may be attributed to the policy intervention. This is fundamentally different from more traditional assessment of research projects and groups that in most cases are carried out by peers from the scientific field, focusing more narrowly on the scientific quality of the project or group.

Opportunities for Data-Driven Research Evaluation

Getting the balance right between top-down steering through policy-driven research programs and curiosity-driven response-mode funding is the hallmark of any well-functioning research system. This does not mean that basic funding is exempt from demonstrating its societal relevance – rather, we need to acknowledge the value of curiosity-driven research (in itself and potentially for society) and that there may be many different pathways from research activities and results to societal outcomes and impacts. Effective use of data can help demonstrate this value for both curiosity-driven and policy-driven evaluation.

The two modes of funding come with different challenges for evaluators. Top-down funding instruments normally address explicit societal goals. The typical procedure for planning such interventions is to establish a program logic that translates the target goals into research questions by mapping societal challenges to concrete knowledge needs and actions that can be stimulated through a call for research proposals. When evaluating a funding instrument with a clearly defined program logic or another type of roadmap, the evaluator will know what type of data to look for through all the phases of the projects. Expected outputs, outcomes, and impacts are typically described explicitly in the call for proposals and should be mirrored in the received proposals. The primary challenge for the evaluator is then to harvest and integrate the data needed to assess program success.

Often, funding councils and sponsors in government will be satisfied when the societal impact of a research funding program has been documented. However, the key mechanisms behind this success may yet remain opaque, preventing us from learning: Why did one project succeed in delivering societal value and not another? We argue that there is potential for using data science more systematically to help answer this deeper question. For example, machine learning could help identify specific characteristics of successful projects, such as types of cooperation, project team qualifications, and properties of the project implementation. This knowledge could in its turn be used to maximize program effectiveness by introducing the same characteristics as project requirements in the next call for proposals.

When it comes to curiosity-driven response-mode funding, many research funders have limited their evaluations to documenting and assessing the scientific quality and impact of the funded research based on bibliometric data, sometimes supported by qualitative data like sample publications and impact case studies. Research councils eager to secure sustained public support for investments in basic research still need to find data and methods that may provide support to the much-repeated claim that investments in basic research will lead to scientific breakthroughs that in turn produce societal benefits. The difficulty is that we do not know where to look for evidence, as there is no specified roadmap for such funding instruments to identify expected societal impact. In the absence of a roadmap, evaluators have to ask the researchers themselves to report on how their research has influenced society. Such self-reporting is common in many large-scale research assessments like the Research Excellence Framework in the UK, contributing substantially to the burden of reporting in Higher Education.[3]

Methods from data science may give research funders better knowledge about how basic research is supporting societal goals without increasing the burden of reporting for project leaders and their institutions. The digitalization of research data, publications, and citations in academic and non-academic media – together with the introduction of transnational persistent identifiers for publications, researchers, and their organizations – presents rich analytic opportunities: for example, research outputs may be linked to societal targets or social changes[4] by processing large corpora of research publications with natural language processing (NLP) (Newman-Griffis et al., 2021a), and social media discussions of research may be used to support narratives of engagement and impact.

As discussed by York and Bamberger (2020) and two other chapters in this volume (York and Bamberger (2025, Ch.3) & Nielsen (2025, Ch.13)), evaluation of public policy may gain both in efficiency and effectiveness by more systematically leveraging the learning opportunities that big data offers. In terms of efficiency, a better use of the vast sources of data already available through the internet or in public archives may reduce the cost of bespoke data collection for singular evaluation projects. Integrating machine learning and other algorithmic methods into evaluations may also increase their effectiveness by enabling data analysis at the population level, making it possible to synthesize a broad set of variables that cover regions, or national and international data. Still, shifting to broader and more diverse data for evaluation and analysis poses significant challenges in implementation and quality assurance but has the potential to drive new and more responsive modes of research analysis.

Explorations of AI and ML in Research Funding and Evaluation

Applications of AI techniques, including ML, are already being explored to inform analysis and decision-making in research funding and evaluation. Most prominently, progress is being made in analytical work linked to the processes of selecting and following up on research projects, as well as the evaluation of published research (Holm et al., 2022). At the same time, new ways are being explored to use AI to help trace impacts in academia as well as in society at large (Rosemberg et al., 2021). Innovations include the introduction of new analytical techniques as well as novel methods for capturing new kinds of data, as much of the data needed to assess societal impact is not readily available. AI is also being explored to assist in the assessment of research quality and the actual grading of research proposals or publications, for instance, in research evaluation (Thelwall et al., 2023; Williams et al., 2023). None of these innovations are yet standard practice, and experiments tend to be characterized by various shortcomings such as limited representativeness of data, lack of transparency into AI operations, and scarcity of guidelines and common standards for the responsible use of AI in the funding context.

There is thus a clear risk that the use and integration of AI into existing processes may become ad hoc, highly dependent on combinations of rare expertise, and inappropriately aligned with strategic goals and funders' public missions (Madaio et al., 2020). Without better, more context-driven guidelines for integrating AI into evaluations, traditional evaluation units may be marginalized because of the latency and costs inherent in tailor-made evaluation design and data collection. On the other hand, machine learning and other algorithmic methods, used with a focus on concrete problem solving, are often blind to the biases inherent in data and lack reflection on how target groups may be affected positively or negatively by choices of methods and data sources (Mehrabi et al., 2021).

It is clear that AI will affect both how research is performed and how it is evaluated. With appropriate design, AI use may be transformative in enabling funding processes to become more agile and responsive to changes in research systems. *We expect a deep-seated awareness of strategic and organizational context to be essential to realizing this transformative change. Approaching AI use as a highly contextualized process will not only help achieve the full potential of the digital revolution but critically will help effectively manage intended as well as unintended consequences and assure transparency and explicability of the algorithms in use.*

To illustrate some of the opportunities and challenges at hand, we will present two cases from the Research Council of Norway that may serve as a starting point for a more general discussion on how research funding bodies could use machine learning and AI to support their mission.

Examples of Using Machine Learning to Link Research to Societal Goals: Two Case Studies

In principle, a research council may work on the alignment of research to societal goals along the whole process of funding and grant management: call formation, selection of projects, grant management, and follow-up of completed projects. Efforts in optimizing the societal impact of investments in research have traditionally been based on research on research and program evaluation. Data science methods offer new opportunities for analysis of a funding portfolio. As a starting point for discussing how research funders may integrate methods from data sciences into their analytical toolbox, we present two examples from the Research Council of Norway.

The first example is related to the programming phase of research funding, asking what types of research will be most effective in supporting a green transition of the Norwegian maritime industries. The second example illustrates that the societal return on investments in research may be documented by mining existing data on funded projects, their publications, and citations (beyond academic venues) using machine learning algorithms.

Case Study 1: Analysing Research Fields in a Call for Proposals to Align to Societal Needs

In 2021, the Research Council of Norway (RCN) was mandated to assess the research knowledge available for developing maritime industries in Norway. The assessment was to serve as input to a new national strategy, laying the grounds for a green transition in these industries. The assessment included the use of clustering methodologies to support broader and more effective discovery of related research. We provide an overview of the study to illustrate many of the steps required for funders to effectively integrate data-driven methods into their work.

Context

In research evaluation, defining the evaluation object is sometimes a challenging task. This may be particularly so if the evaluation object is a field of research that is not included in conventional classifications, i.e., well-established research disciplines.

For example, it is difficult to describe the emergence of new research areas, interdisciplinary fields, or collaborations with such classification, as well as to describe research based on other categorizations (such as UN's sustainable development goals, climate, integration, digitization, and AI).

At the same time, there are advantages to classifications that remain constant over time, and conventional classifications will undoubtedly still be needed in the future. They make it possible to make comparisons over time and to compare with official statistics that use the same classification, such as research resources and personnel. However, for other purposes, this inertia creates various difficulties. We can therefore conclude that they are not sufficient on their own to meet the range of purposes that research evaluation entails.

This combination of clear value from expert-derived standard classifications and the need for flexibility to emerging fields creates a strong value proposition for using data to build on the knowledge embedded in existing classifications. This study explored the value to RCN of expanding on standard classifications using data-driven discovery to explore relevant research.

Methods

Box 7.1: Outline of Case Study 1, examining interdisciplinary assessment of maritime industries knowledge by RCN. We highlight here the motivating context and design decisions informing the use of machine learning (ML) as part of the analysis.

Case Study Outline: Interdisciplinary Assessment of Maritime Industries Knowledge

A typical task for a research council is to provide strategic advice on public investments in research. Such advice often includes an assessment of current research capacity and knowledge gaps. As research policy goals are often defined by targets outside of the research system, the identification of the knowledge needed to reach the targets is not straightforward.

Strategic goal motivating ML integration: Assess research knowledge to support developing maritime industries in Norway, in the context of a new national strategy.

Operational goal to integrate ML into: To assess the available knowledge, the first task is to identify the knowledge that might be relevant for the green transition of the maritime industry.

Limitations of the current (non-ML) process: Traditionally, such identification has been performed through keyword searches. Because this approach depends on authors' awareness of the relevance of their research for the purposes in question, it does not capture the actual connections within the research ecosystems and between research and its areas of applications.

Potential application of ML: Algorithmic methods can help make it possible to map the actual connections within research ecosystems based on broader characteristics of the research itself, rather than specific keywords alone. To discover similarities and connections, clustering methodologies are effective tools.

Data available to inform ML analysis: RCN tested clustering methods based on citation links between publications (bibliographic coupling and co-citations) and textual resemblance (topic modeling).

Opportunity for ML-informed ***categorization*****:** To identify outputs that may not include relevant keywords but can be aligned with known categories, a combination of bibliographic data and textual content could be analyzed to match uncategorized outputs with known exemplars.

Opportunity for ML-informed ***discovery*****:** To discover new, growing interdisciplinary research areas, topic modeling on textual content could be used to find outputs discussing similar topics, methodologies, etc., regardless of known categories.

Box 7.1 describes the design and implementation methodology used in the example case and how these methods were derived from strategic and operational goals within RCN. An initial set of publications to use as a starting point for clustering was constructed by a combination of publications reported from projects funded by a program for maritime research and publications in journals dedicated to maritime research. From a starting publication set of around 1,000 publications, the algorithms (Traag et al., 2019) typically generated a publication set of around 20,000 publications and grouped these into a couple of 100 clusters based on similarity in citation networks (bibliographic coupling and co-citations) or textual content (topic modeling based on Latent Dirichlet Allocation (Blei et al., 2003)).

The primary method in this example was unsupervised learning via clustering (i.e., without aiming to recover specific "true" labels). Still, the unsupervised algorithmic method required prior knowledge of the scientific field to be analyzed in order to be useful. This is because the clustering algorithm used a set of publications known by experts to represent the field in question as its starting point (here: maritime research); experts also reviewed the clusters produced by the algorithms to determine how they informed the overall assessment goal. The validity of the obtained classification is thus dependent on available expert knowledge. This illustrates the importance of differentiating between the knowledge used to inform machine learning directly (e.g., categories and labels) and the knowledge informing the design of the overall analysis. The machine learning in this example was unsupervised, but the overall design and implementation were highly supervised by subject matter expertise.

Findings

The algorithmic method provided clear value beyond the experts' prior knowledge by bringing thousands of potentially relevant publications into the dataset for analysis. Experiments showed that the citation-based methods (bibliographic coupling and co-citations) produced the most promising results by identifying groups of papers that appeared more coherent and easily identifiable than the text-based method. The difficulty in obtaining clear results from text-based methods might be explained by the interdisciplinary nature of maritime research, which combines studies of natural phenomena (and their effect on maritime constructions), various maritime technologies, and the social and political context of human activities on and in the oceans. The lack of a specialized vocabulary across such various fields of research makes it more difficult to determine thematic proximity between research publications based on the linguistic representation of the object of study (Bracken & Oughton, 2006).

Implications for Funder

After the expansion of the dataset and the grouping of publications, the clusters produced by the algorithm were assessed by subject matter experts at RCN to identify groups relevant to maritime research in general and more specifically to the green transition of the maritime industry.

RCN used the resulting dataset for further bibliometric analysis, under the assumption that new and emerging scientific fields relevant to the development of the maritime industry in Norway could be detected by looking at citation chains. It should be noted as a major limitation of this method that there may exist strategically relevant links between disciplines or research themes that are not represented by actual citations. Such links could include the societal context in which maritime technologies are developed or potential environmental effects of maritime technologies. Furthermore, a citation shows that a paper has received some attention; it does not show the specific relevance of the cited paper to the citing paper. In practice, there are various types of references in a research paper, whereas only a subset refers to the subject matter of the study. Sometimes, publications were grouped together based on more general references to common methods or high-level political frameworks, like UN SDGs. Groupings based on such general references were excluded from the analysis.

Despite these limitations, the machine learning approach provided added value to the analysis through its ability to discover potentially relevant topics outside of what is acknowledged as maritime research in RCN programs to date. The limitations of the method were mitigated by using expert knowledge to curate the initial dataset and to assess results in several iterations.

Case Study 2: Analysing Research Outcomes and Impacts to Optimise Societal Benefits

In 2020, RCN commissioned a study to establish a methodology to assess the societal impact of research and research-based innovation across all funding schemes.[5] The methodology, delivered by Technopolis Group (Rosemberg et al. 2021), piloted the use of new data sources and used machine learning for classifying results, outcomes, and impacts according to societal goals. We present the model here in brief, before discussing the advantages and disadvantages of such methods compared to traditional evaluation methods otherwise used by RCN.

Context

Evaluating the societal benefits of research (outcomes and impacts) can be highly challenging for a number of reasons. Many – perhaps even most – societal impacts of research were unintended at the time the research itself was initiated. Identifying impacts and impact pathways of research can be especially challenging when doing so *post hoc* in the absence of an initial roadmap (i.e., a theory of change) for the research. Research is an inherently uncertain endeavor, which limits our ability to decide on particular investments based on estimates of future benefits (Wallace & Rafols, 2015). Evaluation must therefore be flexible to uncertain and unanticipated impacts.

Societal impacts of research can also be very diverse and dependent on the specific research area and application. Some fields – such as biotechnology, genetics, and artificial intelligence – may have a broad reach. Other fields may have impacts that are significant in a more local context. Most fields will see a combination of broad and local impacts. Evaluation must therefore account for different scopes of impact.

Societal impacts often emerge over a long period of time and involve multiple stages, making it difficult to link specific research activities with specific impacts. A wide range of factors, such as culture, politics, and economic conditions, may also influence the societal impacts of research, not the least over longer time frames. Societal impacts may also be positive as well as negative. Evaluation must therefore be able to account for complex relationships between the contextual factors surrounding research over time.

Each of these challenges of uncertainty, scope, and complexity may in principle be mitigated by the well-motivated use of data from multiple sources to create a more holistic picture of research impact. This study explored the value to RCN of integrating a wide range of data sources and using ML to explore unanticipated connections and patterns in evaluating research impact.

Methods

In the pilot study, data on RCN-funded projects and publications were linked across three pillars as shown in Figure 7.1.

- Pillar 1 corresponds to the input or funding element of research.
- Pillar 2 represents the knowledge production/knowledge outputs that result from the funding.
- Pillar 3 tracks early evidence of impact and uptake through the dimensions of technological influence, mass media communication and education, social media, and policy influence.

Algorithmic methods were used to mine the resulting body of evidence in order to assess the contributions made by RCN-funded research to specified societal challenges. A machine learning algorithm (TextRazor[6]) trained on Wikipedia data to tag texts with a controlled vocabulary of media topics delivered by IPTC[7] was used to classify the data of each pillar. The pilot study focused on two societal challenges:

- Achieving better protection/enhancement of natural ecosystems
- Reducing inequalities of opportunity (health, education, economic).

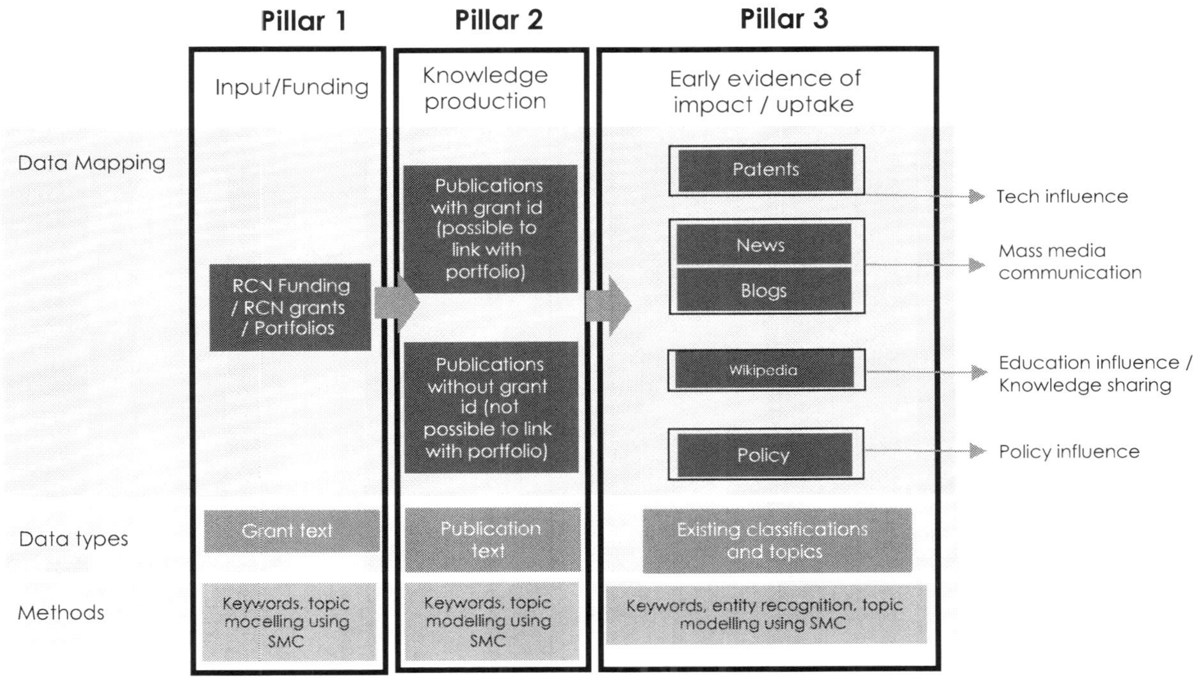

Figure 7.1 Overall methodological strategy and data mapping. This figure illustrates the data sources and methods used to collect an overall picture of research impact.

Source: Rosemberg et al., 2021.

The relevance of projects (pillar 1), publications (pillar 2), and impact (pillar 3) for these two challenges was analyzed in terms of a selection of the IPTC media topics. A selection of existing data sources was used to document societal impact in the pilot (Table 7.1).

Each source was deemed to represent a specific type of impact of a specific research publication, documented by a citation. Among the selected data sources, some are well known and documented in the research policy literature, like citations in patents and policy documents, whereas others, like citations in news media, blog posts, and Wikipedia, have been less used to date, and therefore more difficult to interpret. These new sources, often referred to as Altmetrics,[8] can still be used to shed light on how a piece of research is read, reused, and built upon outside of academia.

There are some important limitations to the use of altmetric citations as indicators of societal impact. *New data sources and novel indicators will embed similar limitations to traditional indicators, with the additional shortcoming of being still relatively unexplored. A publication may be cited and used incorrectly by a news outlet, which would represent an unclear demonstration of impact. Or even worse, a publication can be cited incorrectly to sustain "fake news" arguments in a social media post, which would represent a potential negative impact pathway. Therefore, the indicators we derive from these novel data sources need to be put in context and combined with qualitative checks. But more importantly, the impact pathways that were established by the pilot study*

Table 7.1 Coverage per type of data source used in the analysis

Type of data	*Data source*	*Uptake documents*
Patents	Lens and Dimensions	• ~4,000 patents citing RCN publications in their NPL references
News and blogs	Altmetrics	• ~15,000 news articles and ~5,000 blog posts mentioning RCN-funded publications
Wikipedia and Syllabuses	Altmetrics	• ~1,700 Wikipedia pages mentioning RCN-funded publications
Twitter, Facebook, and Reddit (dropped due to data access and methodological concerns)	Altmetrics	• ~20,000 RCN funded publications with Altmetric scores, which include the number of posts from social media sources
Policy documents	Overton	• ~16,000 policy documents from Norway, ~1,900 documents citing RCN knowledge outputs

Source: Rosemberg et al., 2021.

should be interpreted as early signs of uptake/contribution, rather than as an explicit and causal impact attribution.

Findings

Even with these limitations, the results strongly suggest that algorithmic analysis of these new data sources can be used to gain new knowledge on the societal impact of research funding. As a general rule, descriptive analysis of the data is less affected by the limitations of altmetrics than the direct calculation of simple indicators.

Figure 7.2 provides an example of descriptive analysis. It shows the pathway from RCN-funding via research publications to documented influences of these publications in policy documents relating to the societal challenges selected in the pilot. In addition to just counting citations in relevant policy documents and linking them back to publications and projects, these projects and publications have been classified in disciplines based on the "Field of Research" (FOR) classification used in the publication database Dimensions.[9] The Sykes diagram showcases how a diverse set of topics covered by RCN grants and publications (linked to those grants) then feed into policy documents related to "Protection of ecosystems" and "Inclusive societies." In particular it makes it evident that projects classified in one field may produce publications in a variety of fields, and also that some scientific fields are relevant for

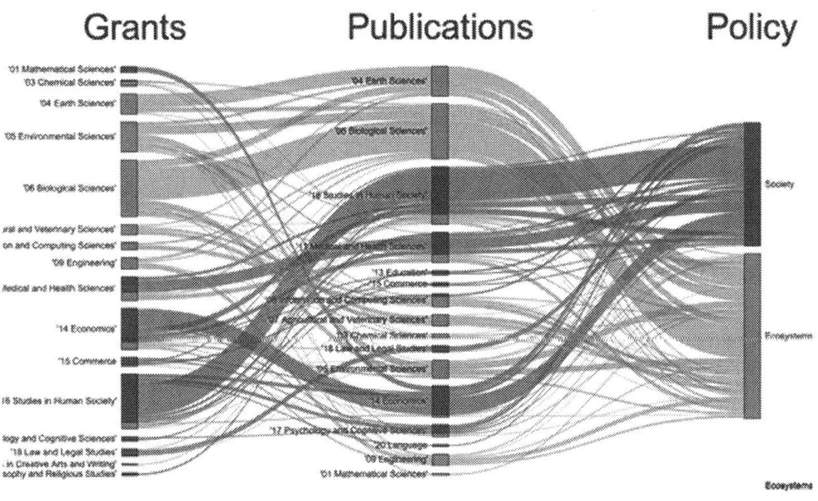

Figure 7.2 Flow diagram illustrating connections between grants, publications, and policy documents.

Source: Rosemberg et al., 2021.

both societal challenges (Medical and Health sciences, Economics, Studies in Human Society).

Implications for Funders

At a strategic level, the *key advantage of pursuing a quantitative approach is its scalability and automation. Once established, the methodology can be scaled, reused, and adapted to future grants and different societal challenges.* Still, the questions of how to integrate machine learning as a standard tool for monitoring and evaluation, and which data sources to use, are not trivial. In the case of RCN, this integration has not yet taken place, mostly due to insufficient in-house competence in data science.

A final consideration in terms of future use of algorithmic data analysis highlighted by this pilot study is the potential to combine this data-driven methodology with complementary methods. The pilot mapped channels of uptake and early signs of impact without providing causal assessments. However, the methodology builds the necessary data infrastructure to undertake a causal analysis. With the implementation of methods such as contribution analysis, process tracing, or comparative case studies, it is possible to assess the degree to which RCN-funded knowledge, through the documented signs of uptake, contributed to effectively tackling a societal challenge. For example, one application would be to trace the grants resulting in publications that were cited in a policy document known to have been instrumental for an important policy change.

Key Considerations in Adapting Data Science and AI Technologies to the Research Funding Context

As our case studies illustrate, data science presents new opportunities for analyzing and optimizing the societal impact of investments in research. However, the use of data science methods must be clearly motivated and put in the context of a funder's strategic and operational goals in research funding and evaluation. In the final part of this chapter, we examine the organizational perspective by discussing key considerations in building capacity for data science within a research funding organization. We take a pragmatic approach, focusing on promoting the responsible use of AI in research funding organizations as a supplement to traditional evaluation methods. This perspective complements the more comprehensive analysis of ethical concerns in the use of AI – including the use of big data, machine learning for information synthesis, and machine learning for decision-making and prediction – found in Greenstein and Cho (2025).

Alignment of Technology Design to Goals and Available Resources

For any individual goal that integrating AI techniques can help a funder achieve, there may be multiple ways to define how those AI technologies will operate and what resources (e.g., data, ML models) can inform their development. In choosing between these strategies, funders must contend not only with technical efficacy but also with alignment with existing processes, resource requirements, and potential risks from AI use.

Consider the example goal of using machine learning to help identify the characteristics of successful projects. We can consider two ways a funder might formulate an AI approach to serve this overall goal: a *holistic* (or data-led) model and a *composite* (or expertise-led) model.

In a holistic, fully data-led approach, the funder aims to have a single machine learning model to analyze a complete project, correlate it with similar projects seen in the past, and estimate its level of success. This is highly flexible as it allows for the emergence of unanticipated features from the data: e.g., particular groupings of research themes, methodologies, and data sources may emerge as more likely to succeed than others. However, in this approach the task to be modeled with AI is also highly complex; it therefore requires large amounts of specialized data (e.g., thousands of projects or more) to deliver a reliable model, and the correlations captured may not reflect what the funder might deem truly informative or even relevant. The use of data-driven approaches may also surface undesirable biases in data as correlated with the outcome of interest (Obermeyer et al., 2019), and without visibility into "black box" AI approaches it is difficult to ensure that models align with organizational principles (Kim et al., 2021).

In a composite, expert-led approach, the overall task is broken down into smaller pieces: for example, (1) categorize authors by expertise area, (2) identify families of research methodologies used, and (3) model types of cooperation. Each of these sub-tasks then becomes an individual target for AI application, with a hierarchical structure to synthesize more specific analyses into broader project judgments. In this approach, as each separate task is smaller in scope and more generalizable, less data is likely to be required to train each machine learning model, and external data sources may be incorporated to inform model development (e.g., informing a "methodology classification" model with auto-generated metadata from scholarly archives). This implies more engineering complexity and is based on expert knowledge about what types of features might be most informative but reduces the resource requirements for AI development and will produce more easily maintained solutions. The composite approach provides more visibility into AI operations and reduces the scope of each step, making it easier to ensure alignment with organizational principles and to monitor for evidence of AI bias (Raji et al., 2020). However, the process is not free of risk, as manual structuring of the process of AI use may enshrine established preferences and omit valuable, unexpected correlations that may emerge from data-driven approaches (Ding & Sterling, 2017).

Of course, the design and implementation of AI systems is a spectrum, on which these two examples are only sample points. Decisions about how AI systems will be designed, implemented, and managed must be driven by alignment with strategic goals as well as pragmatic concerns about available resources and requirements for ongoing management. In other words, the question of how a research funder could make responsible use of data science methods may (and very likely will) find a different answer in each funder depending on its strategic goals, organizational context, and available resources.

This context dependency is key to ensuring that AI use is both effective for its intended purposes and adherent to ethical standards. The use of AI systems must be driven by specific organizational needs to achieve any real difference in practice. At the same time, grounding AI use in specific existing purposes provides valuable scaffolding for assessing AI bias and monitoring to ensure that adding AI into the picture does not harm the delivery of a funder's societal mission.

Defining Goals and Measuring Success

The push to adopt new technologies may originate from various organizational and strategic needs. The introduction of data science and AI must not only be motivated by an explicitly stated goal, but also be aimed at a clear definition of success. It is only when the implementation of data science methods is oriented toward a clear organizational goal (such as improving the alignment of research investments with specific societal needs) that the purposefulness of such methods can be assessed. In our view, the basic assumption that the introduction of artificial intelligence (or other types of automation) will create benefits by replacing more costly and slow manual procedures is insufficient as a primary goal for the use of AI. This is because the introduction of a new technology is likely to affect the outcomes of the process that is automated, not only the efficiency of that process.

The administrative cost of project selection and grant management is a small part of the total budget of a research council (7 percent in the case of RCN).[10] This means that efficiency gains at a research council will only affect total public spending at a rate of 1/20 (5 percent), while an effect of automated procedures on the actual project selection will have an effect of 20/1. In short, it is essential to keep an eye on the direction of travel, not only on the ecometer. If an organization is not stating clearly what type of changes it would expect from the use of data science – and which of these are acceptable or beneficial – it will not be possible to assess whether the implementation of data science technologies is serving its purpose.

When developing machine learning and AI approaches, the primary concern is often how a team may measure a *model's* success (e.g., accuracy, precision, etc.). But even the most accurate model may not make a successful difference if it is poorly integrated into funding or evaluation processes or poorly aligned with strategic goals. Defining how the success of *AI use* will be measured is key to aligning the implementation of AI technologies with a positive impact in the broader funder context.

Co-production Between Different Competencies

While AI is often discussed through an exclusively technical lens, the effectiveness of AI and ML is directly rooted in the alignment between the data to be captured, how it is analyzed, and how that analysis is used in practice. Each of these is a social and organizational process as much as a technical one, and effective design and use of AI in practice, therefore, require interdisciplinary perspectives (Shneiderman, 2016; Newman-Griffis et al., 2021b). Building AI systems that address both strategic goals and public responsibility requires attending to the social origins and limitations of data (e.g., who is and is not represented, and in what ways), how technical design decisions interact with those social understandings of information, and the organizational systems that will make use of AI outputs in operation (Newman-Griffis et al., 2023).

From a practical perspective, it is therefore necessary for research funders to compose interdisciplinary teams to develop AI solutions, representing the diverse competencies needed for technical design, understanding the social context of data, and aligning system use with strategic priorities. These competencies must be represented throughout the AI development process to ensure that each step appropriately accounts for technical efficacy, organizational utility, and social impact. Without these multiple perspectives, it is more likely that AI implementations will replicate or exacerbate existing social biases, lose technical validity, or fail to address the strategic goals that motivated their use in the first place (Ali et al., 2023).

How these different competencies can best be convened and managed in practice is something of an open question. Different approaches will be needed for organizations with centralized data analysis, those with distributed data expertise across different teams, and those who contract out data analytics (Holm et al., 2022). As funders experiment further with integrating AI and ML into their work of funding and evaluation, there are valuable opportunities for mutual learning and the development of shared best practices in how that interdisciplinary integration can best be achieved.[11]

Transparency and Trust

Transparency is a key requirement for building trust in AI applications.[12] This requirement is often linked to the concept of explainability, meaning that it is possible to show how the algorithm arrived at a certain result, for instance, which words are prominent in a class of text in NLP classification. Explainability varies across methods, with more advanced methods (such as BERT [Devlin et al., 2019] or ChatGPT) tending to be less explainable than simpler models (like TF/IDF). While explainability of results from algorithms adds transparency and thus helps build public trust in AI applications, scholars working on questions of ethics in AI have suggested that the requirement for transparency should be extended beyond the purely technical explainability to the whole process of development and application of AI systems (Russo et al., 2023). Russo et al. (2023) suggest a shift in focus from the outcome to the process where the requirement for transparency should be extended to "the actors involved [in developing AI systems] and their expertise: designers, peer experts, the public, institutional stakeholders."

To make its AI solutions trustworthy, a funder should thus document the process of development and implementation of AI solutions. Recent developments of standards such as model cards (Mitchell et al., 2019) and dataset datasheets (Gebru et al., 2021) are mapping out some of this documentation for machine learning components, but funders must go further in reflecting the contexts in which AI systems are developed and used. Similar concerns are also put forward in the EU Artificial Intelligence Act (Madiega, 2023). Relevant items to document may include: analysis of relevant data sources and their biases, reflections on the strengths and weaknesses of alternative methods, assessment of test runs of alternative methods and evaluation of effects on immediate outcomes (funded projects), the research system and society at large. Transparency is of course not an issue of documentation alone: involving relevant stakeholders in the process of developing AI solutions and the evaluation of results may also help foster trust, as well as assure the relevance of the application.

Evaluation and Big Data Analysis

Evaluation of both research and research policy instruments is a well-established part of research council activities. As with any new instrument or intervention, the use of data science and AI methods in the work of research funders must itself be evaluated to ensure both value and societal benefit. The competence already in place in evaluation departments or as part of thematic or disciplinary units is an invaluable resource for a research council embarking on the journey to the promised land of Big Data analysis. As noted above, this evaluation must go beyond the performance-based evaluations that dominate the AI field to encompass the broader domains and processes in which AI is being used.

Although there are certainly technical hurdles to be overcome, the ability to steer the development of AI solutions toward organizational and societal goals, while keeping an eye on ethical issues and managing stakeholders' interests, will help assure the successful and sustainable implementation of AI solutions. The cases presented in this chapter can be taken as examples of the value of big data analysis for tasks performed at a traditional evaluation unit. Of course, traditional evaluation methods may also be used to support and govern the development and implementation of AI solutions.[13] Combining traditional and data-driven methods of evaluation and using established methods to monitor the effectiveness of new interventions will be essential to the continuing development of the evaluation profession.

Conclusion

The long-term shift toward steering research policy in the direction of predefined societal goals, together with increasing drives toward data-driven decision-making across policy spheres, creates significant pressures on research funders. Research activities must be effectively balanced between societally driven and curiosity-driven research, and funding and evaluation decisions often must be demonstrably anchored in data.

Increased use of data science methods, including AI and machine learning, can help funders meet both of these pressures. However, each use of AI or similar methods must be strongly rooted in specific funder and process contexts to deliver on a funder's aims effectively and responsibly. We presented two case studies from the Research Council of Norway, illustrating the processes and decisions involved in the use of AI and machine learning in the funding context, as well as key lessons learned for future applications of these methodologies. We have further highlighted important considerations for connecting the use of AI and machine learning to specific funder contexts and ensuring that this use is as grounded in strategic and societal missions as possible.

Evaluation of public policy may indeed gain both in efficiency and effectiveness by more systematically leveraging the learning opportunities that big data offers (York & Bamberger, 2020). We argue that these gains may only be achieved under a strong understanding of big data contexts and that this understanding may be developed through addressing specific, actionable challenges. Our discussion highlights the need for more sharing of insights and experiences from AI and data science use in research funding, and for the development of new practices to implement and manage data science across competencies and contexts in the research funding ecosystem.

Notes

1 See, e.g., https://www.digg.se/en, https://www.ai.se/en, https://researchonresearch.org/project/funder-data-platform/.

2 ML and AI are related methodologies serving different aims: ML assists in *pattern recognition*, and AI in using knowledge (including patterns identified via ML) to *perform tasks* that typically require human intelligence. AI approaches may use data-driven insights from ML and may also use expert-sourced information such as known patterns and large-scale knowledge resources (e.g., WikiData and Web of Science). ML techniques, and AI more broadly, can be used to ask questions/learn information from data, in a data science paradigm, and/or to inform decision making in more operational contexts. In this chapter, we refer to ML when specifically concerned with data-driven pattern recognition, and AI when concerned with knowledge-driven task performance in general.

3 REF2021 collected 6,781 so-called impact cases from Higher Education Institutions. https://results2021.ref.ac.uk/impact.

4 Several providers of bibliometric analysis have developed algorithms to map scholarly publications to the UN's Sustainable Development Goals (SDGs). For a critical discussion of this service, see Armitage et al. (2020).

5 The presentation of the case is partly based on the text of a report from Technopolis Group (Rosemberg et al. 2021). Passages reproduced verbatim are marked in italics.
6 https://www.textrazor.com/.
7 https://iptc.org/standards/media-topics/.
8 https://nap.nationalacademies.org/content/about-altmetrics.
9 https://plus.dimensions.ai/support/solutions/articles/23000018826-what-is-the-background-behind-the-fields-of-research-for-classification-system-13.
10 Cf. The annual report for the Research council of Norway 2022 (Forskningsrådet 2023) (p. 76–78), https://www.forskningsradet.no/siteassets/publikasjoner/2023/arsrapport-for-forskningsradet-2022---230706.pdf.
11 The Research on Research Institute GRAIL project is an example of ongoing efforts to create this mutual learning and develop best practices: https://researchonresearch.org/project/grail/.
12 https://oecd.ai/en/dashboards/ai-principles/P7.
13 For a more detailed discussion of how traditional evaluation can support data-driven methods, see York and Bamberger (2025).

References

Ali, S. J., Christin, A., Smart, A., & Katila, R. (2023). *Walking the Walk of AI Ethics: Organizational Challenges and the Individualization of Risk among Ethics Entrepreneurs*. Proceedings of the 2023 ACM Conference on Fairness, Accountability, and Transparency, 217–226. https://doi.org/10.1145/3593013.3593990

Armitage, C. S., Lorenz, M., & Mikki, S. (2020). Mapping scholarly publications related to the sustainable development goals: Do independent bibliometric approaches get the same results?. *Quantitative Science Studies*, *1*(3), 1092–1108.

Blei, D. M., Ng, A. Y., & Jordan, M. I. (2003). Latent dirichlet allocation. *Journal of machine Learning research*, *3*(Jan), 993–1022.

Bracken, L. J., & Oughton, E. A. (2006). 'What do you mean?' The importance of language in developing interdisciplinary research. *Transactions of the Institute of British Geographers*, *31*(3), 371–382.

Devlin, J., Chang, M.-W., Lee, K., & Toutanova, K. (2019). BERT: Pre-training of Deep Bidirectional Transformers for Language Understanding. In *Proceedings of the 2019 Conference of the North American Chapter of the Association for Computational Linguistics: Human Language Technologies, Volume 1 (Long and Short Papers)*, (pp. 4171–4186). Minneapolis, Minnesota: Association for Computational Linguistics.

Ding, Y., & Stirling, K. (2017). Data-driven discovery: A new era of exploiting the literature and data. *Journal of Data and Information Science*, *1*(4), 1–9.

Forskningsrådet (2023). *Årsrapport 2022*. Research council of Norway, Oslo. https://www.forskningsradet.no/siteassets/publikasjoner/2023/arsrapport-for-forskningsradet-2022.pdf

Franck, P. G. (1969). *Technology in retrospect and critical events in science-A summary and critique of findings by IIT Research institute*. National Science Foundation, Washington, D.C. 1969.

Gebru, T., Morgenstern, J., Vecchione, B., Vaughan, J.W., Wallach, H., Iii, H.D., & Crawford, K. (2021). Datasheets for datasets. *Communications of the ACM, 64*(12), 86–92.

Gooch, D., Vasalou, A., & Benton, L. (2017). Impact in interdisciplinary and cross-sector research: Opportunities and challenges. *Journal of the Association for Information Science and Technology, 68,* 378–391. https://doi.org/10.1002/asi.23658

Greenstein, N., & Cho, S.-W. (2025). Ethics & Equity in Data Science for Evaluators. In S.B. Nielsen, F. Mazzeo Rinaldi and G.J. Petersson (eds.). *Artificial Intelligence and Evaluation. Emerging Technologies and Their Implications for Evaluation* (pp. 56–77). London: Routledge. https://doi.org/10.4324/9781003512493

Holm, J., Waltman, L., Newman-Griffis, D., & Wilsdon, J. (2022). *Good Practice in the Ue of Machine Learning & AI by Research Funding Organisations: Insights from a Workshop Series*. Research on Research Institute. Report. https://doi.org/10.6084/m9.figshare.21710015.v1

Janssen, M., Brous, P., Estevez, E., Barbosa, L. S., & Janowski, T. (2020). Data governance: Organizing data for trustworthy Artificial Intelligence. *Government Information Quarterly, 37*(3), 101493.

Kim, T.W., Hooker, J., & Donaldson, T. (2021). Taking principles seriously: A hybrid approach to value alignment in artificial intelligence. *Journal of Artificial Intelligence Research, 70,* 871–890.

Madaio, M.A., Stark, L., Wortman Vaughan, J., & Wallach, H. (2020). Co-designing checklists to understand organizational challenges and opportunities around fairness in AI. In *Proceedings of the 2020 CHI Conference on Human Factors in Computing Systems* (pp. 1–14). Association for Computing Machinery, New York.

Madiega, T. (2023). Artificial intelligence act. *European Parliamentary Research Service Briefing: Legislation in Progress*. Retrieved from: https://www.europarl.europa.eu/thinktank/en/document/EPRS_BRI(2021)698792

Mansilla, V.B., Feller, I., & Gardner, H. (2006). Quality assessment in interdisciplinary research and education. *Research Evaluation, 15*(1), 69–74.

Mehrabi, N., Morstatter, F., Saxena, N., Lerman, K., & Galstyan, A. (2021). A survey on bias and fairness in machine learning. *ACM Computing Surveys (CSUR), 54*(6), 1–35.

Ministry of Education and Research. (2018). Long-term plan for research and higher education 2019–2028 — Meld. St. 4 (2018–2019) Report to the Storting (white paper). Retrieved from: https://www.regjeringen.no/en/dokumenter/meld.-st.-4-20182019/id2614131/

Misuraca, G., & Van Noordt, C. (2020). AI Watch - Artificial Intelligence in public services, EUR 30255 EN, Publications Office of the European Union, Luxembourg. ISBN 978-92-76-19540-5, https://doi.org/10.2760/039619, JRC120399.

Mitchell, M., Wu, S., Zaldivar, A., Barnes, P., Vasserman, L., Hutchinson, B., Spitzer, E., Raji, I.D., & Gebru, T. (2019). Model cards for model reporting. In *Proceedings of the Conference on Fairness, Accountability, and Transparency*. Association for Computing Machinery, New York (pp. 220–229).

Newman-Griffis, D., Sivaraman, V., Perer, A., Fosler-Lussier, E., & Hochheiser, H. (2021a). TextEssence: A Tool for Interactive Analysis of Semantic Shifts Between Corpora. In Toutanova K., Rumshisky A., Zettlemoyer L., Hakkani-Tur D., Beltagy I., Bethard S., Cotterell R., Chakraborty T. & Zhou Y. (ed.) *Proceedings of the 2021 Conference of the North American Chapter of the Association for Computational Linguistics: Human Language Technologies* (pp. 106–115). Association for Computational Linguistics.

Newman-Griffis, D., Lehman, J.F., Rosé, C., & Hochheiser, H. (2021b). Translational NLP: A New Paradigm and General Principles for Natural Language Processing Research. In *Proceedings of the 2021 Conference of the North American Chapter of the Association for Computational Linguistics: Human Language Technologies* (pp. 4125–4138). Association for Computational Linguistics.

Newman-Griffis, D., Rauchberg, J. S., Alharbi, R., Hickman, L., & Hochheiser, H. (2023). Definition drives design: Disability models and mechanisms of bias in AI technologies. *First Monday*, 28(1). https://doi.org/10.5210/fm.v28i1.12903

Nielsen, S.B. (2025). The Evaluation Industry and Emerging Technologies. In S.B. Nielsen, F. Mazzeo Rinaldi and G.J. Petersson (eds.). *Artificial Intelligence and Evaluation. Emerging Technologies and Their Implications for Evaluation* (pp. 266–286). London: Routledge. https://doi.org/10.4324/9781003512493

Nobel Prize Outreach AB. (2023, 11 October). *The Nobel Prize in Physiology or Medicine 2023.* NobelPrize.org. https://www.nobelprize.org/prizes/medicine/2023/summary/

Obermeyer, Z., Powers, B., Vogeli, C., & Mullainathan, S. (2019). Dissecting racial bias in an algorithm used to manage the health of populations. *Science*, *366*(6464), 447–453.

Petersson G.J., & Breul, J.D. (2018). *Cyber Society, Big Data, and Evaluation.* Routledge. ISBN 9781138483033.

Raji, I.D., Smart, A., White, R.N., Mitchell, M., Gebru, T., Hutchinson, B., Smith-Loud, J., Theron, D., & Barnes, P. (2020). January. Closing the AI accountability gap: Defining an end-to-end framework for internal algorithmic auditing. In *Proceedings of the 2020 Conference on Fairness, Accountability, and Transparency* (pp. 33–44). Association for Computing Machinery, New York.

Reale, E., Avramov, D., Canhial, K., Donovan, C., Flecha, R., Holm, P., Larkin, C., Lepori, B., Mosoni-Fried, J., Oliver, E., Primeri, E., Puigvert, L., Scharnhorst, A., Schubert, A., Soler, M., Soòs, S., Sordé, T., Travis, C., & Van Horik, R. (2018). A review of literature on evaluating the scientific, social and political impact of social sciences and humanities research. *Research Evaluation*, *27*(4), 298–308.

Rosemberg C., Machado, D., Dijkstal F., Brown N., Nielsen K., Ryd J., A[l]merud M., Arnold M., Melin G. (2021) *Study to establish a methodology to assess the societal impact of research and research-based innovation.* Technopolis group (unpublished).

Russo, F., Schliesser, E., & Wagemans, J. (2023). Connecting ethics and epistemology of AI. *AI & Society,* 1–19.

Schot, J., & Steinmueller, W. E. (2018). Three frames for innovation policy: R&D, systems of innovation and transformative change. *Research Policy*, *47*(9), 1554–1567.

Shneiderman, B. (2016). *The New ABCs of Research: Achieving Breakthrough Collaborations.* Oxford University Press. http://dx.doi.org/10.1080/24751448.2017.1292800

Shneiderman, B. (2018). Twin-win model: A human-centered approach to research success. *Proceedings of the National Academy of Sciences*, *115*(50), 12590–12594.

Smith, T. L. (2022). Demonstrating the value of government investment in science: Developing a data framework to improve science policy. *Harvard Data Science Review*, *4*(2). https://doi.org/10.1162/99608f92.d219b2ce

Stokes, D. E. (1997). *Pasteur's quadrant: Basic science and technological innovation.* Washington, DC: Brookings Institution Press.

Thelwall, M., Kousha, K., Wilson, P., Makita, M., Abdoli, M., Stuart, E., Levitt, J., Knoth, P., & Cancellieri, M. (2023). Predicting article quality scores with machine learning: The U.K. Research Excellence Framework. *Quantitative Science Studies, 4*(2), 547–573. https://doi.org/10.1162/qss_a_00258

Traag, V. A., Waltman, L., & Van Eck, N. J. (2019). From Louvain to Leiden: Guaranteeing well-connected communities. *Scientific Reports*, *9*(1), 5233.

York, P., & Bamberger, M. (2020). *Measuring Results and Impact in the Age of Big Data: The Nexus of Evaluation, Analytics, and Digital Technology*. The Rockefeller Foundation. Retrieved from: https://www.rockefellerfoundation.org/report/measuring-results-and-impact-in-the-age-of-big-data-the-nexus-of-evaluation-analytics-and-digital-technology/

York, P., & Bamberger, M. (2025). The Applications of Big Data to Strengthen Evaluation. In S.B. Nielsen, F. Mazzeo Rinaldi and G.J. Petersson (eds.). *Artificial Intelligence and Evaluation. Emerging Technologies and Their Implications for Evaluation* (pp. 37–55). London: Routledge. https://doi.org/10.4324/9781003512493

Wallace, M.L., & Rafols, I. (2015). Research portfolio analysis in science policy: Moving from financial returns to societal benefits. *Minerva*, *53*, 89–115.

Williams, K., Michalska, S., Cohen, E., Szomszor, M., & Grant, J. (2023). Exploring the application of machine learning to expert evaluation of research impact. *PLoS One*, *18*(8), e0288469.

8 The Use of Quantitative Text Analysis in Evaluations

Laura Gatto and Pirmin Bundi

> Numbers and words can be used together in a variety of ways to produce richer and more insightful analyses of complex phenomena than can be achieved by either one alone.
>
> — Rossman and Wilson (1985)

Introduction

In today's rapidly evolving technological society, data generation and collection offer unprecedented opportunities. While data collection and analysis were largely limited by technological constraints in the past, the digital revolution has catalyzed an explosion of data generation, from user-generated content on social media platforms (Mayrhofer et al., 2020) to sensor data from *Internet of Things* devices (Van Deursen & Mossberger, 2018). The term *big data* has emerged as a descriptor for this vast and complex structured and unstructured data, which has the potential to revolutionize the way we understand and analyze various aspects of our world (Gani et al., 2016). As evaluations aim to analyze the effectiveness and efficiency of programs and policies, the digital revolution has also reached the evaluation community (Petersson & Breul, 2017; Picciotto, 2020; Nielsen et al., 2025, this volume). In the era of evidence-informed policymaking (Head, 2016), decision makers are demanding more information and asking evaluators to incorporate as much information as possible into program evaluations. The rapid pace of technological evolution has challenged evaluators to develop the capacity to process, understand, and extract meaningful insights from big data (Petersson & Breul, 2017; Picciotto, 2020). In recent years, various scholars have been using big data to conduct evaluations (see, for example, Barrados, 2017; Nielsen et al., 2017; Goyal & Howlett, 2019; Leeuw, 2017; Wilkins, 2017; Nielsen, 2025, this volume).

As the accessibility of information grows, so too does the challenge of managing and making sense of this wealth of information. Information overload has become a widespread problem, with individuals and institutions being inundated with more data than they can effectively process. According to Mayne (2007: 97), an infinite number of possible measures and evaluative information can be created for any given program, exacerbating the ability to deal with the information. This paradoxical situation of having easier access to extensive information along with limited cognitive resources to deal with it can lead to reduced

DOI: 10.4324/9781003512493-8

evaluation quality due to an inability to distinguish relevant from irrelevant data (Weiss, 1988). Thus, evaluators often struggle with the problem of having more information than they can process. At best, information is often condensed so that evaluators can process it, but sometimes documents are ignored due to a lack of resources to analyze them (Kouame, 2010).

In response to this new information, new technologies have emerged to help evaluators navigate big data. While big data includes data generation and its ecosystems, data analytics in particular offers evaluators new avenues (Bruce et al., 2025, this volume; Nielsen, 2025, this volume; York & Bamberger, 2024). According to Cintron and Montrosse-Moorhead (2022: 414), data analytics "requires that data science, data analysis, data visualization, and evaluation methods [are] combined to answer evaluation questions." However, the use of these analytical techniques is still very limited (Højlund et al., 2017). This chapter aims to introduce one such tool, quantitative text analysis, to the evaluation community. Quantitative text analysis has emerged as an important tool for measuring the (policy) positions of stakeholders, such as interest groups (Bunea & Ibenskas, 2015; Klüver, 2009, 2015), governmental (Wratil et al., 2022), and non-governmental organizations (James et al., 2021; J. C. Lam et al., 2019), but has rarely been used in program evaluations.[1] In doing so, these studies use techniques that allow for the extraction of policy preferences using text as data (Benoit & Herzog, 2017). Therefore, we explain how this approach can help evaluators to process, understand, and analyze large amounts of documents. The chapter also addresses the question of what skills evaluators need in order to use quantitative text analysis.

This chapter aims to address these issues through the following structure. First, we will provide an introduction to quantitative text analysis and the intersections of evaluation, discussing evaluation models that focus on stakeholders (e.g., democratic evaluation and participatory evaluation) and may need to analyze their positions or preferences (Bundi & Pattyn, 2022). Second, we will present a quantitative textual analysis of a stakeholder consultation in which interest groups and political parties made proposals on financial regulation. We will then describe how we mined the text, which methods can be used to analyze the documents, and which computer content analysis software can be used to identify the preferences of different stakeholders. Specifically, we use an unsupervised scaling technique called Wordfish, which aims to infer actors' preferences from the vocabulary used in political texts (speeches, party manifestos, and position papers). According to this approach, the position is therefore considered latent and is retrieved without the need for a reference text or a word dictionary (Slapin & Proksch, 2008). The final section concludes the chapter and discusses the use of quantitative text analysis for evaluation research.

Text Analysis and Evaluation

In principle, text analysis is "any systematic reeducation a flow of text (or other symbols) to a standard set of statistically manipulable symbols representing the presence, the intensity, or the frequency of some characteristics relevant to social sciences" (Shapiro & Markoff, 1997: 14). The approach of text analysis[2] has a long tradition in social science research. During World War II, the Allied governments launched a series of projects to analyze the content of Nazi propaganda (Krippendorff, 2018: 8). In the following 20th century, the analysis of texts was driven by technological innovations. The proliferation of personal computers in the 1970s and the introduction of the internet have marked important milestones in the systematic analysis of documents (Mehl, 2006). According to Krippendorff (2018), several qualitative approaches have evolved from the early version of text analysis: discourse analysis (Foucault, 1971; Johnstone, 2017), social constructivist analysis (Gergen, 1992), rhetorical analysis (Jamieson, 1988), ethnographic content analysis (Altheide, 1987), and conversation analysis (Goodwin & Heritage, 1990).

There are several conceptual and empirical contributions that text analysis can make to evaluation. According to Bundi and Pattyn (2022), evaluation can be conducted in different ways and focus on different aspects. Two types of evaluation models may be of interest to text analysts. First, theoretical approaches aim to evaluate the design, implementation, and expected outcome of a public action, such as CIPP evaluation (Stufflebeam & Zhang, 2017), theory-driven evaluation (Chen, 2014), and realistic evaluation (Pawson & Tilley, 1997). A systematic analysis of the documents that form the evaluation's object helps evaluators to understand and illustrate the underlying program theory. Second, stakeholder-oriented evaluation models focus on the question of whether stakeholders' needs are satisfied. In doing so, they can focus on different stakeholder groups, i.e., responsive evaluation (Stake, 2003), democratic evaluation (MacDonald, 1976; Picciotto, 2015), participatory evaluation (Cousins & Whitmore, 1998), and empowerment evaluation (Fetterman, 2001). Preskill and Jones (2009) argue that stakeholder involvement is important to ensure that evaluations reflect their expectations, experiences, and insights. As stakeholders are potential users of evaluation findings, their perspective is essential to maximize the benefits of an evaluation. Not surprisingly, classic text analysis has also found its way into the evaluation community. Text analysis has been used to identify program theories (Fujita-Conrads et al., 2023; Leeuw, 2003), but also to measure stakeholder preferences (Christie & Rose, 2003; Cintron & Montrosse-Moorhead, 2022; Jacobson et al., 2013; Stevahn & King, 2016).

However, modern technologies allow evaluators to analyze a larger number of documents without spending more time than with classical text analysis tools. More recently, Cintron and Montrosse-Moorhead (2022) provide an example of topic models, a specific type of quantitative text analysis. In these topic models, the goal is to extract themes that define large amounts of textual data in a short period of time. In doing so, the topic models allow an evaluation team that wants to involve stakeholders but is limited by budget to collect information from many stakeholders by analyzing open-ended questions. Alternatively, evaluators faced with the task of synthesizing a large volume of existing program or policy reports can use quantitative text analysis to identify themes in the literature by mapping the key concepts embedded in the reports. Finally, the technique allows evaluators to assess an intervention that has moved entirely to an online format due to the COVID-19 pandemic (e.g., a professional development program). The use of topic modeling could help uncover dominant themes within these postings. Thus, Cintron and Montrosse-Moorhead (2022) argue that quantitative text analysis, such as topic models, is a valuable contribution to the evaluation community (see Mazzeo Rinaldi et al., 2025, this volume).

In order to analyze text with quantitative text analysis, users, i.e., evaluators, must transform text into quantifiable data. According to Benoit (2020: 463), all forms of text contain information that can be treated as a form of data because they communicate a message. Thus, these messages can be captured and treated as data, but this requires a process in which the characteristics of textual data are abstracted from the acts of communication. The essence of treating text as data is to "transform it into more structured, summarized, and quantitative data to make it amenable to the familiar tools of data analysis" (Benoit, 2020: 463). In the following chapter, we will provide an introduction to quantitative text analysis and explain what evaluators need to do to transform text into data.

Quantitative Text Analysis: Classification and Approaches

At the intersection of linguistics, data science, and social research, quantitative text analysis has emerged as a powerful toolkit for systematically unraveling the hidden patterns, sentiments, and knowledge embedded in large textual datasets. According to Mehl (2006: 144–145), approaches to quantitative text analysis can be distinguished along a number of different dimensions (see also Popping, 1999): (1) aim, (2) approach, (3) breadth of scope, and (4) focus. In the following, we present the different types of approaches to quantitative text analysis.

Aim: In a very general sense, methods of text analysis differ according to whether their aim is representational or instrumental. The representational approach seeks to decode the message as closely as possible to the intended meaning of the message, which is why it focuses on the manifest content of a text. In comparison, and more often, instrumental analysis focuses mainly on latent content, i.e., the text is analyzed independently of the author's intention for the occurrence of themes. For example, the manifest content of the word "blue" refers to a color, while the latent content may be used in the quarterback's cadence, which indicates the timing or protection scheme of an American football offense (Popping, 2012: 89).

Approach: The second dimension concerns the extent to which textual analysis exclusively identifies themes or the relationship between them (Roberts, 1997). According to Mehl (2006), until the 1980s, almost all text analysis was thematic. These approaches map the occurrence of a set of concepts in a text, i.e., they count the frequency of certain words and phrases. There are several techniques for implementing thematic models that aim to derive themes that characterize large amounts of text data in a short period of time, such as *latent Dirichlet allocation* (LDA) (see Figgou & Pavlopoulos, 2015). In comparison, semantic analysis attempts to extract information in the context of the topic. For example, Mehl (2006: 144) argues that for the topic of killing, it is important to identify whether it occurs in the context of "self" or "other people." These approaches solve this problem by defining the concrete nature of the relationships between themes. While this process has been top-down by the investigator, newer models such as Latent Semantic Analysis (LSA) use a bottom-up approach, where information about the semantic similarity of words is generated by analyzing a large body of text, i.e., bottom-up (Campbell & Pennebaker, 2003). These models used to rely on human coders to parse large amounts of text, but nowadays these models can also be unsupervised thanks to machine learning algorithms (Berry et al., 2020; Celebi & Aydin, 2016).

Breadth of scope: Text analytics also differ in their bandwidth, or the number of variables they seek to examine. While some studies focus exclusively on some word uses, they ignore other potentially relevant information. In doing so, they tend to have a stronger theoretical background, while broad approaches tend to be more data-driven. According to Mehl (2006), broader approaches provide a broad linguistic profile of a text, which often offers more flexibility.

Focus: Finally, quantitative text analyses differ in whether they analyze context or style (Groom & Pennebaker, 2002). According to Mehl (2006), this distinction is based on the difference between adaptive and stylistic aspects of behavior. While the former serves a purpose in a given context, the latter is not contextualized and serves an expressive rather than instrumental function. In particular, how someone says something has been studied extensively in the psychological literature (Pennebaker et al., 2003), but it is also popular in the political science literature, for example, in the study of populism (Pauwels, 2011; Storz & Bernauer, 2018).

There are many different techniques that can be used in quantitative text analysis. Grimmer and Stewart (2013) provide an overview of text analysis methods (see Figure 8.1). They distinguish between classification and ideological scaling approaches. While the former organizes texts into a set of categories (using dictionaries or some kind of supervision), the latter estimates the location of actors in a policy space, i.e., they produce a scale. One of these approaches, Wordfish, infers actors' preferences from the language found in political texts such as speeches, party manifestos, and position papers. In this approach, the position is treated as hidden and can be extracted without relying on a reference text or a

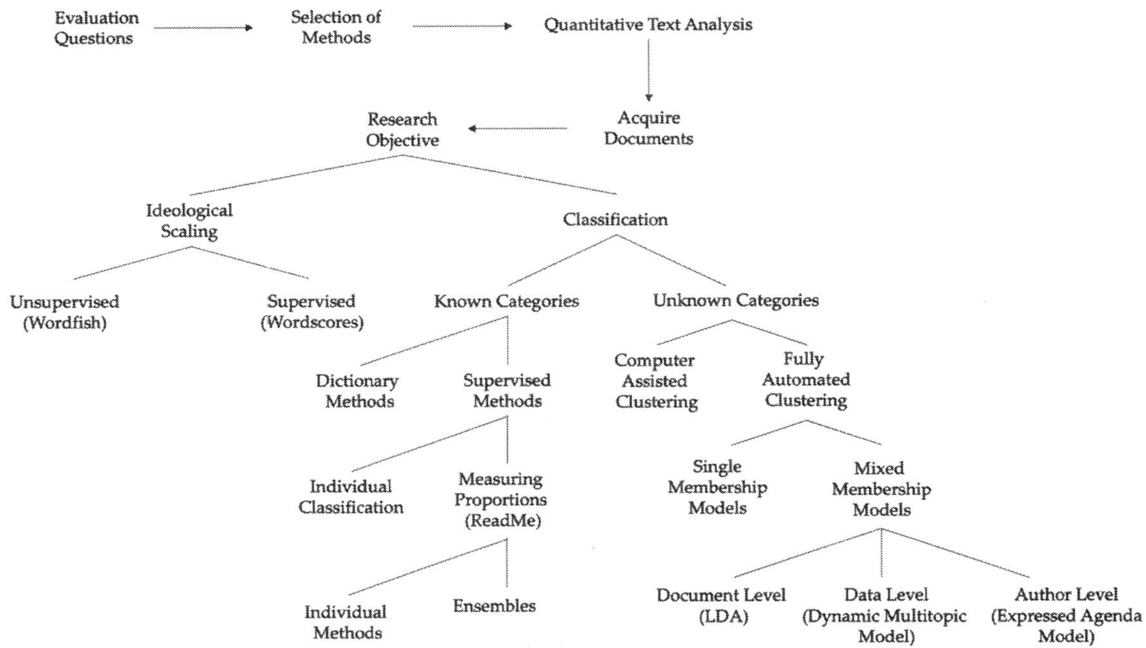

Figure 8.1 Overview of quantitative text analysis methods.

(adapted from Grimmer & Stewart 2013).

word dictionary (Slapin & Proksch, 2008). In the following, we argue that this method is particularly suitable for identifying stakeholder preferences. On the one hand, word scores rely on reference texts to locate political actors in space. In doing so, they are a special case of dictionary methods. On the other hand, the Wordfish method argues that ideologies or preferences affect word usage, which is why preferences can be derived from texts. Since program evaluations often lack reference texts, it may be difficult to identify a similar stakeholder group from another policy. Thus, we argue that the Wordfish method is particularly interesting for evaluations.

The Wordfish Method

In the following paragraphs, we will explain the main features of this scaling technique. Wordfish discovers words that distinguish locations on a political spectrum (Grimmer & Stewart, 2013: 292). Various approaches have emerged for gauging policy preferences, such as expert surveys, manual analysis of party manifestos, and computational coding methods (Gross & Jankowski, 2020). Over the past few years, computer-based coding has garnered increased attention, primarily owing to its capacity to effectively process substantial data volumes. Particularly when applied to textual data, this approach can yield more profound insights into how actors align themselves ideologically with specific policies. A major advantage of using quantitative textual analysis is its potential to bring objectivity to the research. Rather than relying solely on perceived information, this approach aims to examine the actual positions of stakeholders.

Wordfish is an unsupervised scaling technique that plays a crucial role in data analysis by allowing us to uncover trends and patterns in data without the need for pre-existing information or labeled data (Grimmer & Stewart, 2013). Moreover, Wordfish has been widely used to assess the preferences of political parties through party manifestos (Slapin & Proksch, 2008; Proksch, Slapin & Thies 2011; Gross & Jankowski, 2020) or parliamentary debates (Lauderdale & Herzog, 2016; Frid-Nielsen, 2018; Vignoli et al., 2022).

In short, Wordfish tries to infer the latent positions of actors from the language used in different types of text such as speeches, party manifestos, and position papers. It assumes that these positions are not directly observable but can be inferred from the words chosen by these stakeholders. Unlike some other techniques, such as Wordscores,[3] Wordfish does not require predefined reference points to define ideological categories. Instead, it is a frequency-based scaling method that assumes that word frequencies in texts follow a Poisson distribution. This means that the number of times a word occurs in a speech is influenced by the context in which the speech is delivered. In terms of breadth of scope, Wordfish attempts to study a "bag of words" and thus has a fairly broad reach that pays less attention to written style. The bag-of-words approach used by Wordfish assumes that words are related but not interdependent. Specifically, Wordfish relies on the following formula:

$$y_{ijt} \sim poisson\left(\lambda_{ijt}\right)$$

$$\lambda_{ijt} = exp\left(\alpha_{it} + \psi_{j} + \beta_{j} * \omega_{it}\right) \tag{8.1}$$

where λ_{ijt} is the count of word j for every text i. α stands for the stakeholder fixed effects, while ψ accounts for the text fixed effects. β is an estimate of a word-specific weight capturing the importance of word j in discriminating between positions, and ω is the estimate of the stakeholder's position (Proksch & Slapin, 2009).

In the next section, we will present a practical case that applies this model in order to analyze the preferences of political parties and interest groups concerning financial regulation.

A Case Study of a Quantitative Text Analysis

The development of a policy and its subsequent implementation is directly linked to the preferences of its stakeholders. Examining policy preferences can provide a comprehensive and well-rounded understanding of the potential course a particular policy issue might take. Namely, it can enhance our understanding of how a policy is designed and in which direction it will be implemented (Knill et al., 2012). It also sheds light on the evolution of preferences over time. This case study[4] discusses the use of quantitative text analysis, specifically Wordfish, as a method for estimating the policy positions of both political parties and interest groups. It highlights its application in the Swiss context, the advantages of this approach, and its mathematical underpinnings.

Stakeholders' Preferences in the Swiss Context

The case study, based on Gatto & Bundi[5] (2023), analyzes the position of two stakeholder groups – parliament and interest groups – regarding financial regulation. First, representatives are among the most important stakeholders in public policy. They initiate debates and decide on political issues. As such, it is important to have an understanding of the functioning and structure of the Swiss parliamentary arena. The Swiss Parliament (Federal Assembly) is based on a perfect egalitarian bicameral system, where both the National Council and the Council of States have equal powers (Varone & Ingold, 2023: 3). As a working parliament,[6] parliamentary committees play a particularly important role during the parliamentary phase. In particular, committee representatives discuss the bill in detail and determine the key parameters of the final legislation. In practice, the committee's proposals for debate in the plenary session of parliament strongly anticipate the final decisions (Pilotti, 2012). This makes parliamentary committees, and their members, key players in the parliamentary phase of policy making (Bellanger, 2006). By being part of a committee, representatives refine their knowledge by becoming much more precise and subject-matter specific (Bellanger, 2006). It is also important to highlight the secrecy surrounding the closed-door meetings of the parliamentary committees, which are however available for research purposes and can provide a more authentic and reliable data source, free from external pressures such as media and public influence. To analyze the general preferences of representatives with Wordfish, we selected two legislative committees: the Economic Affairs and Taxation Committees (EATC).

Second, interest groups enjoy a strong position in Switzerland, which is traditionally considered a neo-corporatist country, i.e., in which interest groups exert strong influence (Christiansen et al., 2018; Lijphart & Crepaz, 1991). Based on Article 147, they benefit from the constitutional freedom to express their opinions. According to Christiansen et al. (2018), interest groups, particularly those with competence in technical subjects, are crucial for policymakers because they provide specific expertise that the political actors seldom have. Hence, in smaller countries, the collaboration among parties representing diverse interests enables the fusion of political stability and economic adaptability, as noted by Sciarini (2014). Specifically, interest groups can clearly express their positions in the pre-parliamentary phase through the consultation process and extra-parliamentary committees (Mach et al., 2020).

Operationalization and Data Mining

The utilization of parliamentary committee data constitutes a novel and enriching approach. Within these committee settings, members are granted the opportunity to articulate their political perspectives, free from the apprehension that their opinions will be subjected to immediate public scrutiny in the press. On the other hand, to measure the preferences of interest groups, we used their position papers. These were systematically collected during the federal consultation phase and several consultations of the Banking Commission, which became the Swiss Financial Market Supervisory Authority in 2009 (CFB, 2007). With respect to the selected interest groups, we considered both business interest groups at the federal level (associations and companies) and public interest groups. In this study, public interest groups are defined as follows: "individuals who focus on the attainment and protection of common goods" (see Gava et al., 2017: 79), but they also represent a particular notion and worldview of a selected group.

An inherent challenge in analyzing this dataset arises from its bilingual nature, encompassing both German and French languages. To mitigate this issue, we opted to translate the French texts into German, given that 75% of the texts are originally in German. According to Proksch and Slapin (2009: 32), German works particularly well for word-based analyses: "In contrast to English compound words, which are separated by spaces or hyphens, German allows the concatenation of nouns to form a long word, and theoretically there is no limit to the number of nouns that can be compounded." Another challenge arises from ensuring the comparability of documents. Position papers tend to be longer than parliamentary speeches. Furthermore, within committee debates, there may be differences in the level of participation of different representatives, with some speaking more frequently than others. This could lead to a problem of comparability. To tackle this issue, we have incorporated the recommendation put forth by Gross and Jankowski (2020), which involves the inclusion of words that manifest in a minimum of two documents and occur at least ten times.

As a first step, the speeches of representative *y* at time *t* for policy *x* and the position papers of the interest group *b* at time *t* for policy *x* are merged. We created individual texts and then checked the language with Word. Next, we applied the procedure developed by Slapin and Proksch (2008), using the Quanteda package (Benoit et al., 2021). For the committee minutes, we examined all speeches and eliminated those that contained only a few sentences, resulting in a total of 441 documents on financial regulation. We repeated the process for the position papers, resulting in a total of 246 documents.

When using this technique, the availability of a substantial corpus of text data proves essential, even though the precise threshold for the minimum required text length to ensure the efficient operation of Wordfish remains an area of ongoing investigation (Slapin & Proksch, 2008). The text was then preprocessed. Specifically, we removed punctuation, spaces, numbers, and the so-called "stop words,"[7] while stripping the stem and converting it to lowercase. Finally, we utilized the Quanteda package to remove names and numbers from the position papers, and we manually deleted any address-related information pertaining to the interest groups.[8]

Finally, as explicated by Proksch and Slapin (2009), for each policy, we selected two position papers that represent polarized viewpoints at opposite ends of the political spectrum. To achieve this contrast, we specifically chose two representatives, one from the Swiss Socialist Party (SP) and another from the Swiss People's Party (SPP), each aligning with divergent ideologies. In the case of interest groups, our selection encompasses both business and public interests, featuring the "Union suisse des arts et métiers" (USAM) representing business interests and the Swiss Trade Union Federation (USS) advocating for public interests.[9] We then used Wordfish to estimate the policy preferences of both interest groups and the representatives. Once the representatives' preferences

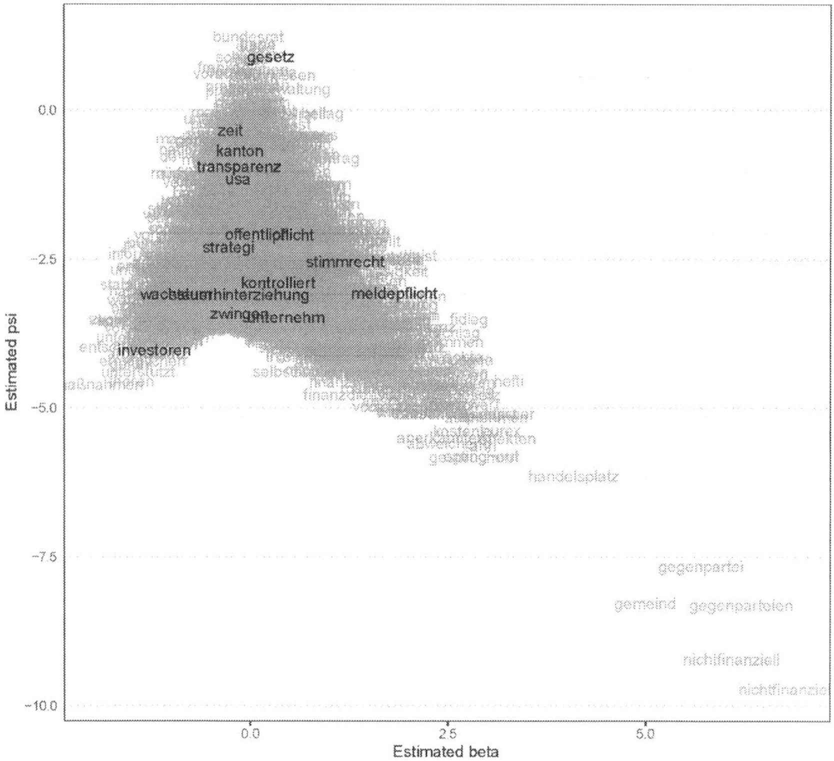

Figure 8.2 Words Position of Representatives.

were obtained, they were also aggregated by party affiliation to see the general trend over time.

Analysis

Prior to delving into the elucidation of the actors' preferences, we need to establish the credibility of our corpus selection and the veracity of our findings. We initiate this validation process by analyzing the position of words, as visually presented in the following figures, denoted as Figures 8.2 and 8.3.

The estimated psi (ψ) measures the occurrence of the word (fixed effect), while the estimated β takes into account the weight of the words, thus reflecting the ideological spectrum. The tokens at the top of the graph are the most frequent, while the tokens located at the bottom of the so-called "Tour Eiffel," moving away from the center, are less frequently used and can thus better represent the position of the document.

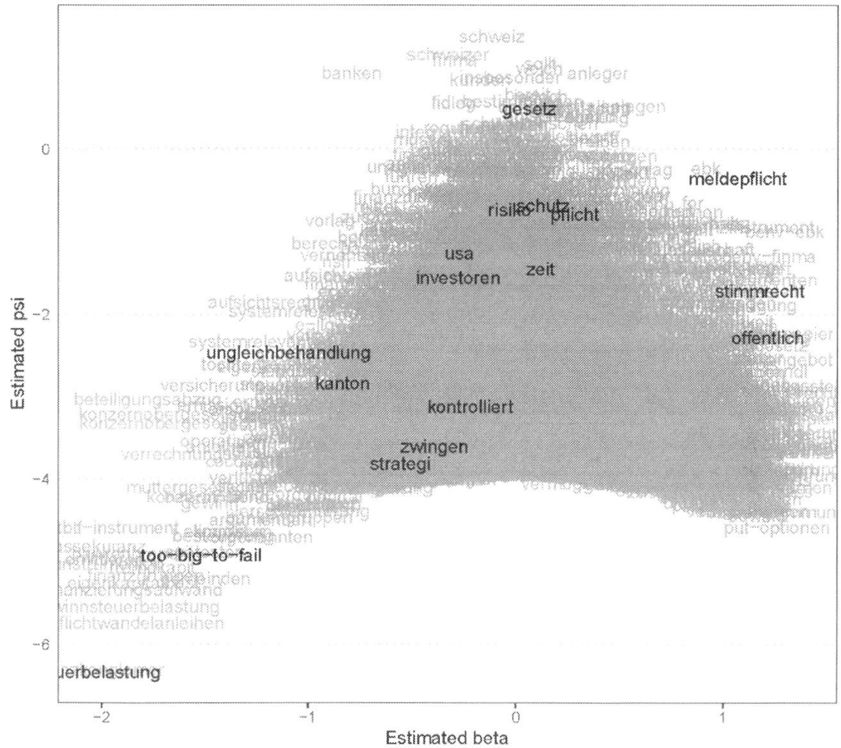

Figure 8.3 Words Position of Interest Groups.

Moreover, the values of *j* weight define the real function of each word; negative values of *j* correlate with low values of ω (document position), while positive values of *j* correspond to higher values of this parameter, which determines the placement of each text. It is also important to note that, according to the method, words associated with USAM and SPP (used as a reference point in the document selection) are more likely to be positioned on the left side of the "Tour Eiffel" graph. In addition, both figures indicate that terms such as "gesetz" (law), "zeit" (time), and "USA" are located at the top of the figure. In both groups, these are recurring words that are less likely to influence the weight of the words. Figure 8.2 shows that tokens associated with right-wing ideology are found on the left: "wachstum" (growth), "investoren" (investors). Similarly, terms like "steuerbelastung" (tax burden) and "too-big-to-fail", which may be associated with the same ideology, are also located on the left side of the interest groups figure. Interestingly, in comparison with Figure 8.3 the term "investoren" is placed on the outskirts rather than in the center.

Looking at the right side of both figures, it is possible to observe identical tokens "meldepflich" (reporting obligation) and "stimmrecht" (right to vote). These words are associated with the banking secrecy regulation, since the first term (reporting obligation) is highly essential in the sphere of financial regulation. However, the use of words related to the notions of obligation and reporting, which can improve transparency standards, can be associated with left-wing ideologies. Nevertheless, there are also some minor differences between these two sections of the graph. For example, the word "offentlich" (public) is more centered for representatives, while it is significantly more to the right side (positive side) for interest groups. As a result, tokens reflecting left-wing beliefs can be found on the left flanks of both figures, and vice versa in the opposite half of the figure. This corresponds to the directory used for the Wordfish estimation, where both the USAM's position paper and the minutes of an SPP representative were forced to be more negative than the other two documents. This suggests that in general the words used by left-wing parties and public interest groups are generally in the same direction. The same should be true for business interest groups and parties further to the right. There are clearly differences that could be explained by the type of document.

To continue the analysis of preferences, Figure 8.4 shows the position of the documents by estimating the variable ω. As explained earlier, we took into account business interest groups and public interest groups. Points above zero on the y-axis are associated with a preference for more regulation, while points below zero refer to the opposite. First, the figure shows that there is a distinction between business interest groups and public interest groups, with the former generally being less inclined to regulate. The interesting findings are that the preferences of business interest groups seem to be skewed toward less regulation, especially before the crisis. Conversely, they seem to be more in favor during the subsequent new wave of regulation that hit the country in its wake. Notably, both lines have risen since 2010, indicating a preference for more regulation. However, it is important to acknowledge that there are differences between both business interest groups and public interest groups. Thus, in a period of increased issue salience, such as the post-global financial crisis era, business interest groups may thus have found it imperative to recalibrate public opinion by adopting preferences in favor of more stringent regulatory measures.

To compare interest groups and representatives, we group the latter into their different parties (Figure 8.5). Similar to the previous graph on interest groups' preferences, the preferences of representatives seem to follow a relatively similar pattern. Interestingly, similar to the previous graph, there seems to be a desire for more regulation in the aftermath of the financial crisis, or at least the statements point in that direction. It should be noted, however, that the lack of variance across parties is counterintuitive, but this could be due to significant changes in the framing over time or the consensual environment within the committees. Another important point to note is the spike in the figure between

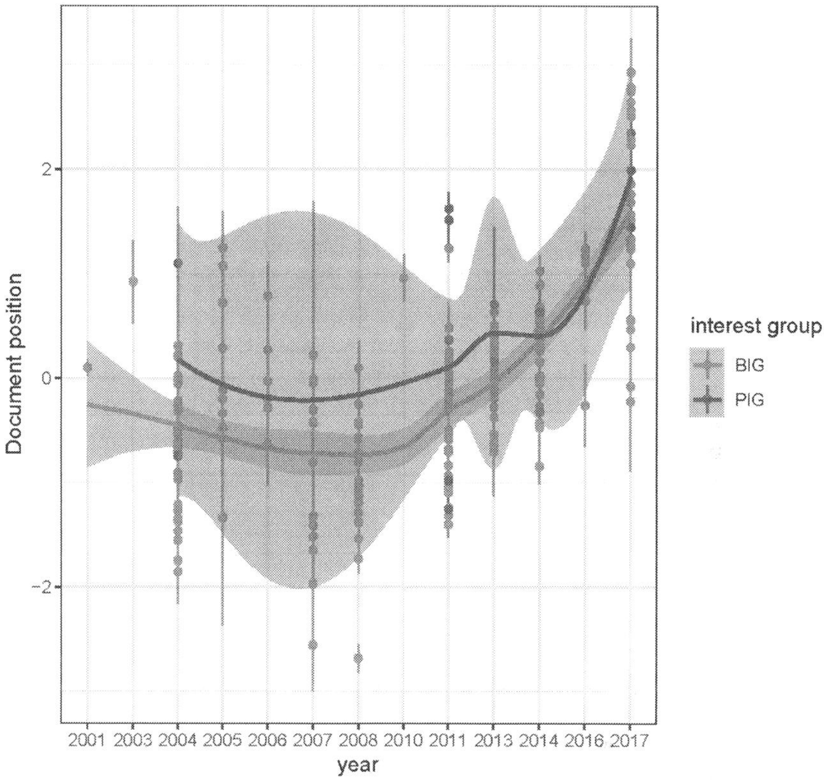

Figure 8.4 Interest groups' preferences.

2005 and 2007, which could be attributed to initial pressure for more regulation, taking into account the money laundering regulation. These results are particularly important as it appears that both actors are likely to prefer more regulation. Finally, it should be emphasized that while a certain tendency can be observed, a direct causality with the financial crisis is not yet possible.

Discussion

This chapter has discussed how quantitative text analysis can be used for program evaluation. This technique, particularly the Wordfish method, presents a multifaceted landscape of advantages and disadvantages that evaluators must navigate carefully. The advantages of this approach lie in its ability to extract objective insights from large amounts of textual data and to provide a systematic and replicable analysis process. Wordfish and similar techniques offer a way to

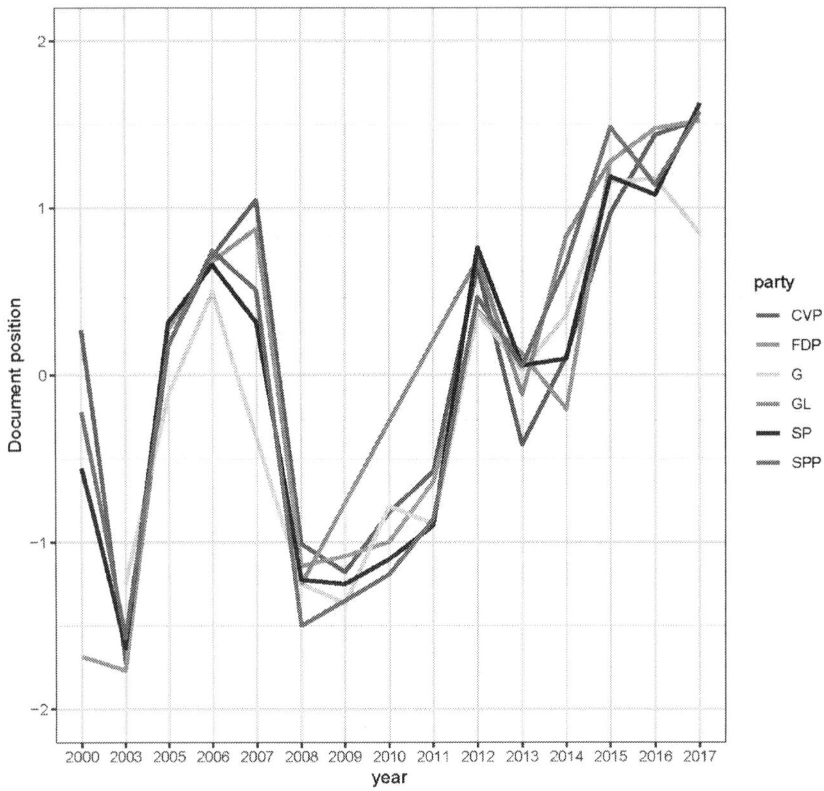

Figure 8.5 Party preferences.

uncover latent patterns, ideological stances, and the use of particular vocabulary within textual content, thereby enriching the depth of program evaluations. Specifically, it is an effective tool for identifying stakeholder preferences from various types of documents, such as position papers and political speeches, which is less resource-intensive than conducting interviews, while considering a more extensive dataset. In addition, different groups can be compared with each other, indicating different preferences for programs. However, these advantages must be weighed against the limitations of the method, such as the potential oversimplification of complex human language (see Grimmer et al., 2022: 37), sensitivity to preprocessing decisions (Proksch et al., 2019; Yano et al., 2021), and the risk of missing context (Eichstaedt et al., 2021; Kučera & Mehl, 2022). Importantly, scaling techniques tend to aggregate texts with a high co-occurrence of terms into a single policy dimension (Grimmer & Stewart, 2013).

Therefore, for this technique to be effective, it is not only important that actors with different positions use different vocabularies, but also that similar issues are identified and focused on when measuring preferences over time (Slapin & Proksch, 2008; Ferrara, 2020). Consequently, a comprehensive understanding of the texts used is paramount in order to identify the specific language patterns employed. Finally, this method is highly dependent on the existence of documents describing preferences.

To effectively master the power of quantitative text analysis, evaluators need to acquire a variety of skills. Knowledge of both basic statistical methods and programming languages is essential for the effective use of Wordfish. However, there are now good open-access text analysis tools in both R and Python that can be particularly helpful to the user in supporting the analysis. Specifically, in R, the "readtext" and "Quanteda" packages (Benoit et al., 2021) are particularly useful. The "readtext" package helps to import and read text data into the R environment. Meanwhile, the "Quanteda" package plays a crucial role in generating the corpus needed to construct the document term matrix (DTM). This DTM is an essential component for the successful application of both unsupervised and supervised techniques in the analysis process. A solid understanding of linguistic nuances and domain-specific knowledge is also essential to accurately interpret results and avoid misinterpretations. Therefore, a critical eye is essential for fine-tuning parameters and ensuring the quality of the final analysis.

The application of quantitative text analysis in program evaluation is extensive and promising. While previously scholars have emphasized the importance of text analysis, notably context analysis for evaluations (Christie & Fleischer, 2010; McKibben et al., 2020; Stemler, 2000), quantitative text analysis provides a new tool to open up the unlimited possibilities of document analyses. This approach can help evaluators decipher public perceptions, political dynamics, and stakeholder preferences to inform evidence-based decision making. By integrating quantitative text analysis into evaluation frameworks, practitioners can enrich traditional methods with insights from the vast digital landscape of textual data (Thomas, 2006). This integration could lead to more comprehensive and nuanced program evaluations, improving accountability and transparency across sectors, and balancing out biases in survey methods, such as misreporting and social desirability (Bundi et al., 2018; Johanson et al., 1993; Lam & Bengo, 2003; Stanton, 2004).

Looking ahead, the prospects for evaluation research in quantitative text analysis remain vibrant. The field is poised for continued advancement, with potential areas of exploration including refining methods to mitigate inherent biases, developing hybrid models that combine quantitative and qualitative approaches (Andreotta et al., 2019; Eichstaedt et al., 2021; Parks & Peters, 2023), and adapting existing techniques to evolving forms of communication, such as social media platforms (Driss et al., 2019). In addition, the ethical implications of quantitative text analysis warrant a thorough examination, particularly in terms of privacy concerns, algorithmic fairness, and potential societal impacts (Dolata et al., 2022).

Conclusion

In conclusion, quantitative text analysis, as exemplified by Wordfish, represents a valuable toolset for evaluators seeking to delve into the intricacies of textual data. While challenges remain, the acquisition of the necessary skills, coupled with a discerning approach to implementation, can maximize the benefits of this method in program evaluation. As the field continues to evolve, the fusion of quantitative text analysis with established evaluation practices holds the promise of advancing our understanding of complex human communication and taking evidence-based decision making to unprecedented heights.

Notes

1 For a rare exception, see Cintron & Montrosse-Moorhead (2022).
2 Text analysis is often referred to as content analysis. We will use text analysis to make a link to quantitative text analysis (Krippendorff, 2018).
3 As explained by Gross & Jankowski (2020:17), Wordscores estimates the policy-specific position of a document by comparing the relative word frequency of "reference texts" to the word distribution of "virgin texts."
4 When we use the term "case study," we intend to convey "the detailed analysis of either a particular case or more cases that aim to shed light on a larger population of cases" (Gerring & Cojocaru, 2016: 394).
5 In this research paper, Gatto & Bundi (2023) conducted a comparative analysis of the positions of business interest groups and their affiliated representatives over time, with the primary objective of examining potential changes in trends related to three different economic policies. In this case, we take advantage of our extensive dataset and narrow our scope to analyze all interest groups and representatives solely in the context of only one policy, namely, financial regulation. This approach will allow us to observe and assess the evolving trends in stakeholder preferences.
6 Working parliaments are characterized by a strong committee system, which mainly drafts and decides on bills. In contrast, the majority of bills are dealt with in plenary sessions in the so-called speech parliaments (Dann, 2003).
7 Highly common terms (e.g., and, or, the, etc.).
8 The final number of features is 3,722 (92.32% sparse) for fiscal policy and 3,424 features (92.86% sparse) for Financial Regulation.
9 Two documents are selected, the first of which is constrained to have a more negative value than the second. For example, dir = c(1,5) would constrain ω1 to be less than ω5 (Proksch & Slapin, 2009: 340).

References

Altheide, D. L. (1987). Reflections: Ethnographic content analysis. *Qualitative Sociology*, *10*(1), 65–77. https://doi.org/10.1007/BF00988269

Andreotta, M., Nugroho, R., Hurlstone, M. J., Boschetti, F., Farrell, S., Walker, I., & Paris, C. (2019). Analyzing social media data: A mixed-methods framework combining computational and qualitative text analysis. *Behavior Research Methods, 51*, 1766–1781.

Barrados, M. (2017). Getting Started with Big Data: The Promises and Challenges of Evaluating Health-Care Quality. In G. J. Petersson and J. D. Breul (eds.). *Cyber Society, Big Data, and Evaluation*. Routledge.

Bellanger, F. (2006). Parlement et administration en Suisse. *Annuaire européen d'administration publique, 29*, 337–369.

Benoit, K. (2020). Text as Data: An Overview. In L. Curini and R. Franzese (eds.). *The SAGE Handbook of Research Methods in Political Science and International Relations* (pp. 461–497). SAGE Publications Ltd. https://doi.org/10.4135/9781526486387

Benoit, K., & Herzog, A. (2017). Text Analysis: Estimating Policy Preferences from Written and Spoken Words. In J. Bachner, K. H. Wagner and B. Ginsberg (eds). Analytics, Policy and Governance (pp. 137—159). Yale University Press.

Benoit, K., Watanabe, K., Wang, H., Lua, J. W., Kuha, J., & Benoit, M. K. (2021). Package 'quanteda. textstats.' *Research Bulletin, 27*(2), 37–54.

Berry, M. W., Mohamed, A., & Yap, B. W. (Eds.). (2020). *Supervised and Unsupervised Learning for Data Science*. Springer International Publishing. https://doi.org/10.1007/978-3-030-22475-2

Bruce, K., Vandelanotte, J., & Gandhi, V. (2025). Emerging Technology and Evaluation in International Development. In S. B. Nielsen, F. Mazzeo Rinaldi and G. J. Petersson (eds.). *Artificial Intelligence and Evaluation. Emerging Technologies and Their Implications for Evaluation* (pp. 13–36). Routledge. https://doi.org/10.4324/9781003512493

Bundi, P., & Pattyn, V. (2022). Citizens and evaluation: A review of evaluation models. *American Journal of Evaluation*, 10982140211047219. https://journals.sagepub.com/doi/full/10.1177/10982140211047219

Bundi, P., Varone, F., Gava, R., & Widmer, T. (2018). Self-selection and misreporting in legislative surveys. *Political Science Research and Methods, 6*(4), 771–789.

Bunea, A., & Ibenskas, R. (2015). Quantitative text analysis and the study of EU lobbying and interest groups. *European Union Politics, 16*(3), 429–455.

Campbell, R. S., & Pennebaker, J. W. (2003). The secret life of pronouns: Flexibility in writing style and physical health. *Psychological Science, 14*(1), 60–65. https://doi.org/10.1111/1467-9280.01419

Celebi, M. E., & Aydin, K. (Eds.). (2016). *Unsupervised Learning Algorithms*. Springer International Publishing. https://doi.org/10.1007/978-3-319-24211-8

Chen, H. T. (2014). *Practical Program Evaluation: Theory-Driven Evaluation and the Integrated Evaluation Perspective*. SAGE Publications.

Christiansen, P. M., Mach, A., & Varone, F. (2018). How corporatist institutions shape the access of citizen groups to policy-makers: Evidence from Denmark and Switzerland. *Journal of European Public Policy, 25*(4), 526–545. https://doi.org/10.1080/13501763.2016.1268194

Christie, C. A., & Fleischer, D. N. (2010). Insight into evaluation practice: A content analysis of designs and methods used in evaluation studies published in North American evaluation-focused journals. *American Journal of Evaluation, 31*(3), 326–346.

Christie, C. A., & Rose, M. (2003). The language of evaluation theory: Insights gained from an empirical study of evaluation theory and practice. *Canadian Journal of Program Evaluation*, *18*(2), 33–45. https://doi.org/10.3138/cjpe.18.002

Cintron, D. W., & Montrosse-Moorhead, B. (2022). Integrating big data into evaluation: R code for topic identification and modeling. *American Journal of Evaluation*, *43*(3), 412–436. https://doi.org/10.1177/10982140211031640

Commission fédérale des banques (CFB). Rapport de gestion 2007. Commission fédérale des banques, 2007.

Cousins, J. B., & Whitmore, E. (1998). Framing participatory evaluation. *New Directions for Evaluation*, *1998*(80), 5–23. https://doi.org/10.1002/ev.1114

Dann, P. (2003). European parliament and executive federalism: Approaching a parliament in a semi-parliamentary democracy. *European Law Journal*, *9*(5), 549–574.

Dolata, M., Feuerriegel, S., & Schwabe, G. (2022). A sociotechnical view of algorithmic fairness. *Information Systems Journal*, *32*(4), 754–818.

Driss, O. B., Mellouli, S., & Trabelsi, Z. (2019). From citizens to government policy-makers: Social media data analysis. *Government Information Quarterly*, *36*(3), 560–570. https://doi.org/10.1016/j.giq.2019.05.002

Eichstaedt, J. C., Kern, M. L., Yaden, D. B., Schwartz, H. A., Giorgi, S., Park, G., Hagan, C. A., Tobolsky, V. A., Smith, L. K., & Buffone, A. (2021). Closed-and open-vocabulary approaches to text analysis: A review, quantitative comparison, and recommendations. *Psychological Methods*, *26*(4), 398.

Ferrara, F. M. (2020). The battle of ideas on the euro crisis: Evidence from ECB inter-meeting speeches. *Journal of European Public Policy*, *27*(10), 1463–1486. https://doi.org/10.1080/13501763.2019.1670231.

Fetterman, D. M. (2001). *Foundations of Empowerment Evaluation*. SAGE.

Figgou, L., & Pavlopoulos, V. (2015). Social psychology: Research methods. *International Encyclopedia of the Social & Behavioral Sciences*, *22*, 544–552.

Foucault, M. (1971). Orders of discourse. *Social Science Information*, *10*(2), 7–30. https://doi.org/10.1177/053901847101000201

Frid-Nielsen, S. S. (2018). Human rights or security? Positions on asylum in European Parliament speeches. *European Union Politics*, *19*(2), 344–362. https://doi.org/10.1177/1465116518755954

Fujita-Conrads, E., Christie, C. A., & FastHorse, E. (2023). Evaluation Policy as a Bridge between Evaluation Theory and Practice. In M. C. Alkin and C. A. Christie (eds.). *Evaluation Roots: Theory Influencing Practice* (pp. 292–299). Guilford Publications.

Gani, A., Siddiqa, A., Shamshirband, S., & Hanum, F. (2016). A survey on indexing techniques for big data: Taxonomy and performance evaluation. *Knowledge and Information Systems*, *46*, 241–284.

Gatto, L., & Bundi, P. (2023). *Till crisis us do part: Business Interest Groups, Representatives and Business Regulation* [Unpublished manuscript].

Gava, R., Varone, F., Mach, A., Eichenberger, S., Christe, J., & Chao-Blanco, C. (2017). Interests groups in parliament: Exploring MP s' interest affiliations (2000-2011). *Swiss Political Science Review*, *23*(1), 77–94.

Gergen, K. J. (1992). The social constructionist movement in modern psychology. In *The Restoration of Dialogue: Readings in the Philosophy of Clinical Psychology* (pp. 556–569). American Psychological Association. https://doi.org/10.1037/10112-044

Gerring, J., & Cojocaru, L. (2016). Selecting cases for intensive analysis: A diversity of goals and methods. *Sociological Methods & Research*, *45*(3), 392–423. https://doi.org/10.1177/0049124116631692

Goodwin, C., & Heritage, J. (1990). Conversation analysis. *Annual Review of Anthropology*, *19*(1), 283–307. https://doi.org/10.1146/annurev.an.19.100190.001435

Goyal, N., & Howlett, M. (2019). Combining internal and external evaluations within a multilevel evaluation framework: Computational text analysis of lessons from the Asian Development Bank. *Evaluation*, *25*(3), 366–380. https://doi.org/10.1177/1356389019827035

Grimmer, J., & Stewart, B. M. (2013). Text as data: The promise and pitfalls of automatic content analysis methods for political texts. *Political Analysis*, *21*(3), 267–297. https://doi.org/10.1093/pan/mps028

Grimmer, J., Roberts, M. E., & Stewart, B. M. (2022). *Text as data: A new framework for machine learning and the social sciences*. Princeton University Press.

Groom, C. J., & Pennebaker, J. W. (2002). Words. *Journal of Research in Personality*, *36*(6), 615–621. https://doi.org/10.1016/S0092-6566(02)00512-3

Gross, M., & Jankowski, M. (2020). Dimensions of political conflict and party positions in multi-level democracies: Evidence from the local Manifesto Project. *West European Politics*, *43*(1), 74–101. https://doi.org/10.1080/01402382.2019.1602816

Head, B. W. (2016). Toward more "evidence-informed" policy making? *Public Administration Review*, *76*(3), 472–484.

Højlund, S., Olejniczak, K., Petersson, G. J., & Rok, J. (2017). The current use of big data in evaluation. In G. J. Petersson, F. Leeuw, & K. Olejniczak (Eds.), Cyber society, big data, and evaluation (pp. 35–60). Routledge.

Jacobson, M. R., Azzam, T., & Baez, J. G. (2013). The nature and frequency of inclusion of people with disabilities in program evaluation. *American Journal of Evaluation*, *34*(1), 23–44. https://doi.org/10.1177/1098214012461558

James, S., Pagliari, S., & Young, K. L. (2021). The internationalization of European financial networks: A quantitative text analysis of EU consultation responses. *Review of International Political Economy*, *28*(4), 898–925. https://doi.org/10.1080/09692290.2020.1779781

Jamieson, K. H. (1988). *Eloquence in an Electronic Age: The Transformation of Political Speechmaking*. Oxford University Press.

Johanson, G. A., Gips, C. J., & Rich, C. E. (1993). " If you can't say something nice" A variation on the social desirability response set. *Evaluation Review*, *17*(1), 116–122.

Johnstone, B. (2017). *Discourse Analysis*. John Wiley & Sons.

Klüver, H. (2009). Measuring interest group influence using quantitative text analysis. *European Union Politics*, *10*(4), 535–549.

Klüver, H. (2015). The promises of quantitative text analysis in interest group research: A reply to Bunea and Ibenskas. *European Union Politics*, *16*(3), 456–466. https://doi.org/10.1177/1465116515581669

Knill, C., Schulze, K., & Tosun, J. (2012). Regulatory policy outputs and impacts: Exploring a complex relationship. *Regulation & Governance*, *6*(4), 427–444.

Kouame, J. B. (2010). Using readability tests to improve the accuracy of evaluation documents intended for low-literate participants. *Journal of MultiDisciplinary Evaluation*, *6*(14), 132–139.

Krippendorff, K. (2018). *Content analysis: An introduction to its methodology*. Sage Publications.

Kučera, D., & Mehl, M. R. (2022). Beyond English: Considering language and culture in psychological text analysis. *Frontiers in Psychology*, *13*, 819543.

Lam, J. C., Cheung, L. Y., Wang, S., & Li, V. O. (2019). Stakeholder concerns of air pollution in Hong Kong and policy implications: A big-data computational text analysis approach. *Environmental Science & Policy*, *101*, 374–382. https://doi.org/10.1016/j.envsci.2019.07.007

Lam, T. C., & Bengo, P. (2003). A comparison of three retrospective self-reporting methods of measuring change in instructional practice. *The American Journal of Evaluation*, *24*(1), 65–80.

Lauderdale, B. E., & Herzog, A. (2016). Measuring political positions from legislative speech. *Political Analysis*, *24*(3), 374–394.

Leeuw, F. L. (2003). Reconstructing program theories: Methods available and problems to be solved. *The American Journal of Evaluation*, *24*(1), 5–20. https://doi.org/10.1016/S1098-2140(02)00271-0

Leeuw, H. B. M. (2017). Using Big Data to Study Digital Piracy and the Copyright Alert System. In G. J. Petersson and J. D. Breul (eds.). *Cyber Society, Big Data, and Evaluation*. Routledge.

Lijphart, A., & Crepaz, M. M. (1991). Corporatism and consensus democracy in eighteen countries: Conceptual and empirical linkages. *British Journal of Political Science*, *21*(2), 235–246.

Mach, A., Varone, F., Eichenberger, S. (2020). Transformations of Swiss neo-corporatism: From pre-parliamentary negotiations towards privileged pluralism in the parliamentary venue. In: Careja, R., Emmenegger, P., Giger, N. (eds) The European Social Model under Pressure. Springer VS, Wiesbaden. https://doi.org/10.1007/978-3-658-27043-8_4

MacDonald, B. (1976). Evaluation and the Control of Education. In D. A. Tawney (ed.). *Curriculum Evaluation Today: Trends and Implications* (pp. 125–136). MacMilland Education.

Mayne, J. (2007). Challenges and lessons in implementing results-based management. *Evaluation*, *13*(1), 87–109.

Mayrhofer, M., Matthes, J., Einwiller, S., & Naderer, B. (2020). User generated content presenting brands on social media increases young adults' purchase intention. *International Journal of Advertising*, *39*(1), 166–186. https://doi.org/10.1080/02650487.2019.1596447

Mazzeo Rinaldi, F.., Celardi, E., Miracula, V., & Picone, A. (2025). Artificial Intelligence and Text Analysis in Evaluating Complex Social Phenomena. The Russia-Ukraine Conflict. In S. B. Nielsen, F. Mazzeo Rinaldi and G. J. Petersson (eds.). *Digital Era Evaluation. Emerging Technologies and Their Implications for Evaluation* (pp. 168–195). Routledge. https://doi.org/10.4324/9781003512493

McKibben, W. B., Cade, R., Purgason, L. L., & Wahesh, E. (2020). How to conduct a deductive content analysis in counseling research. *Counseling Outcome Research and Evaluation*, *13*(2), 156–168. https://doi.org/10.1080/21501378.2020.1846992

Mehl, M. R. (2006). Quantitative Text Analysis. In M. Eid and E. Diener (eds.). *Handbook of Multimethod Measurement in Psychology* (pp. 141–156). American Psychological Association. https://doi.org/10.1037/11383-011

Nielsen, S. B., Ejler, N., & Schretzman, M. (2017). Exploring Big (Data) Opportuninities: The Case of the Center for Innovation through Data Intelligence (CIDI), New York City. In G. J. Petersson and J. D. Breul (eds.). *Cyber Society, Big Data, and Evaluation.* Routledge.

Nielsen, S. B. (2025). The Evaluation Industry and Emerging Technologies. In S. B. Nielsen, F. Mazzeo Rinaldi and G. J. Petersson (eds.). *Artificial Intelligence and Evaluation. Emerging Technologies and Their Implications for Evaluation* (pp. 266–286). Routledge. https://doi.org/10.4324/9781003512493

Nielsen, S. B., Mazzeo Rinaldi, F., & Petersson, G. J. (2025). Evaluation in the Era of Artificial Intelligence. In S. -B. Nielsen, F. Mazzeo Rinaldi and G. J. Petersson (eds.). *Artificial Intelligence and Evaluation. Emerging Technologies and Their Implications for Evaluation* (pp. 1–12). Routledge. https://doi.org/10.4324/9781003512493

Parks, L., & Peters, W. (2023). Natural language processing in mixed-methods text analysis: A workflow approach. *International Journal of Social Research Methodology, 26*(4), 377–389. https://doi.org/10.1080/13645579.2021.2018905

Pauwels, T. (2011). Measuring populism: A quantitative text analysis of party literature in Belgium. *Journal of Elections, Public Opinion and Parties, 21*(1), 97–119. https://doi.org/10.1080/17457289.2011.539483

Pawson, R., & Tilley, N. (1997). *Realistic Evaluation.* SAGE.

Pennebaker, J. W., Mehl, M. R., & Niederhoffer, K. G. (2003). Psychological aspects of natural language use: Our words, our selves. *Annual review of psychology, 54*(1), 547–577. https://doi.org/10.1146/annurev.psych.54.101601.145041

Petersson, G. J., & Breul, J. D. (2017). *Cyber Society, Big Data, and Evaluation.* Routledge.

Picciotto, R. (2015). Democratic evaluation for the 21st century. *Evaluation, 21*(2), 150–166. https://doi.org/10.1177/1356389015577511

Picciotto, R. (2020). Evaluation and the big data challenge. *American Journal of Evaluation, 41*(2), 166–181.

Pilotti, A. (2012). *Les parlementaires suisses entre démocratisation et professionnalisation (1910-2010). Biographie collective des élus fédéraux et réformes du Parlement helvétique* (Doctoral dissertation, Université de Lausanne, Faculté des sciences sociales et politiques).

Popping, R. (2000). *Computer-Assisted Text Analysis.* SAGE.

Popping, R. (2012). Qualitative decisions in quantitative text analysis research. *Sociological Methodology, 42*(1), 88–90. https://doi.org/10.1177/0081175012460854

Preskill, H., & Jones, N. (2009). *A Practical Guide for Engaging Stakeholders in Developing Evaluation Questions.* https://folio.iupui.edu/handle/10244/683

Proksch, S. O., & Slapin, J. B. (2009). How to avoid pitfalls in statistical analysis of political texts: The case of Germany. *German Politics, 18*(3), 323–344. https://doi.org/10.1080/09644000903055799

Proksch, S. O., Slapin, J. B., & Thies, M. F. (2011). Party system dynamics in post-war Japan: A quantitative content analysis of electoral pledges. *Electoral Studies, 30*(1), 114–124. https://doi.org/10.1016/j.electstud.2010.09.015

Proksch, S.-O., Wratil, C., & Wäckerle, J. (2019). Testing the validity of automatic speech recognition for political text analysis. *Political Analysis, 27*(3), 339–359.

Roberts, C. W. (1997). A generic semantic grammar for quantitative text analysis: Applications to East and West Berlin Radio News content from 1979. *Sociological Methodology*, *27*(1), 89–129. https://doi.org/10.1111/1467-9531.271020

Rossman, G. B., & Wilson, B. L. (1985). Numbers and words: Combining quantitative and qualitative methods in a single large-scale evaluation study. *Evaluation Review*, *9*(5), 627–643.

Sciarini, P., Nai, A., & Tresch, A. (2014). *Analyse de la votation fédérale du 9 février 2014*. gfs. bern et Université de Genève.

Shapiro, G., & Markoff, J. (1997). A matter of definition. In C. W. Roberts (Ed.), Text analysis for the social sciences (1st ed., pp. 24). Routledge.

Slapin, J. B., & Proksch, S.-O. (2008). A scaling model for estimating time-series party positions from texts. *American Journal of Political Science*, *52*(3), 705–722. https://doi.org/10.1111/j.1540-5907.2008.00338.x

Stake, R. E. (2003). *Standards-Based and Responsive Evaluation*. SAGE Publications.

Stanton, C. K. (2004). Methodological issues in the measurement of birth preparedness in support of safe motherhood. *Evaluation Review*, *28*(3), 179–200.

Stemler, S. (2000). An overview of content analysis. *Practical Assessment, Research, and Evaluation*, *7*(1), 17.

Stevahn, L., & King, J. A. (2016). Facilitating interactive evaluation practice: Engaging stakeholders constructively. *New Directions for Evaluation*, *2016*(149), 67–80. https://doi.org/10.1002/ev.20180

Storz, A., & Bernauer, J. (2018). Supply and demand of populism: A quantitative text analysis of cantonal SVP manifestos. *Swiss Political Science Review*, *24*(4), 525–544. https://doi.org/10.1111/spsr.12332

Stufflebeam, D. L., & Zhang, G. (2017). *The CIPP Evaluation Model: How to Evaluate for Improvement and Accountability*. Guilford Publications.

Thomas, D. R. (2006). A general inductive approach for analyzing qualitative evaluation data. *American Journal of Evaluation*, *27*(2), 237–246.

Van Deursen, A. J., & Mossberger, K. (2018). Any thing for anyone? A new digital divide in internet-of-things skills. *Policy & Internet*, *10*(2), 122–140.

Varone, F., Ingold, K. (2023). Switzerland, Public Policy in. In: van Gerven, M., Rothmayr Allison, C., Schubert, K. (eds) Encyclopedia of Public Policy. Springer, Cham. https://doi.org/10.1007/978-3-030-90434-0_54-1

Vignoli, V., Ostermann, F., & Wagner, W. (2022). Ideological talk, strategic vote: German parties' positions on the military intervention in Afghanistan in Parliament. *German Politics*, 1–23. https://doi.org/10.1080/09644008.2022.2137497

Weiss, C. H. (1988). If program decisions hinged only on information: A response to Patton. *Evaluation Practice*, *9*(3), 15–28.

Wilkins, P. (2017). Keeping Traffic and Transit Passengers Moving—The Use of Big Data. In G. J. Petersson and J. D. Breul (eds.). *Cyber Society, Big Data, and Evaluation*. Routledge.

Wratil, C., Waeckerle, J., & Proksch, S.-O. (2022). Government rhetoric and the representation of public opinion in international negotiations. *American Political Science Review*, 1–18.

Yano, K., Endo, S., Kimura, S., & Oishi, K. (2021). Effective coping strategies employed by university students in three sensitivity groups: A quantitative text analysis. *Cogent Psychology*, *8*(1), 1988193.

York, P., & Bamberger, M. (2024). The Applications of Big Data to Strengthen Evaluation. In S. B. Nielsen, F. Mazzeo Rinaldi and G. J. Petersson (eds.). *Digital Era Evaluation. Emerging Technologies and Their Implications for Evaluation* (pp. 37–55). London: Routledge. https://doi.org/10.4324/9781003512493

9 Artificial Intelligence and Text Analysis in Evaluating Complex Social Phenomena

The Russia–Ukraine Conflict

Francesco Mazzeo Rinaldi, Elvira Celardi, Vincenzo Miracula, and Antonio Picone

Introduction

The social issues that plague postmodern and globalized society are complex and affect various dimensions of people's lives. In the past three years, moreover, the spread of the COVID-19 pandemic has not only changed the social reality we were used to but also confirmed how data is one of the most valuable resources, challenging the tools and techniques that social scientists, in general, and evaluators have used to conduct their studies. This calls for reflection not only on the tools and techniques used to predict and evaluate the outcomes of interventions but also on the skills and, more generally, on the working practices that have so far characterized the evaluator's role and profession.

In today's "hyper-digitized" social reality, individuals and groups constantly leave traces of their behaviors. Big Data (BD) can be collected and analyzed using interconnected data platforms to reveal hidden patterns and trends of great use in many decision-making contexts (see Petersson & Breul, 2017; York & Bamberger in this book). Extracting and analyzing this data can further help to understand the workings of the complex social systems in which social programs and projects are shaped.

Across the world, even the public sector has become acutely aware of the exponential growth in data, an unstoppable deluge. Numerous public agencies and organizations have recognized the immense potential lying within the torrent of data originating from sources such as smartphones, sensors, satellites, and digital transformation endeavors (see Bruce et al., in this book). This potential is harnessed when coupled with artificial intelligence (AI) systems and, in particular, machine learning (ML) algorithms, which possess the capability to unearth valuable insights from this vast sea of information (see Nielsen et al., 2025, this volume).

DOI: 10.4324/9781003512493-9

Hand in hand with these advances, various professions have exhibited diverse responses. Social scientists, for example, have recognized and embraced the new opportunities presented by emerging technologies (see Grossmann et al., 2023). In this context, evaluators seem slow to respond to these new challenges (Petersson & Breul, 2017; Picciotto, 2020; Nielsen, 2023; Nielsen, 2025, this volume). How evaluation practices can be adapted to harness the power of AI systems does not seem to be a question that evaluators have asked in the recent past. Today, it appears that this is, albeit slowly, changing.

For instance, the vast success of ML applications in the past decade has inspired its adoption in several evaluation contexts (in this volume, Næss et al., 2025; Holtermann & Engebretsen, 2025; Gatto & Bundi, 2025; Ziulu et al., 2025; Holm et al., 2025).

This chapter aims to highlight the value such technologies can add to analyzing and evaluating complex social phenomena, offering a *venue* for methodological innovations and for the development of tools geared toward understanding and assessing large-scale complex social events. We investigate the potential of using ML and text analysis tools by presenting a case study on the Ukraine conflict. The Russia–Ukraine war today represents a phenomenon that is changing geopolitical scenarios globally, one that policymakers need to consider when developing policies and programs (national and international). Through the case study, we also explore how such technologies enable us to understand people's reactions to the imposed sanctions against Russia, focusing on (1) perceptions of the Ukraine conflict in the digital context, (2) emotional responses to the application of sanctions, and (3) applicability/reliability of ICT in macro contexts.

We will address the following questions: What features allow these tools to collect and analyze data? What kinds of information do they allow us to capture that more traditional tools cannot? How may they help identify the response mechanisms people use under specific circumstances?

Evaluation and Digital Context

The social science and evaluation literature offers numerous definitions of context (Fitzpatrick, 2012; Pawson, 2013), many of which are in apparent conflict. One of the tasks of evaluation research is to explain how and why context shapes the mechanisms through which an intervention (a policy, a program, a project, or, more generally, a political strategy) works and thus explain why it might work differently in different contexts (Greenhalgh & Manzano, 2022). In recent years, evaluators and researchers have given more consideration to addressing contextual challenges in their causal explanations, for instance, in theory-driven evaluation studies (Shaw et al., 2018; Coldwell, 2019; Nielsen et al., 2022).[1]

In the Realistic evaluation field, in more detail, the context has a broad scope (Pawson, 2013; Goicolea & Kermode, 2018; Kerr et al., 2018; Ebenso et al., 2019; Nielsen et al., 2022). It is omnipresent, complex, dynamic, constantly changing, relational, agentic-creating, and not simply moderating change (Coldwell, 2019, p. 109). It can be perceived as something "physical or a non-physical construct" (Pfadenhauer et al., 2015, p. 106). In the first case, Pfadenhauer and colleagues highlighted the observable features (like space, place, people, things, etc.) that triggered or blocked the intervention, assuming that context operates at one moment and sets a chain reaction of events in motion. In the second case, context is understood as the relational and dynamic features that shape the intervention's mechanisms, assuming that context operates dynamically over time at multiple social system levels (Greenhalgh & Manzano, 2022).

In the latter sense, context is no longer reduced to a set of things (a list of tangible, material facts, and inputs, which often include consumables and staff) but as systems of interactions: meaning, rules, and sets of relationships that shape stakeholders' reasoning in response to program resources and, consequently, influence program outcomes. Ray Pawson notes that "contexts have multiple levels" (Pawson, 2016). In other words, contexts operate at all levels of social systems, and the different levels interact and influence each other. They are multi-layered entities operating in borderless micro–meso–macro systems (Greenhalgh & Manzano, 2022, p. 19). So, a given policy or measure can activate multiple mechanisms with divergent outcomes.

In developing evaluation plans, the complexity of the contextual conditions in which the interventions are inserted cannot be ignored (Rogers, 2008). They should be understood as complex social systems embedded in realities that are themselves complex (Pawson, 2013). The formulation of any intervention today is part of a global governance system (Rosenau & Czempiel, 1992; Hooghe & Marks, 2001) that is exercised in different places with the participation of different actors (Raffini, 2011). The national government (within whose borders political decisions take shape and manifest their effects) becomes an actor in a more complex and layered power system.

Today, more than ever before, evaluating the impact of interventions in response to large-scale phenomena requires contextual knowledge that extends beyond physical places to encompass interpersonal and social relationships within digital contexts.

Connecting (or being always connected) online, exchanging likes or comments on social media, purchasing products online, downloading an app, updating one's virtual profile, exchanging emails, SMS, or WhatsApp messages, creating short videos on TikTok, and seeking information on the web are just a few of the countless activities and daily gestures that have become part of the lives of billions of people, altering access to information, economic opportunities, the shape of social relationships, and the processes of identity construction. The digital realm profoundly impacts our lives and daily routines. The invasion of online platforms has occurred across various fields in a society where social and economic interaction increasingly occurs through a global and highly interconnected digital infrastructure. The growth of digital platforms is hailed as the engine of economic progress and technological innovation. Individuals can reap significant benefits from this transformation, enabling them to initiate activities, exchange goods, and share information online.

As Van Dijck and colleagues argue, the use of digital platforms today increasingly influences sectors of fundamental importance in society, such as healthcare, education, and public transportation, assuming a role of growing significance to the point of gradually converging with the institutions (offline, traditional) and practices that structure democratic societies from an organizational standpoint. In the "platform society," platforms are not an external factor; they do not merely reflect society but produce the social structures of our everyday lives. Online platforms disseminate content, values, and culture worldwide. They can contribute to global cultural homogenization or, conversely, facilitate cultural diversity by sharing unique cultural experiences. They influence user behavior through interface design, recommendation algorithms, and content personalization.

Furthermore, social media platforms can shape social dynamics and user behavior by counting "likes" and shares, influencing social conformity behavior and the desire for approval. They promote trends and contribute to the creation of new cultural forms. They provide spaces where users can create profiles, share personal information, interests, and opinions, and participate in online communities. These spaces influence identity formation and self-representation. Similarly, platforms can play a role in constructing collective identities, such as social movements and political groups. Online platforms have changed the nature of participation, which has gradually become more and more online. The increasing interaction and interdependence between the real and the virtual contribute to creating a new social environment for the individual, characterized by belonging to multiple networks of physical and non-physical relationships (Wellman, 2001). Today, millions of people join one or more online communities to meet their need for communication, information, and entertainment. In this context, social media has gained popularity and created a virtual reality where people can express their thoughts and feelings about products, services, brands, individuals, personalities, or others. The proliferation of digital communities has sparked a heated debate between those who believe that technology can be a valuable tool to facilitate social relationships and expand the boundaries of community and those who argue the exact opposite, emphasizing the dangers of this type of community. It becomes crucial for social scientists in general, and evaluators in particular, to consider what tools can be used to understand contexts when the relationships established within them occur not in a physical but in a digital space.

The now unstoppable process of globalization is moving simultaneously and triggering a process of separating social experience from the physical boundaries of the territory. Territories in a global society (Castells, 2004) are profoundly transformed by the flows crossing them (Appadurai, 2001). These flows involve the movement of people, goods, information, messages, and symbols (Raffini, 2011). In this context, social life is no longer defined by spatial logic but is (re)defined in a plurality of spaces unrelated to territory and spatial contiguity. Advances in technology have made it easier for individuals to communicate and connect with each other. The rise of social media, instant messaging, and other forms of digital communication have made it possible for people to interact with each other across vast distances. People are no longer merely consumers of information but have become prosumers of content shared in real-time globally[2] (Pearce & Rodgers, 2020).

The emerging literature on digitalization (Coleman & Blumler, 2009; Svensson, 2014; Parycek et al., 2017; Reis et al., 2020) highlighted the creation of a "*digital agora*," a new electronic public sphere that can be seen as a symbol of a more efficient and more emotionally rewarding way to connect citizens and stakeholders (Van Dijk, 2020; Jing, 2022) ICTs are a stimulus toward a more participatory society and provide support for decision-making. At the same time, new challenges emerge, such as new types of privilege in networked societies (Svensson, 2014). One of the most critical and well-known aspects of power distribution in ICTs (Parycek et al., 2017) is, in fact, that of the digital divide, most notably between groups of different social status, migration backgrounds, or gender (Van Dijk, 2012; Lythreatis et al., 2022). It can exacerbate existing inequalities and limit opportunities for those on the wrong side of the divide.

Recent studies also highlight a dark and less explored side of ICT (Bishop, 2018; Fourcade & Johns, 2020) related to information control and, especially, to the possibility of enabling the guiding or control of people's behavior through reactive, cybernetic feedback loops that operate in real-time. These elements have important implications for how people ultimately perceive themselves and how social identities are formed (Burrell & Fourcade, 2021).

Digital platforms (like Twitter, Facebook, YouTube, Instagram, Reddit, TikTok, etc.) extend relationships over time that might otherwise have dwindled or even never existed. Still, as some scholars point out, through psychological and economic techniques, they *have built themselves into a de facto global public sphere with near-monopoly power over the social distribution of attention* (Burrell & Fourcade, 2021, p. 230).

In "*The Society of Algorithms*," Burrell and Fourcade write about it:

> Public debate, knowledge circulation, affirmative pursuits, and reportage have all become intimately dependent upon social media intermediaries and their secretive algorithms. But the sheer abundance of information, which people are supposed to parse through on their own, "often provoke[s] paranoid and otherwise speculative forms of public knowledge and participation" (Hong, 2020, p. 8). Established actors have been displaced by skilled or well-funded activist upstarts, coordinated online mobs, and clickbait producers. The spirits of collective mobilization and counter-mobilization are easily

> overwhelmed in the unequal struggle over the means of digital production (Schradie, 2019). It is instead the old demons of conspiracy, belief, and gossip that have ascended from their graves, called to the surface by rapacious algorithmic spells.
>
> (Burrell & Fourcade, 2021)

Contextual knowledge is crucial to evaluating interventions' impact in response to large-scale phenomena. But what is contextually significant may relate not only to a physical place but also to systems of interpersonal and social relationships that nowadays are widespread and developed in digital contexts. From that perspective, evaluation must gear up with tools that enable it to investigate, in real-time, the dynamics unfolding in these new and as yet unexplored digital contexts.

We think using these new technologies in evaluation designs might allow us to capture the complexity of the generally unexplored constellations of circumstances that characterize digital contexts. This is a necessary step to intercept the nonlinear cause-and-effect mechanisms resulting from participating in debates within the digital *agorà.*

Materials and Methods

The Russia–Ukraine war is an ongoing conflict that began in 2014 when Russia annexed Crimea from Ukraine. The conflict escalated with the involvement of pro-Russian separatists in eastern Ukraine and a military intervention by Russia. The war caused significant damage to infrastructure and resulted in the deaths of thousands of people, causing a massive humanitarian crisis – almost seven million Ukrainians have fled the country. Attempts at resolving the conflict have been made through diplomatic negotiations and various ceasefire agreements, but the fighting has continued. The war has also considerably impacted international relations, with many countries imposing sanctions on Russia for its actions in Ukraine. The conflict remains unresolved, and its impact on the region and the world continues to be significant.

For several reasons, gaining insight into the public's perception of the Russia–Ukraine conflict might be necessary. Understanding how people feel about the conflict and the parties involved is crucial for policymakers and governments, who can better understand its impact on individuals, communities, and the broader region and identify areas of common ground and disagreement between different groups. Understanding public attitudes and opinions can help design strategies for building initiatives that promote dialogue and reconciliation between different parties involved in the conflict. Understanding how different the public perceives narratives and messages can help in crafting more effective messaging and communication strategies. Public opinion can also play a role in holding governments and other actors accountable for their actions. By understanding public perceptions of the conflict, it is possible to identify areas where accountability is needed and work toward ensuring that those responsible for violence and human rights abuses are held accountable.

In this context, social media has emerged as a critical platform for shaping public opinion on the conflict, with a range of voices and perspectives amplified on these platforms. However, social media can also facilitate the spread of misinformation and create echo chambers that polarize public opinion. One way social media may shape public opinion on the conflict is by amplifying voices that may not have been heard otherwise. Social media platforms such as Twitter, Facebook, and Instagram have allowed people on the ground in Ukraine and those with political or personal connections to the conflict to share their perspectives and experiences. This has provided a more diverse and nuanced understanding of the conflict than traditional media outlets.

However, social media can also be a breeding ground for misinformation and propaganda. False information and conspiracy theories can spread quickly on social media, leading to polarized and entrenched positions in the conflict. This has been particularly true in the Russia–Ukraine conflict, with both sides using social media to spread false information about the actions of the other. Furthermore, social media algorithms can create echo chambers where people only see content supporting their opinions. This can reinforce existing beliefs and make finding common ground more difficult and working toward a resolution. Social media can also provide real-time updates on the conflict, allowing people to stay informed about the latest developments.

Thus, social media plays a complex and multi-faceted role in shaping public perception of the conflict. Understanding these dynamics is crucial for policymakers and researchers seeking to make sense of the conflict and its effects on the region.

In general, gaining insight into the public's perception of the Russia–Ukraine conflict requires a multi-faceted approach that combines different methods. One way to gain insight into the public's perception of this conflict is through the use of Google Trends and X (formerly known as Twitter).

To understand these complex dynamics, we observed the flow of information about the conflict to better understand the role of social media in shaping opinions and spreading information (firstly, search the hashtag #RussiaUkraineWar). To this end, we conducted a social network analysis (SNA) (Figure 9.1) that allowed us to identify key countries, track the spread of information, and examine the structure of connected publics. SNA is a method for analyzing social structures through networks and graph theory. In the context of social media, SNA can be used to analyze the connections between users, groups, and content to gain insights into how information is disseminated and how opinions are formed (Xu & Li, 2013; Kapoor et al., 2018). SNA can also identify the different groups and communities on social media platforms (Giuffrida et al., 2018, 2019; Freire et al., 2023).

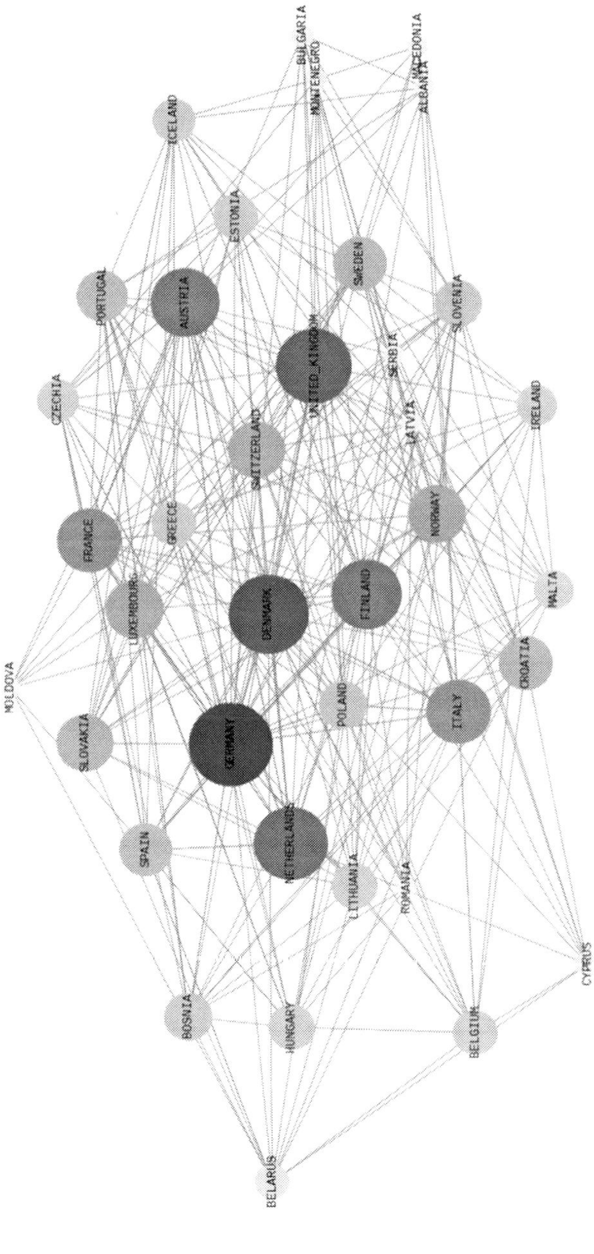

Figure 9.1 Centrality in SNA.

According to the SNA analysis conducted for this research, Germany, Denmark, and the United Kingdom are the nations that participate the most in the Twitter discussion. It is important to note that these countries are members of the Atlantic Alliance (NATO). However, despite not being members of NATO, Finland and Austria also play a prominent role.

Their active participation in these dynamics underscores the importance of international collaboration and building bridges between different political and geographical entities. The SNA thus provides a detailed picture of the complexity of the social interactions and information networks at play. The results show that both NATO members and some actors outside this alliance play a role in building a solid foundation for possible ongoing conflict resolution.

These groups may be based on shared interests, beliefs, or geographic location. By analyzing the connections between these groups, researchers can gain insights into how information is disseminated within and between different communities, identifying patterns in disseminating information on social media. This can include identifying the types of content shared most frequently, the timing and frequency of posts, and the use of hashtags and other metadata. Analyzing these patterns makes it possible to better understand how information is shared and how opinions are formed.

Once all the relationships within the digital context were mapped, we used Google Trends to select which keywords were linked to the hashtag #RussiaUkraineWar. Google Trends is a powerful tool for analyzing search behavior over time and can provide insights into the popularity of specific keywords and phrases. The added value of using Google Trends with hashtags is that it allows one to track the popularity of a hashtag over time, analyze geographic trends, and identify related search queries.

Starting from the search term "Russia-Ukraine war," we looked at the most searched words ("trend"). "Sanctions" and "Nuclear Threat" were then selected. After that, we used these hashtags/keywords to collect tweets related to them. Lastly, Power BI was used to build a dynamic dashboard to find insights about our data. The following section will describe the tools used to conduct the research.

Google Trends

Google Trends allows users to see the relative popularity of search terms over time, providing insight into the public's interest in a particular topic. When combined with hashtags (Miracula & Celardi, 2023), Google Trends may provide a quantitative analysis of the popularity and interest of any specific hashtag over time and across different regions, and how this popularity relates to events and trends. Researchers can quickly identify the most relevant tweets and understand how people discuss a particular issue by searching for keywords or hashtags. This can provide insight into the public's interest in the conflict and how it may have evolved. Additionally, by looking at the geographic location of searches, it is possible to understand how conflict is perceived in different regions.

Google search volume is used in several research areas where it is essential to have information about individual concerns, interests, and perspectives. In medicine, for example, examining search terms related to flu symptoms has been shown to predict flu activity (Ginsberg et al., 2009). Search volume can predict economic indicators (Da et al., 2011; Choi & Varian, 2012). Finally, during the COVID-19 pandemic, several studies analyzed the pandemic using search volume (Pan et al., 2020; Walker et al., 2020).

Google Trends provides a time series index of the volume of queries users enter into Google. The maximum share of queries in a given period is normalized to 100. Queries such as "Nuclear Threat" are counted in the calculation of the query index for "Nuclear." Note that the Google Trends data are calculated using a sampling method, so the results vary by a few percentage points daily. Thus, as shown in several research fields, the search volume analysis can reveal insights into individuals' search for information. However, its use comes with several important considerations. Researchers should exercise caution when utilizing Google Trends data, as it has limitations and potential biases.[3]

Data were extracted for individual countries and then re-aggregated according to their membership or non-membership in NATO, as shown in Table 9.1 , to better understand how the topic of war is researched in European countries. This resulted in a dataset with 730 observations for each country (365 observations for 2021 and 2022).

Twitter and Power BI

Based on what we observed in Google Trends, we tried to understand how information about a given event spreads across social networks. We decided to collect textual data from a specific social network, Twitter. The reason behind this choice is simple. Twitter is a social network from which one can quickly get data, as it provides a regular API, a kind of API (Application Programming Interface) designed to exchange data over the Internet.[4]

Table 9.1 European Ccountries analyszed

Status NATO	*Country*
Members	Belgium, Bulgaria, Czechia, Denmark, Estonia, France, Germany, Greece, Hungary, Iceland, Italy, Latvia, Lithuania, Luxembourg, Netherlands, Norway, Poland, Portugal, Romania, Slovakia, Slovenia, Spain, United Kingdom
Associated	Albania, Austria, Croatia, Finland, Moldova, North Macedonia, Sweden, Switzerland
Not Members	Belarus, Bosnia, Cyprus, Ireland, Malta, Montenegro, Serbia

We used Tweepy, a library that can be used through Python code to get tweets. It is an open-source library that has gained popularity among Python developers due to its ease of use and versatility. Tweepy provides an easy-to-use interface for accessing the Twitter API. The library enables developers to write Python scripts that interact with Twitter in various ways, such as searching for tweets, posting tweets, and streaming real-time tweets.

To get data for this work, tweets were collected using a combination of keywords and hashtags related to the 2022 Russia–Ukraine war, such as "Russia-Ukraine war" and "Nuclear Threat." We collected half a million (500K) tweets from European countries filtered by location, language, and hashtag from February 2021 to October 2022.[5] Then, we built a Power BI dashboard for data analytics purposes. Power BI is a powerful data visualization tool that can be used to analyze large datasets. It has features such as data modeling, transformation, and visualization that make it easy to analyze and visualize data using the drag-and-drop interface of Power BI Desktop (see Figure 9.2). Power BI also integrates with many different data sources, transforming and cleaning data using Power Query.

Once the information of the digital context has been extracted, we applied machine learning techniques for text mining.

NLP Techniques Applied to Artificial Intelligence

The use of language, from the earliest times, has been an essential tool for humans. It has allowed the transmission of important information, first through its oral form and later through writing. This tool immediately proved to be an effective way of disseminating knowledge. Over time, the means and language have changed, but the instrument has remained the same. Today's information is vast; countless texts are stored in libraries worldwide. The advent of technology and its rapid spread in daily life have contributed to creating unimaginable quantities of texts. It was soon realized that machine evolution could play a role in solving this problem: natural language processing is a hybrid discipline involving computer science and linguistics for studying texts in an automated way through computers (Chowdhary & Chowdhary, 2020).

Numerous approaches to natural language processing (NLP) can generally be categorized into four types.

1. *Rule-based approaches*: where linguistic rules and patterns are defined by experts or linguists to identify and extract relevant information from text. These approaches are typically used for sentiment analysis, information extraction, and text classification tasks.
2. *Statistical approaches*: use mathematical models and algorithms to learn patterns and relationships in language data. These approaches are widely used in text classification, machine translation, and speech recognition tasks.

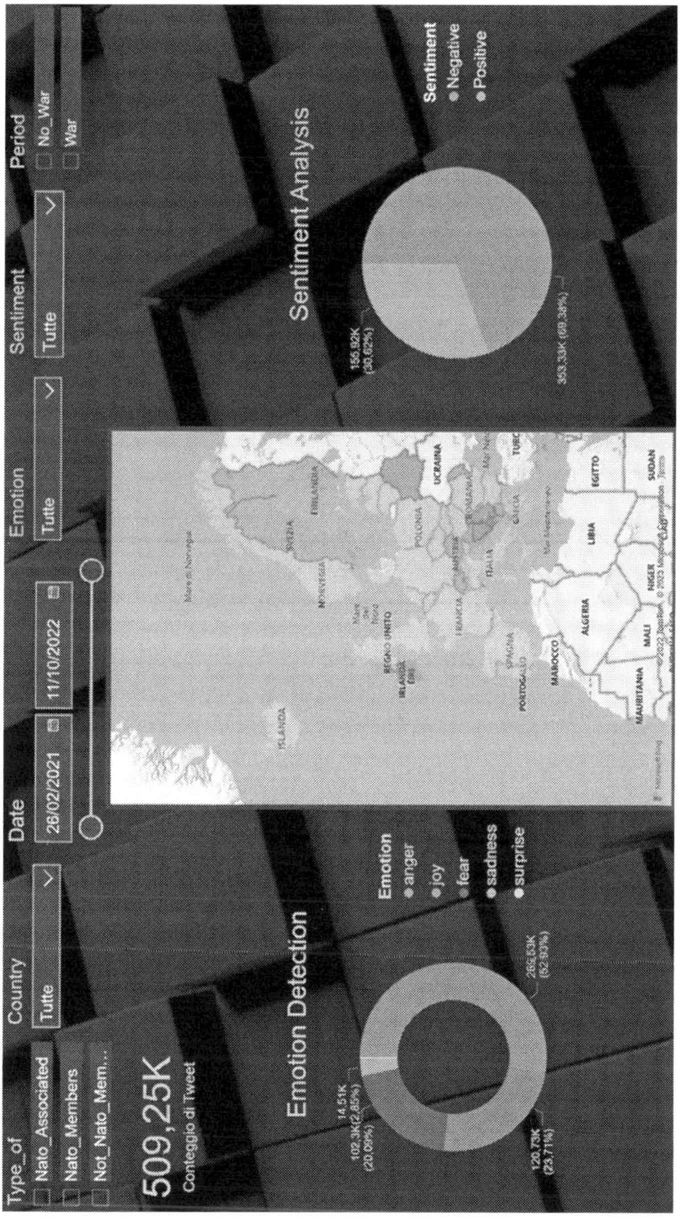

Figure 9.2 Power BI dashboard.

3. *Neural network-based approaches*: use deep learning models, such as recurrent neural networks (RNNs) and convolutional neural networks (CNNs), to learn the patterns and structures in language data. These approaches benefit sentiment analysis, machine translation, and natural language generation tasks.
4. *Hybrid approaches*: combine two or more of the above approaches to improve the accuracy and performance of NLP models.

To extract meta-textual data from our corpus, we had to pre-process it,[6] before moving on to the next step: sentiment and emotion detection – two typical use cases of AI applied to text analysis.

Sentiment Analysis and Emotion Detection

Sentiment Analysis is a field of study that deals with the computational treatment of opinions, sentiments, and emotions expressed in text. It is an area of NLP that aims to determine the attitude of a speaker or writer toward a particular topic, product, or service. Sentiment Analysis has become an important research area due to the exponential growth of digital data and the increasing demand for insights into public opinions and customer feedback (Sahayak et al., 2015; Birjali et al., 2021; Parveen et al., 2023).

The purpose of Sentiment Analysis is to automatically classify the polarity of a given text as positive, negative, or neutral. Positive polarity indicates a favorable or supportive opinion, negative polarity represents an unfavorable or critical opinion, and neutral polarity suggests the absence of strong feelings or opinions. Sentiment Analysis aims to extract meaningful information from text data and use it to conclude public opinions, brand reputation, and consumer behavior.

Emotion Detection, also known as Affective Computing, is a field of study that deals with recognizing and analyzing emotions in human behavior, including facial expressions, speech patterns, and body movements. It is an area of AI and computer science that aims to develop computer systems that can identify and respond to human emotions naturally and intuitively. Emotion Detection has become an important research area due to the growing need for more personalized and human-centered interactions in various domains, such as education, entertainment, health, and customer service (Garcia-Garcia et al., 2017; Nandwani & Verma, 2021; Kusal et al., 2023).

The purpose of Emotion Detection is to automatically identify and categorize emotions a person expresses based on their behavioral cues. Emotions can be categorized into basic emotions like *happiness*, *sadness*, *anger*, *fear*, and *surprise*, or more complex emotions like *joy*, *frustration*, *excitement*, and *boredom*. Emotion Detection aims to extract meaningful information from behavioral data and use it to improve human–computer interactions, enhance customer experience, and promote emotional well-being.

Results

The data collected from Twitter shows that the public reaction to the events of the 2022 Russia–Ukraine war was diverse, with a large proportion of tweets expressing negative sentiment (~70%) toward the events. Many tweets expressed concern and condemnation of the actions of the Russian government, while others showed support for Ukraine's territorial integrity. On the emotional side, we can see that the analysis of tweets results in anger (~50%), fear (~20%), and, surprisingly, joy (~25%) (Figure 9.3).

We conducted sentiment and emotion analysis on the war phenomenon in general. We found that in the pre-war period, the primary emotion is fear (~47%), in line with expectations. Only in the first two weeks of the war is there a considerable percentage of surprise (~52%), which decreases in the following months in favor of the emotions of anger and fear, which stand at similar values of ~30% and ~29%, respectively. In the same period, unexpectedly, there is also a significant percentage of joy (~20%). This figure attracted our attention. Therefore, we explored the tweets labeled "joy" individually to understand the motivations behind this emotion.

Insights on Emotions

Joy: In the cases observed, regardless of the territorial origin of the tweet-joy, most of them referred to the practices of welcoming and supporting Ukrainian citizens fleeing the conflict. In other cases, the interpretation of the tweet required more in-depth analysis. It emerged, for example, that joy was not related to welcoming in the classical sense but that, instead, there was sarcasm and irony present in some tweets that the tool failed to recognize. For example, in cases where the tweets referred to the reception of "beautiful girls" ("*bunnies*") from war zones (see Figure 9.4), an evident example showing the risk of inaccurate results.

Anger was the dominant emotion expressed in most content related to the conflict. This suggests that people were highly frustrated and angry about the

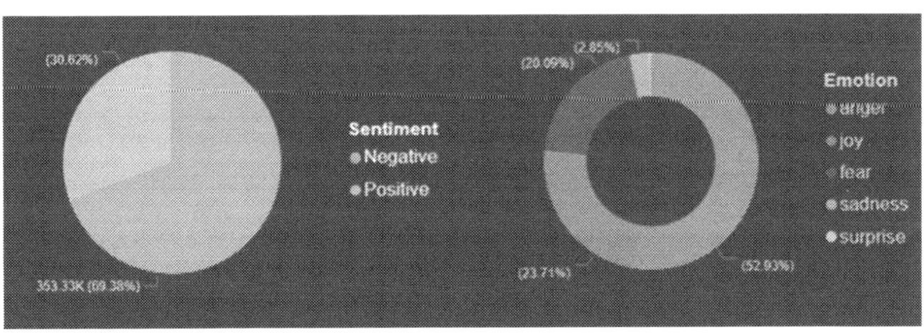

Figure 9.3 Sentiment and emotion for the Russia–Ukraine conflict.

Figure 9.4 Joy in Russia–Ukraine war.

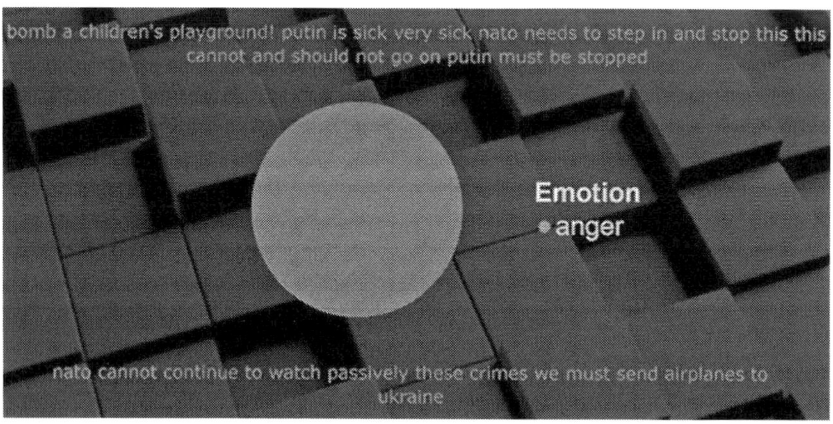

Figure 9.5 Anger during Russia–Ukraine war.

situation, possibly due to the violence, political tensions, and human rights abuses associated with the conflict (see Figure 9.5). This suggests that while people were upset about the situation, they were primarily focused on expressing their anger and frustration.

In the first phase of the conflict (and throughout the observation period), anger was referred to Russia, especially Putin. In particular, negative feelings were directed at the bombing of civilians and the effects the war would have on children. As time passed, the topic of sanction measures targeting specific sectors of the economy of the Russian Federation, initially not much debated, occupied more space until it peaked after the summer (2022), when the effects of the high cost of living affected the lives of individuals and families (particularly regarding the cost of electricity), causing feelings of anger to grow. This occurred mainly in southern European countries rather than in northern European countries.

These findings provide valuable insights into the emotional nature of the Russia–Ukraine conflict and the impact it has been having on people's lives and perceptions. By understanding the underlying emotions driving the conflict, policymakers and other stakeholders can better address the root causes of the conflict in the search of *sustainable* solutions.

Fear: War causes deaths and injuries to military personnel and civilians. The fear of losing loved ones, friends, or oneself can be traumatic and terrifying for all parties involved. The conflict may impact the economy of both Russia and Ukraine and send ripple effects across the world. Fear of an economic downturn could leave people uncertain about their future. Also, war causes displacement of people from their homes and can lead to humanitarian needs. The thought of being displaced, living in a refugee camp, or being unable to return home can be distressing. Lastly, the fear of nuclear weapons: Russia has nuclear weapons, and the use of these weapons could cause mass disruptions, environmental damage, and loss of life. The thought of the use of nuclear weapons could be terrifying.

Even for Fear (see Figure 9.6), emotion detection allows us to identify different causal roots among the countries examined. In northern European countries (Denmark, Sweden, and Finland), fear is predominantly associated with the unpredictable escalation of the conflict, with the risks of both direct NATO involvement and Russia's use of tactical nuclear weapons.

Figure 9.6 Fear in RussiaUkraine war.

In other countries, especially the Mediterranean ones, fear is predominantly associated with the economic impact of the war. The Russian invasion of Ukraine is significantly affecting the EU economy. The conflict and resulting sanctions have disrupted exports from the region for commodities like metals, food, oil, and gas, pushing inflation to levels not seen in decades. Further trade disruptions or increased economic sanctions could plunge the European economy into recession. The slowdown in growth is particularly evident in countries close to Ukraine, like Poland and Hungary. Italy, some other Mediterranean countries, and Germany, heavily dependent on Russian oil and gas, are also feeling the pressure, and, consequently, the fear associated with prolonging the conflict and its consequences.

Surprise: During emotion detection in collected tweets, we noticed a significant amount of expressions of surprise (Figure 9.7). We decided to investigate the reason behind this, and we found that most of the tweets expressing this emotion were related to the peaceful socio-political context that the West had been experiencing for several years. Some people see similarities with dynamics from previous wars. Some are surprised by Russia's deceitful choices, while others are surprised because, for the first time, the war is also being fought on digital and social media.

The president of Ukraine decided to communicate via social media, and the surveillance camera video showing the first Russian vehicles entering Ukrainian territory will always be remembered in history. This series of events has caused a lot of shock, disbelief, and, indeed, surprise in many people belonging to different generations who use media in different ways.

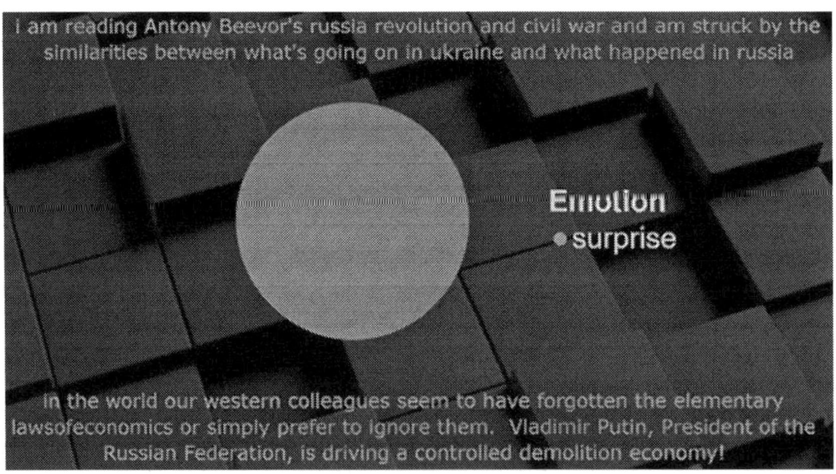

Figure 9.7 Surprise in Russia–Ukraine war.

Table 9.2 Contexts, mechanisms, and emotional responses

Context			*Mechanism hypothesis*	*Emotional response*	
Type of speech (topic)	*Conflict stage*	*Territorial areas*		*Emotion*	*Sentiment (+/-)*
War event (disruption of peace, bombings and war actions, contempt for Putin, possible involvement)	From the beginning of the conflict to about three months later	All	Conflict's unexpected nature triggers fear and concerns about its potential economic, political, and humanitarian implications.	Surprise	-
	The entire observation period	All	Engagement in war causes victims among both the military and civilians. The deliberate targeting of civilians through bombings and the profound impact of the conflict on children contribute to a sense of injustice and cruelty.	Anger	-
	The entire period with the most significant peak in the early phase	Northern Europe	The conflict's unpredictable escalation and the potential use of nuclear weapons by Russia trigger fear among neighboring nations, who fear immediate and severe consequences for regional security and stability.	Fear	-
Economic situation (increasing cost of living)	The entire observation period	Southern Europe	The Russian invasion of Ukraine significantly disrupted the region's exports of critical commodities leading to high inflation. Further trade disruptions or heightened economic sanctions could plunge the European economy into recession.	Fear	-

(*Continued*)

Table 9.2 Continued

Context			*Mechanism hypothesis*	*Emotional response*	
Type of speech (topic)	*Conflict stage*	*Territorial areas*		*Emotion*	*Sentiment (+/-)*
	June to October 2022	Southern Europe	The slowdown is particularly evident in Ukraine's neighboring countries, such as Poland and Hungary. Italy, along with some other Mediterranean countries and Germany, heavily reliant on Russian oil and gas, are also feeling the pressure and, consequently, the anger associated with the prolonged conflict and its economic repercussions	Anger	-
Welcoming and humanitarian actions	The entire observation period	All	Positive reception and assistance provided to refugees can evoke feelings of compassion and generosity. Joy may result from the perception of positively impacting the lives of those fleeing war, offering refuge and support. The humanitarian aspect of welcome and empathy toward those facing challenging situations can activate the mechanism of joy.	Joy	+

Some Realistic Considerations

The power of ML, sentiment analysis, and emotion detection to quickly process large datasets was critical in identifying emerging emotional patterns during the different phases of the conflict. This allowed for a comprehensive exploration of the multiple factors contributing to emotional shifts, enhancing the depth of analysis required in a realistic assessment framework (Leeuw, 2025, this volume).

While the strict application of the realist approach is not feasible due to the inherent nature of the event analyzed, taking a cue from it, we sought to explore the emotional ramifications of the war within the digital realm of Twitter. The dynamic nature of the unfolding events allowed us to discern three distinct "discourse contexts" that defined the broader Twitter landscape. Each context is distinguished by thematic elements encapsulating the emotional reactions conveyed through the dissemination of information, communication, and interaction on Twitter in response to the war. The different territorial areas (Northern and Southern Europe) and conflict phases during the data collection also characterize the contexts.

Table 9.2 shows the contexts, hypothesized mechanisms, and the emotional responses elicited from Twitter users.

In exploring realistic evaluation within political decision-making, it becomes apparent that ML faces inherent limitations. ML's strength lies in pattern detection within data, yet it struggles with the intricate nature of the human context emphasized by realistic evaluation. The latter highlights the importance of understanding the context in political interventions, where ML often falls short in interpreting the complexities of human interactions. The subtle nuances of social relationships, cultural intricacies, and evolving dynamics within communities often elude algorithms that tend to oversimplify the richness of these interactions.

Moreover, realistic evaluation recognizes the dynamic nature of political decisions and societal contexts over time. In contrast, ML systems typically remain static once trained, lacking the adaptability to keep pace with evolving situations. This limitation hampers ML's capacity to offer insights into the long-term impacts of policies, impeding its effectiveness in meeting the adaptability requirements essential for realistic evaluation. As political landscapes and societal dynamics continually shift, the static nature of ML systems proves insufficient in capturing the nuanced, ever-changing reality of political interventions.

Conclusion

The spread of digital technologies in an increasing number of institutions and practices has reconfigured relationships and social dynamics in numerous domains of society. In a scenario where technologies, devices, and digital data are woven into the fabric of society, the digital and the social are evolving into a socio-digital world where it is increasingly difficult to separate sociality and materiality (Amaturo & De Falco, 2022). Digital transformation has thus opened up scenarios that require new conceptual and methodological models to be studied and interpreted.

In these pages, we showed how ML and text analysis tools might provide helpful support to evaluators in understanding the contextual conditions in which complex social phenomena develop. These techniques can be used to analyze large amounts of text data, such as news articles, social media posts, and other sources of information, to extract insights about the context in which they were produced. ML algorithms, in particular, can be used to analyze the sentiment of text data to identify positive, negative, or neutral sentiments. This can provide insights into the attitudes and emotions of those who produced the text data.

The case presented here does not have an evaluative purpose in the strict sense. We did not evaluate interventions such as policies, programs, or projects to judge their effectiveness. Starting from the idea that it is difficult to explain the outcomes of complex social interventions without being related to the context in which they are activated, we tested the potential of ML and text analysis tools in contextual analysis. In particular, our idea is that the study of digital contexts or digital agoras should be considered when faced with measures taken in response to global or otherwise international events. In some cases, such as the one presented here, it might be the only way to quickly acquire information in response to a complex event.

Despite the inherent limitations of ML discussed above, these tools can provide valuable insights into what collective responses may be triggered following the enactment of measures that significantly impact the lives and consumption of individuals and households (e.g., EU sanctions against Russia and the long-term impact they will have on the cost of living). By gauging the sentiment and emotions of individuals, governments and policymakers can better understand how the public perceives these initiatives.

We showed how this analysis might serve several essential purposes. It allows for an assessment of public perception. By categorizing sentiments and emotions, it becomes possible to determine whether a specific measure is viewed favorably or critically. Second, this analysis can identify common concerns and issues the public raises. If a substantial portion of the population expresses negative sentiments, this may indicate shortcomings in the measure that must be addressed.

Moreover, international comparisons offer insights into how contextual factors influence the perception of the same measure. Tracking sentiment changes over time can inform how public opinion evolves as an intervention develops. Governments can use this data to make necessary adjustments or address issues promptly (Mazzeo Rinaldi et al., 2017).

Unpopular measures/phenomena of global significance may trigger anger and fear, resulting in boycotts of policy measures or leading people to seek reassurance in populist instances. From an evaluation point of view, therefore, for decision-makers, having real-time information on widespread sentiment and prevailing emotions might be crucial. Real-time information about people's sentiments can help policymakers identify emerging issues and concerns that may require a response. For example, suppose social media posts or news articles indicate that a particular measure (e.g., sanction) is not working as expected or is causing people's reactions that were not anticipated. In that case, decision-makers can use this information to change their strategies.

These tools are helpful for evaluative research because they allow for the reconstruction – through the collection of large amounts of data – of the digital and globalized contexts in which information spreads, and emotions and feelings arise. That way, trends can be observed, and reactions and behaviors can be speculated faster than traditional evaluation research tools.

To understand what mechanisms are triggered by certain contextual conditions, leading individuals to put specific behavioral responses in place, it is considered necessary to supplement the information found online with other information related to the historical, cultural, social, technological, etc. dimensions of the real contexts. It would also be necessary to include in the analysis the particular point of view of the social actors who populate the real contexts. This element escapes the analysis of BD.

Like all social networks, Twitter analysis is subject to various biases that can affect the accuracy of insights. Selection bias arises from the non-representative demographics of Twitter users, who are predominantly younger and tech-savvy. Volunteer bias emerges as users self-select what they share, leading to a skewed dataset. Confirmation bias occurs as users engage with content that aligns with their beliefs, fostering echo chambers. Temporal bias also results from the platform's rapid trends and event-driven nature. One potential issue is the risk of bias in the algorithms used, which can lead to inaccurate or skewed results (Greenstein & Cho, 2025, this volume).

Additionally, there may be concerns about the transparency and accountability of AI-powered evaluations, particularly in cases where the data used is sensitive or controversial. To mitigate these pitfalls, evaluators should employ diverse data sources, employ appropriate statistical methods, and acknowledge data limitations to achieve more accurate Twitter analyses. Finally, it is crucial to recognize that while AI tools can be powerful aids to evaluators, they should not replace the critical thinking and judgment required to conduct high-quality evaluation research.

Notes

1 Pawson and Tilley (1997), in their application of scientific realism for evaluation research, consider that the mechanisms through which programs work will only operate if certain contextual circumstances are present. In a realistic theoretical

framework, while it is not possible to make generalizations about what constitutes "context" in isolation, it is possible to form generalizable, middle-range causal explanations (Merton, 1968) about how contexts interact with mechanisms to produce outcomes.

2 Today's digital landscape is far from a one-way street; individuals are active contributors who shape it. In contrast to the past, when the internet was a platform for passive content consumption, the emergence of social media has altered the power dynamic. Platforms like Facebook and Twitter enable "prosumers" to connect globally, offering spaces to share experiences and expertise. Prosumers influence public opinion, drive conversations, and impact policy through their online presence. However, this shift isn't without challenges. The abundance of user-generated content blurs the lines between fact and fiction, leading to misinformation and eroding trust in online sources. Critical evaluation and authenticity verification are now crucial for users.

3 However, its use comes with several important considerations. Researchers should exercise caution when utilizing Google Trends data, as it has limitations and potential biases. One key limitation is that Google Trends lacks context. It provides data on search query volumes and patterns but does not explain causation or reflect user intent accurately. Consequently, it should not be the sole basis for drawing definitive scientific conclusions. Another concern is the bias and representativeness of Google Trends data. It primarily captures the interests of internet users, excluding those without internet access and those who use non-Google search engines. Additionally, search trends can be influenced by media coverage and external factors, which may not align with genuine public sentiment.

4 From February 9, 2023, Twitter will no longer support free access to its API – both v2 and v1.1. "A paid basic tier will be available instead. Over the years, hundreds of millions of people have sent over a trillion tweets, with billions more every week. Twitter data are among the world's most powerful data sets. We're committed to enabling fast and comprehensive access so you can continue to build with us," the Twitter Dev account posted at the beginning of February 2023. While the official account assured that further details would be provided soon, it is not yet clear how much these new paid tiers would cost. New owner Elon Musk, however, tweeted that "just approximately $100 per month for API access with ID verification" would help ward off bot scammers and opinion manipulators (Mitra, 2023) T

5 Hashtags are not translated, as they were extracted already in English. Only the tweets themselves were in their original language.

6 We apply the following techniques to extract meta-textual data from our corpus: (1) Tokenization, for breaking down the text into smaller units or tokens, which can be words, characters, or sub-words; (2) Stemming, for reducing inflected words to their root form; (3) Lemmatization, to find the base word form called a lemma; (4) Part of Speech Tagging, to assign a grammatical tag to each word in a sentence based on its role and function in the sentence; (5) Named Entity Recognition, which uses the tokens to recognize in a text an entity (e.g., the name of a company, numbers, values, places, an organization, and a person).

References

Amaturo, E., & De Falco, C. C. (2022). Traces and Algorithms as Socio-digital Objects. In F. Comunello, F. Martire, and L. Sabetta (eds.). *What People Leave Behind.* Frontiers in Sociology and Social Research, vol 7. Springer. https://doi.org/10.1007/978-3-031-11756-5_18

Appadurai, A. (2001). Deep democracy: Urban governmentality and the horizon of politics. *Environment and urbanization, 13*(2), 23–43. https://doi.org/10.1177/095624780101300203

Birjali, M., Kasri, M., & Beni-Hssane, A. (2021). A comprehensive survey on sentiment analysis: Approaches, challenges and trends. *Knowledge-Based Systems, 226*, 107134. https://doi.org/10.1016/j.knosys.2021.107134

Bishop, J. (2018). Evaluating the Risk of Digital Addiction in Blended Learning Environments: Considering ICT Intensity, Learning Style, and Architecture. In B. Bozoglan (ed.). *Psychological, Social, and Cultural Aspects of Internet Addiction* (pp. 169–185). IGI Global.

Burrell, J., & Fourcade, M. (2021). The society of algorithms. *Annual Review of Sociology, 47*, 213–237. https://doi.org/10.1146/annurev-soc-090820-020800

Castells, M. (2004). Informationalism, networks, and the network society: a theoretical blueprint. In M. Castells (ed). *The Network Society: A Cross-cultural Perspective* (pp. 3–45). Egaronline.

Choi, H., & Varian, H. (2012). Predicting the present with Google Trends. *Economic Record, 88*(1), 2–9. https://doi.org/10.1111/j.1475-4932.2012.00809.x

Chowdhary, K., & Chowdhary, K. R. (2020). Natural language processing. *Fundamentals of Artificial Intelligence,* 603–649. https://doi.org/10.1007/978-81-322-3972-7_19

Coldwell, M. (2019). Reconsidering context: Six underlying features of context to improve learning from evaluation. *Evaluation, 25*(1), 99–117. https://doi.org/10.1177/1356389018803234

Coleman, S., & Blumler, J. G. (2009). *The Internet and Democratic Citizenship: Theory, Practice and Policy*. Cambridge University Press.

Da, Z., Engelberg, J., & Gao, P. (2011). In search of attention. *Journal of Finance*, 66(5), 1461–1499. https://doi.org/10.1111/j.1540-6261.2011.01679.x

Ebenso, B., Manzano, A., Uzochukwu, B., Etiaba, E., Huss, R., Ensor, T., & Mirzoev, T. (2019). Dealing with context in logic model development: reflections from a realist evaluation of a community health worker programme in Nigeria. *Evaluation and Program Planning, 73*, 97–110. https://doi.org/10.1016/j.evalprogplan.2018.12.002

Fitzpatrick, J.L. (2012). An introduction to context and its role in evaluation practice. *New Directions for Evaluation* 135: 7–24. https://doi.org/10.1002/ev.20024

Fourcade, M., & Johns, F. (2020). Loops, ladders and links: The recursivity of social and machine learning. *Theory and Society, 49*, 803–832. https://doi.org/10.1007/s11186-020-09409-x

Freire, M., Antunes, F., & Costa, J. P. (2023). Enhancing decision-making support by mining social media data with social network analysis. *Social Network Analysis and Mining, 13*(86), 1–15 https://doi.org/10.1007/s13278-023-01089-6

Garcia-Garcia, J. M., Penichet, V. M., & Lozano, M. D. (2017). Emotion Detection: A Technology Review. In *Proceedings of the XVIII International Conference on Human Computer Interaction* (pp. 1-8). https://doi.org/10.1145/3123818.3123852

Gatto, L., & Bundi, P. (2025). The Use of Quantitative Text Analysis in Evaluations. In S. B. Nielsen, F. Mazzeo Rinaldi and G. J. Petersson (eds.). *Artificial Intelligence and Evaluation. Emerging Technologies and Their Implications for Evaluation* (pp. 144–167). Routledge. https://doi.org/10.4324/9781003512493

Ginsberg, J., Mohebbi, M. H., Patel, R. S., Brammer, L., Smolinski, M. S., & Brilliant, L. (2009). Detecting influenza epidemics using search engine query data. *Nature, 457*(7232), 1012–1014. https://doi.org/10.1038/nature07634

Giuffrida, G., Gozzo S., Mazzeo Rinaldi, F., & Tomaselli, V. (2018). Big Data and Network Analysis: A Combined Approach to Model Online News. In F. Greselin, L. Deldossi, L. Bagnato, and M. Vichi (eds.). *Statistical Learning of Complex Data* (pp. 183–191). Springer.

Giuffrida, G., Gozzo, S., Mazzeo Rinaldi, F., & Tomaselli, V. (2019). Building decision-making indicators through network analysis of big data. *Social Indicators Research.* https://doi.org/10.1007/s11205-020-02363-2

Greenhalgh, J., & Manzano, A. (2022). Understanding 'context' in realist evaluation and synthesis. *International Journal of Social Research Methodology, 25*(5), 583–595. https://doi.10.1080/13645579.2021.1918484

Greenstein, N., & Cho, S.-W. (2025). Ethics & Equity in Data Science for Evaluators. In S. B. Nielsen, F. Mazzeo Rinaldi and G. J. Petersson (eds.). *Artificial Intelligence and Evaluation. Emerging Technologies and Their Implications for Evaluation* (pp. 56–77). Routledge. https://doi.org/10.4324/9781003512493

Grossmann, I., Feinberg, M., Parker, D. C., Christakis, N. A., Tetlock, P. E., & Cunningham, W. A. (2023). AI and the transformation of social science research. *Science, 380*(6650), 1108–1109. https://doi.org/10.1126/science.adi1778

Holm, J., Newman-Griffis, D., & Petersson, G. J. (2025). Big Data for Big Investments: Making Responsible and Effective Use of Data Science and AI in Research Councils. In S. B. Nielsen, F. Mazzeo Rinaldi and G. J. Petersson (eds.). *Artificial Intelligence and Evaluation. Emerging Technologies and Their Implications for Evaluation* (pp. 120–143). Routledge. https://doi.org/10.4324/9781003512493

Hong, S. (2020). *Technologies of Speculation: The Limits of Knowledge in a Data-Driven Society New York*: NYU Press.

Hooghe, L., & Marks, G. (2001). *Multi-level Governance and European Integration.* Rowman & Littlefield.

Jing, F. S. (2022). Our digital agora: Politics without privacy. *Techné: Research in Philosophy and Technology, 26*(3), 529–533. https://10.5840/techne2022263167

Kapoor, K. K., Tamilmani, K., Rana, N. P., Patil, P., Dwivedi, Y. K., & Nerur, S. (2018). Advances in social media research: Past, present and future. *Information Systems Frontiers, 20*, 531–558. https://doi.org/10.1007/s10796-017-9810-y

Kerr, H., Price, J., Nicholl, H., & O'Halloran, P. (2018). Facilitating transition from children's to adult services for young adults with life-limiting conditions (TASYL): Programme theory developed from a mixed methods realist evaluation. *International Journal of Nursing Studies, 86,* 125–138. https://doi.org/10.1016/j.ijnurstu.2018.06.015

Kusal, S., Patil, S., Choudrie, J., Kotecha, K., Vora, D., & Pappas, I. (2023). A systematic review of applications of natural language processing and future challenges with special emphasis in text-based emotion detection. *Artificial Intelligence Review*, 1–87. https://doi.org/10.1007/s10462-023-10509-0

Leeuw, F. L. (2025). The Algorithmization of Policy and Society: The Need for a Realist Evaluation Approach. In S. B. Nielsen, F. Mazzeo Rinaldi and G. J. Petersson (eds.). *Artificial Intelligence and Evaluation. Emerging Technologies and Their Implications for Evaluation* (pp. 242–265). Routledge. https://doi.org/10.4324/9781003512493

Lythreatis, S., Singh, S. K., & El-Kassar, A. N. (2022). The digital divide: A review and future research agenda. *Technological Forecasting and Social Change*, *175*, 121359. https://10.1016/j.techfore.2021.121359

Mathias, K., Mathias, J., Goicolea, I., & Kermode, M. (2018). Strengthening community mental health competence - A realist informed case study from Dehradun, North India. *Health & Social Care in the Community*, *26*(1), e179–e190. https://doi.org/10.1111/hsc.12498

Mazzeo Rinaldi, F., Giuffrida, G., & Negrete, T. (2017). Real-time monitoring and evaluation–- Emerging news as predictive process using big data based approach. In G. Petersson and J. D. Breul (eds.). *Cyber Society, Big Data and Evaluation* (vol. 24, pp. 191–214). Routledge.

Merton R. K. (1968). *Social Theory and Social Structure*. New York: The Free Press, Extended edition.

Miracula, V., & Celardi, E. (2023). *The Role of Twitter and Google Trends in Identifying the Perception of Russia-Ukraine Wars*. 5th International Conference on Advanced Research Methods and Analytics (CARMA 2023) (pp. 71–80). Editorial Universitat Politècnica de València.

Mitra, A. (2023). Twitter to stop free API access – what does it mean, and will it affect users? *Mint*, 04 February 2023, https://www.livemint.com/companies/news/twitter-to-stop-free-api-access-what-does-this-mean-and-will-it-affect-users-11675516987841.html

Næss, T., Prabhu, C., Mjaaland, M., Holtermann, H., & Engebretsen, L. S. (2025). Text Mining and Machine Learning in an Evaluation of Police Handling of Cybercrime in Norway. In S. B. Nielsen, F. Mazzeo Rinaldi and G. J. Petersson (eds.). *Artificial Intelligence and Evaluation. Emerging Technologies and Their Implications for Evaluation* (pp. 103–119). Routledge. https://doi.org/10.4324/9781003512493

Nandwani, P., & Verma, R. (2021). A review on sentiment analysis and emotion detection from text. *Social Network Analysis and Mining*, *11*(1), 81. https://doi.org/10.1007/s13278-021-00776-6

Nielsen, S. B., Lemire, S., & Tangsig, S. (2022). Unpacking context in realist evaluations: Findings from a comprehensive review. *Evaluation*, *28*(1), 91–112. https://doi.org/10.1177/13563890211053032

Nielsen, S. -B. (2023). Disrupting evaluation? Emerging technologies and their implications for the evaluation industry. *New Directions for Evaluation, 178–179*, 45–57. https://doi.org/10.1002/ev.20558

Nielsen, S. B., Rinaldi, F. M., & Petersson, G. J. (2025). Evaluation in the Era of Artificial Intelligence. In S. B. Nielsen, F. Mazzeo Rinaldi and G. J. Petersson (eds.). *Artificial Intelligence and Evaluation. Emerging Technologies and Their Implications for Evaluation* (pp. 1–12). Routledge. https://doi.org/10.4324/9781003512493

Pan, Z., Nguyen, H. L., Abu-Gellban, H., & Zhang, Y. (2020). *Google trends analysis of covid-19 pandemic*. In 2020 IEEE International Conference on Big Data (Big Data) (pp. 3438–3446). IEEE.

Parveen, N., Chakrabarti, P., Hung, B. T., et al. (2023). Twitter sentiment analysis using hybrid gated attention recurrent network. *Journal of Big Data, 10*(50). https://doi.org/10.1186/s40537-023-00726-3

Parycek, P., Rinnerbauer, B., & Schossböck, J. (2017). Democracy in the digital age: Digital agora or dystopia. *International Journal of Electronic Governance*, *9*(3–4), 185–209. https://doi.org/10.1504/IJEG.2017.088224

Pawson, R. (2013). *The Science of Evaluation: A Realist Manifesto*. Sage.

Pawson, R. (2016). The ersatz realism of critical realism: A reply to Porter. *Evaluation*, *22*(1), 49–57. https://doi.org/10.1177/1356389015605

Pawson, R., & Tilley, N. (1997). An introduction to scientific realist evaluation. In E. Chelimsky & W. R. Shadish (eds.). Evaluation for the 21st century: A handbook (pp. 405-418). Sage Publications, Inc.

Pearce, S. C., & Rodgers, J. (2020). Social media as public journalism? Protest reporting in the digital era. *Sociology Compass*, *14*(12), 1–14. https://doi.org/10.1111/soc4.12823

Petersson, G. J., & Breul, J. D. (eds.). (2017). *Cyber Society, Big Data, and Evaluation*. Routledge.

Pfadenhauer, L. M., Mozygemba, K., Gerhardus, A., Hofmann, B., Booth, A., Lysdahl, K. B., & Rehfuess, E. A. (2015). Context and implementation: A concept analysis towards conceptual maturity. *Zeitschrift für Evidenz, Fortbildung und Qualität im Gesundheitswesen*, *109*(2), 103–114. https://doi.org/10.1016/j.zefq.2015.01.004

Picciotto, R. (2020). Evaluation and the big data challenge. *American Journal of Evaluation*, *41*(2), 166–181. https://doi.org/10.1177/10982140198503

Raffini, L. (2011). Partecipazione. In G. Bettin Lattes and L. Raffini (eds.). *Manuale di sociologia*, *2* (pp. 709–728). CEDAM.

Reis, J., Amorim, M., Melão, N., Cohen, Y., & Rodrigues, M. (2020). Digitalization: A Literature Review and Research Agenda. In Z. Anisic, B. Lalic, and D. Gracanin (eds.). *Proceedings on 25th International Joint Conference on Industrial Engineering and Operations Management – IJCIEOM. IJCIEOM 2019. Lecture Notes on Multidisciplinary Industrial Engineering*. Springer.

Rogers, P. J. (2008). Using Programme Theory to Evaluate Complicated and Complex Aspects of Interventions. *Evaluation*, *14*(1), 29-48. https://doi.org/10.1177/1356389007084674

Rosenau, J. N., & Czempiel, E. O. (eds.). (1992). *Governance without Government: Order and Change in World Politics* (No. 20). Cambridge University Press.

Sahayak, V., Shete, V., & Pathan, A. (2015). Sentiment analysis on twitter data. *International Journal of Innovative Research in Advanced Engineering (IJIRAE)*, *2*(1), 178–183.

Schradie, J. (2019). *The Revolution that Wasn't: How Digital Activism Favors Conservatives* Cambridge, MA. Harvard Univ. Press.

Shaw, R. L., Hiles, D. R., West, K., Holland, C., & Gwyther, H. (2018). From mixing methods to the logic (s) of inquiry: Taking a fresh look at developing mixed design studies. *Health Psychology and Behavioral Medicine*, *6*(1), 226–244. https://doi.org/10.1080/21642850.2018.1515016

Svensson, M. (2014). Voice, power and connectivity in China's microblogosphere: Digital divides on SinaWeibo. *China Information*, *28*(2), 168–188. https://doi.org/10.1177/0920203X14540082

Van Dijk, J. (2020). *The Digital Divide*. John Wiley & Sons.

Van Dijk, J. A. (2012). The Evolution of the Digital Divide - The Digital Divide Turns to Inequality of Skills and Usage. In J. Bus, M. Crompton, M. Hildebrandt and G. Metakides (eds.). *Digital Enlightenment Yearbook 2012* (pp. 57–75). IOS Press.

Walker, A., Hopkins, C., & Surda, P. (2020). Use of Google trends to investigate loss of smell–related searches during the COVID-19 outbreak. *International Forum of Allergy & Rhinology, 10*(7), 839–847. https://doi.org/10.1002/alr.22580

Wellman, B. (2001). Physical Place and Cyberplace: The Rise of Personalized Networking. In*ternational Journal of Urban and Regional Research*. 25. 227–252. https://doi.org/10.1111/1468-2427.00309.

Xu, G., & Li, L. (eds.). (2013). *Social Media Mining and Social Network Analysis: Emerging Research.* Information Science Reference.

York, P., & Bamberger, M. (2025). The Applications of Big Data to Strengthen Evaluation. In S. B. Nielsen, F. Mazzeo Rinaldi and G. J. Petersson (eds.). *Artificial Intelligence and Evaluation. Emerging Technologies and Their Implications for Evaluation* (pp. 37–55). Routledge. https://doi.org/10.4324/9781003512493

Ziulu, V., Anuj, H., Hagh, A., Raimondo, E., & Vaessen, J. (2025). Extracting Meaning from Textual Data for Evaluation. Lessons from Recent Practice at the Independent Evaluation Group of the World Bank. In S. B. Nielsen, F. Mazzeo Rinaldi and G. J. Petersson (eds.). *Artificial Intelligence and Evaluation. Emerging Technologies and Their Implications for Evaluation* (pp. 78–102). Routledge. https://doi.org/10.4324/9781003512493

10 Harnessing Geospatial Approaches to Strengthen Evaluative Evidence

Anupam Anand, Geeta Batra, and Juha I. Uitto

Introduction

Integration of spatial thinking[1] and its applied uses in the social sciences and humanities has been an ongoing process over the past few centuries. Still, the development of geospatial tools for handling spatial data began only in the 1960s (Goodchild & Janelle, 2010). Since then, geospatial techniques and tools have been widely used to map and monitor changes because of their ability to provide spatially explicit, synoptic, time-series data for various earth system processes (Awange & Kyalo Kiema, 2013; Melesse et al., 2007). Evaluation, in comparison, is transdisciplinary and draws on methods from other fields to deal with both real-world evaluation constraints and methodological challenges (Patton, 2018; Scriven, 1993). These evaluation methods have their strengths and weaknesses, and selecting the appropriate methods and techniques is guided by evaluation questions, available resources, and the context of the interventions being evaluated (Vaessen et al., 2020). More than a decade ago, the lead editorial in *Nature* urged that all observations in the environmental sciences be georeferenced, given the tremendous advances in data availability and geospatial technologies (Nature, 2008). Currently, there are more than 2,500 operating satellites in space, of which 446 are used for earth observation (World Economic Forum, 2020). The use of geospatial data and methods in the development sector has increased immensely as data and tools become readily available, along with a growing need for good data for monitoring, reporting, learning, and generating quantitative evidence on the success or failures of projects (Uitto, 2016).

Geospatial approaches can be used to analyze both biophysical and environmental conditions and the distribution and changes in settlement patterns and infrastructure, as well as socioeconomic development. As discussed in the chapters by Bruce, K., Vandelanotte, J., and Gandhi, V.J. (2025) and York, P., and Bamberger, M. (2025) in this book, digital tools, data, and analytics can significantly improve the performance of monitoring, evaluation, and reporting systems, increasing efficiency and cutting down costs of data collection, analysis, and storage while increasing coverage and quality (Bamberger et al., 2016). Similarly, using geospatial techniques in conjunction with other methods and

DOI: 10.4324/9781003512493-10

data sources – such as literature research, surveys, on-the-ground investigations, etc. – contributes to strengthening evaluative evidence, allows for enhanced triangulation between various sources of information, and increases the accuracy and transparency of evidence (Lech et al., 2018).

Evaluation as a discipline can also contribute to geospatial science in several ways, including providing feedback on improving the methods and quality of data, insights into effectiveness and utility, and generating evidence for future research and application areas. Evaluations assess the relevance, value, performance, and durability of interventions. They also highlight methodological challenges and data gaps in tracking progress and measuring the outcomes of initiatives to tackle major global issues such as climate change or changes in land use. For example, until recently, there was limited data to track progress toward emission reduction – a significant data gap, but GHGSat[2] pioneered the satellite monitoring of greenhouse gases in high resolution. Through rigorous evaluation methods, evaluators can assess the effectiveness and utility of geospatial data and help data providers and scientists identify areas for enhancements and future applications. Geospatial scientists can use evaluative evidence to develop tailor-made solutions for evaluation and to improve the outcomes of spatial analysis and methods. By working together, evaluators and data scientists can help ensure that development interventions are based on solid evidence and produce meaningful impact. Development initiatives happen somewhere on earth, and Geographical Information System (GIS) is a spatial science – evaluations through robust evidence on what works (or does not), where, and why can add to expanding the knowledge and applications of geospatial science.

In this chapter, we draw on the experience of the Independent Evaluation Office (IEO) of the Global Environment Facility (GEF) to explore how spatial science has enhanced the IEO's ability to strengthen its evaluative evidence and demonstrate the global impact of GEF-supported environmental projects. We employ case studies to spotlight the tangible benefits of geospatial science, including integrating satellite and drone data in evaluations alongside traditional evaluation methods.

The chapter begins with an introduction to geospatial analysis in evaluation, discusses the surge in its uptake and use by different evaluation entities, and elaborates a few key considerations on skills, resource needs, and privacy, legal, and ethical issues. The later section highlights how geospatial analysis strengthened the evaluative evidence in various GEF IEO evaluations. Following a discussion on the challenges and opportunities, the chapter concludes by highlighting the need for addressing current challenges and the need for collaborations to set up guidelines and standards for integrating geospatial work.

Evaluative Insights Through Geospatial Analysis

Geographic Information Systems (GIS) science effectively integrates space, place, and temporal elements – all of which are important for considering critical humanitarian and environmental issues and for interpreting human well-being and the changing environment. GIS integrates many types of data, providing deeper insights into patterns, interlinkages, and context and thereby helping evaluate development interventions. In its earlier applications in evaluation, GIS was mainly used for data visualization using two-dimensional maps. Highlighting its lack of use in evaluation, Renger et al. (2002) demonstrated how GIS can be used for data visualization, change detection, and presenting results in a user-friendly manner. Assessing change over time is of great interest to evaluators. The evaluation literature has discussed using geospatial data for measuring baselines and outputs and how these can be used to enhance evaluation practice (Lech et al., 2018; Azzam, 2013; Azzam & Robinson, 2013). Quasi-experimental designs leveraging geospatial data were used to conduct impact evaluations in forestry and biodiversity interventions. Geospatial analysis has also been used in randomized control trials (Ferraro & Pattanayak, 2006; Andam et al., 2008). A recent systematic review of 437 evaluation studies that used big data to measure or evaluate development outcomes found that satellite data was used in over seventy percent of the measurement studies and over eighty percent of the impact evaluations (Rathinam et al., 2021).

Further, the role of geospatial science is increasingly being recognized by several major environmental and development policy conventions and institutions as countries move toward more evidence-based policy decisions and practices. For example, the United Nations Convention to Combat Desertification (UNCCD) has endorsed the use of indicators obtained from remote sensing to monitor progress toward reversing and halting the degradation and desertification of land (Anand & Batra, 2022). The United Nations Framework Convention on Climate Change (UNFCCC) and the Convention on Biological Diversity (CBD) also endorse the use of objective indicators, many of which are derived through geospatial methods. The Independent Evaluation Office (IEO) of the Global Environment Facility (GEF) has been at the forefront of pioneering geospatial approaches for environmental interventions in various domains, including land degradation, climate change, international waters, and biodiversity (Anand & Batra, 2021). In a specific evaluation of GEF's support for protected areas, we used satellite data equivalent to billions of observations (pixels) of forest data for 37,000 protected areas in 147 countries averaging about 400 sq. km each (GEF IEO, 2016). The satellite data-driven analysis enabled us to assess the relevance and effectiveness of GEF-supported protected areas compared to areas (buffer areas, other protected areas) that did not receive GEF support. In the land degradation evaluation, we applied machine learning algorithms driven by geospatial data and econometric analysis, which allowed us to work with the high volume of data, measure environmental changes in biophysical indicators, draw insights into the factors associated with the outcomes, and estimate the co-benefits in terms of one ecosystem service, carbon sequestration (GEF IEO, 2017). This global analysis was complemented by fieldwork, beneficiary surveys, and the collection of GIS data using smartphones. We have also used geospatial analysis and ecological forecasting methodologies to quantify land cover change, estimate above-ground carbon stock, and evaluate ecosystem services provided by GEF-supported protected areas ex-ante in Kenya (Thieme et al., 2020).

Evaluation offices of several multilateral organizations, such as the World Bank, the Food and Agriculture Organization of the United Nations (FAO), and the International Fund for Agricultural Development (IFAD), are also utilizing GIS and geospatial data to improve the quality and effectiveness of their evaluations. These entities have combined various methods, such as randomized controlled trials, quasi-experimental designs, and case studies, with geospatial approaches to assess the results of development projects and programs. The World Bank's Independent Evaluation Group (IEG) has been using geospatial methods and data in a variety of evaluations, including those related to urban growth, disaster risk management, and socio-economic development. For example, in its Managing Urban Spatial Growth evaluation,[3] IEG used a combination of machine learning techniques and econometrics to understand the change in urban growth. Similarly, IEG, through its Country Program Evaluations,[4] is using geospatial analysis to ascertain the relevance of development programs and identify areas where assistance is most needed. IFAD's Independent Office of Evaluation (IOE) published a manual to provide practical guidance on using GIS in the monitoring and evaluation of rural development projects (IFAD, 2022). Besides evaluation, the manual also discusses the utility of GIS for improving monitoring and reporting and provides geospatial data standards and quality checklists for project designers and managers. Similarly, the FAO's Office of Evaluation (OED) has used geospatial data in several evaluations, including in fragile and crisis-affected countries (FAO, 2017).

As geospatial approaches continue to unveil new possibilities for evaluations, we delve into the critical considerations that may influence their adoption in the evaluation process.

Key Considerations for Integrating Geospatial Approaches

Technology Infrastructure Requirements

Geospatial analysis can be understood as a component of a decision process and support infrastructure. It is a multifaceted system that includes software programs (GIS) and hardware, data from sensors, disciplinary expertise, and spatial thinking and analytical skills. Geospatial data is collected using various technologies such as GPS, satellites, drones, and mobile devices (see Table 10.1 for definitions of common terms). Besides asking the questions,

Table 10.1 The differences between GIS, satellite data, and drones, along with their limitations and advantages

	Geographic Information System (GIS)	*Satellite data*	*Drones*
Definition	A computer-based system for capturing, storing, analyzing, and displaying geographically referenced data	Data collected by satellites orbiting the Earth	Uncrewed aerial vehicles (UAVs) used for collecting data
Advantages	Provides a wide range of spatial analysis tools, can integrate various types of data, and enables decision-making based on geographical context	It covers large areas, captures data at regular intervals, and can be used to monitor changes over time	Can capture high-resolution data in real time and can be used to access hard-to-reach areas
Limitations	Requires a high level of technical expertise, can be expensive to implement and maintain, and may be limited by the quality of the input data	Limited to what can be captured from space, may be affected by cloud cover and other atmospheric conditions, and may require significant post-processing to be useful	Limited flight time and range, may be affected by weather conditions, and requires a skilled operator
General applications	Urban planning, natural resource management, emergency response, transportation planning, agriculture	Environmental monitoring, weather forecasting, disaster management, national security, urban planning	Environmental monitoring, precision agriculture, infrastructure inspection, search and rescue, disaster response
Applications in M&E	Change detection, experimental design such as RCT, Quasi-experimental methods such as Difference in Difference	Data could be used for time series analysis, change detection, pattern analysis, and as an input to econometric models and experimental design. Applicable at multiple geographic scales	Visualization is challenging to use for change detection or time series analysis and econometric models. Applicable mostly at a smaller scale.

judging the suitability of geospatial techniques for the evaluation objective, and identifying relevant satellite-based indicators or proxy indicators, successful integration of geospatial methods in evaluation requires a robust technical infrastructure. This includes data storage and management systems, software tools for data processing and analysis, and platforms for data visualization and dissemination (Figure 10.1). The development and maintenance of such infrastructure require collaboration between technical experts, domain experts, and end-users.

Data storage and management include the development of a database system capable of storing and managing large volumes of geospatial data and providing access to users in a secure and efficient manner. This requires the use of appropriate hardware and software systems, such as cloud-based solutions, database management systems (DBMS), and geographic information systems (GIS). Software tools for processing and analyzing geospatial data and satellite imagery are popular tools, such as ENVI or ERDAS Imagine, as well as GIS software, such as ArcGIS and the widely used open-source tool QGIS. Additionally, programming languages such as Python and R can be used for data processing and analysis. Lastly, data visualization and dissemination involve the development of tools and platforms, including web-based mapping tools, such as Google Maps and OpenStreetMap, as well as custom-built web-based platforms for specific applications for visualizing and disseminating results to the end-users. Also, data visualization tools, such as Tableau and D3.js, can be used for visualization and communication. Lately, analysis using large data sets and computational power is being performed on cloud-based solutions such as Amazon Web Services and Google Earth Engine.

Skills

Application of geospatial science, data, and methods is at a nascent stage in evaluation, and evaluators are generally not trained in data science and even less so in spatial science. Hence, it is imperative to collaborate within multidisciplinary teams, where domain experts play a vital role in interpreting geospatial analyses, establishing clear linkages with evaluation criteria, and crafting messages tailored for non-technical audiences. In the early stages of the IEO, we partnered with external experts who were integrated into the evaluation team. However, as time progressed, we expanded our team by bringing in staff members equipped with specialized training and expertise in spatial science. The number of such staff has grown over the last few years. In addition, partnerships with institutions specializing in specific types of analysis have been beneficial.

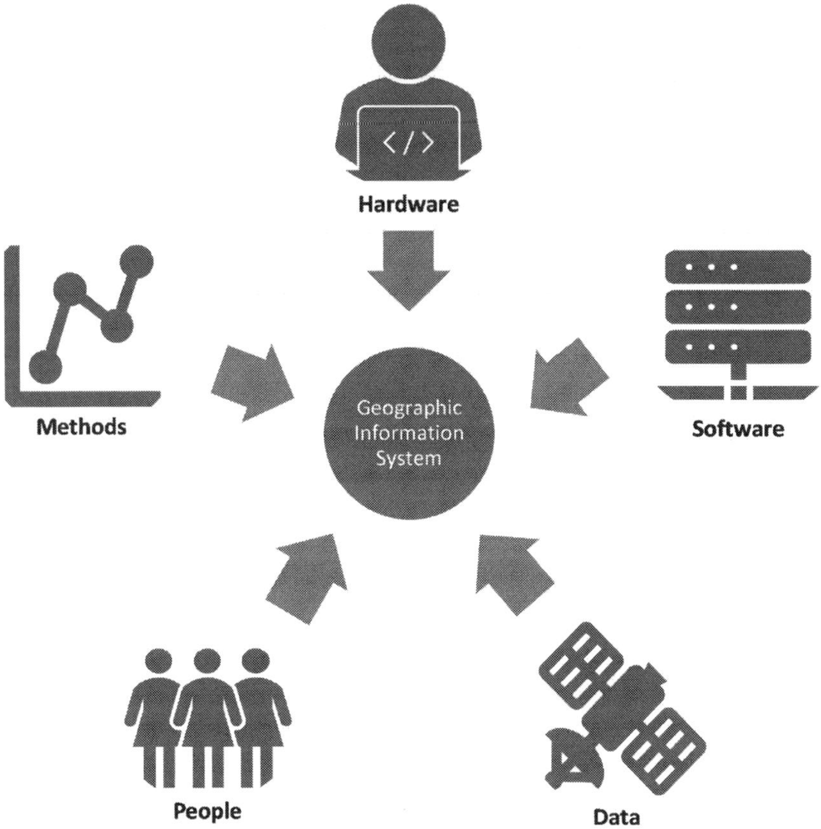

Figure 10.1 Various components of the Geographic Information System.

Aspects Related to Privacy, Ethics, Legality, and Security

The use of geospatial data raises several concerns related to privacy, ethics, legal, and security aspects that are outlined, along with potential solutions, in Table 10.2. To address these concerns, it is important to follow best practices and guidelines for the collection, processing, and sharing of geospatial data; these are discussed in detail in Greenstein, N., Cho, S-W. (2025). Additionally, importance should be given to staying up-to-date with the latest regulations and laws related to geospatial data. The third section provides examples of how we addressed some of these concerns in our own evaluations.

Insights from IEO Case Studies: Integrating Geospatial Information with Diverse

Methods and Data Sources

This section showcases the diverse applications of geospatial science within various environmental domains by GEF IEO. It also demonstrates the effective use of geospatial data and techniques to combine quantitative and qualitative data from a variety of data sources, managing data inconsistencies and quality issues, examining the factors influencing environmental outcomes, estimating results,

Table 10.2 Privacy, ethics, legal, and security issues associated with geospatial data[5]

Privacy, ethical, legal, and security issues	*Risks and challenges*	*Potential solutions*
Privacy issues	Potential for geospatial data to reveal sensitive information such as location, behavior, and activities; use for surveillance and violation of privacy rights	Measures can be taken, such as anonymization of data, restricting access to sensitive information, and implementing robust data protection policies and purpose limitation
Ethical issues	Data accuracy, bias, and accessibility may lead to incorrect or unfair decisions being made based on that data; accessibility issues may lead to inequitable outcomes	Recognize and address biases in data collection, processing, and analysis; use unbiased algorithms in decision-making; promote fair use and transparency in data collection, storage, and usage
Legal issues	Ownership and intellectual property; national security issues due to unauthorized use for creating maps, identifying military targets, and monitoring borders	Through compliance with relevant laws and regulations related to data protection and privacy, as well as protection through various forms of intellectual property rights, including patents, trademarks, and copyrights
Data security	Access or use of personnel or sensitive data (on groups of people, such as ethnic minorities or political dissidents) by unauthorized individuals or organizations; vulnerability to cyber-attacks or hacking	Access control, data encryption, backups, and responsible data governance

and conducting on-field validations. Furthermore, this section delves into the application of geospatial approaches to tackle the challenges presented by remote or difficult-to-access regions. It also underscores the utility of geospatial methodologies in addressing key questions aligned with the evaluation criteria established by the Organization for Economic Co-operation and Development's Development Assistance Committee (OECD DAC). Consequently, it illustrates the application of geospatial techniques in addressing questions regarding the relevance, effectiveness, and overall impact of intervention efforts.

Integrating Geospatial Data with Field Visits

The GEF IEO has effectively integrated geospatial data and methodologies with other complementary evaluation approaches. In the case of the Evaluation of the Small Island Developing State (GEFIEO, 2018), geospatial analysis and field visits were strategically combined to enhance the accuracy and reliability of the findings. Geospatial analyses were specifically employed to illustrate the project's relevance and its initial outcomes. This project, initiated in 2015, aimed to enhance the efficient management and sustainable utilization of the natural resources on the northeastern coast of Saint Lucia, located in the Eastern Caribbean. It sought to generate numerous global environmental benefits. Notably, this region encompasses the Iyanola dry forests, which hold the classification of Key Biodiversity Areas (KBA) and are also designated as Important Bird Areas (IBA).

Geospatial analysis served as a valuable complement to the field visits, contributing to the triangulation of results. It facilitated the evaluation of the project's relevance and its initial impact, including the assessment of progress achieved at specific restoration sites (as depicted in Figures 10.2, 10.3, and 10.4).

Throughout the field visits, the IEO evaluation team harnessed handheld GPS devices and Unmanned Aerial Vehicles (UAVs), commonly known as drones, to gather ground truth data. These GPS devices and drones were pivotal in providing essential information regarding deforested areas and the restoration sites within the project area. This data served as a valuable input for geospatial analysis. This case exemplifies how integrating two distinct methodologies, geospatial and field-based, contributed to the triangulation of evaluative evidence. The geospatial analysis consisted of (i) forest change analysis to examine the long-term trends of forest loss in the protected area and its surrounding areas and (ii) the long-term vegetation productivity trend analysis within the protected area and the selected restoration sites visited by the evaluation team (Figure 10.2). Field visits also corroborated that besides forest loss, forest degradation is a major environmental factor affecting the health of the ecosystem in the region.

The analyses of forest loss and vegetation productivity served to corroborate the preexisting challenges of forest loss and degradation within the Iyanola ecosystem, predating the initiation of the project. This evaluative evidence also provided valuable insights into the significance of GEF support in addressing the delicate Iyanola forest ecosystem through a comprehensive approach aimed at tackling the root causes of ecosystem degradation. This approach encompassed both national-level planning and regulatory adjustments alongside site-specific activities. Nonetheless, during the course of this assessment, the team encountered several challenges. These included grappling with data gaps and inconsistencies, all the while maintaining a keen awareness of ethical and legal considerations related to geospatial data. These considerations encompassed respecting privacy and adhering to designated flight zones while operating the drones.

Resolving Data Inconsistencies and Validation of Field Observations

The IEO attempted to leverage the existing World Database on Protected Areas (WDPA) and satellite data to retrieve data for geospatial analysis. However, the boundary data and the satellite products were inconsistent. This is a major issue with small island nations as the global datasets are generally unavailable at a fine resolution. Therefore, the data available from the Ministry of Environment, Government of St. Lucia, was used for the boundary delineation, and additional satellite data products were generated by the IEO for geospatial analysis. Further

Figure 10.2 A forest restoration site inside the Iyanola National Park.

analysis showed that the global database of satellite data products for SIDS is relatively less accurate.

Therefore, additional dense time series vegetation productivity was carried out to analyze and highlight the long-term trends of vegetation health. Sixteen-day Moderate Resolution Imaging Spectroradiometer (MODIS) was used to derive the normalized difference vegetation index (NDVI), a widely used proxy for vegetation health. This additional analysis also helped deal with the inconsistencies in the global forest data for smaller areas such as small islands. This dense time series vegetation productivity analysis helped assess the spatial and temporal extent of vegetation trends. The results showed that overall, there had been a minor increase in vegetation productivity since 2018, despite the precipitation showing a downward trend (Figures 10.3 and 10.4).

Given the precipitous nature of the terrain and the logistical challenge of visiting dense forests, the IEO used a drone to collect deforestation data samples to train[6] the landcover classification algorithm. Some of these samples were also used to independently validate the accuracy of the satellite-driven classified landcover map.

While drones have gained popularity, it is essential to prioritize compliance with regulatory requirements when using them. The IEO team diligently secured the requisite government permissions and meticulously adhered to drone operation regulations, including altitude limits and flight time restrictions. To address

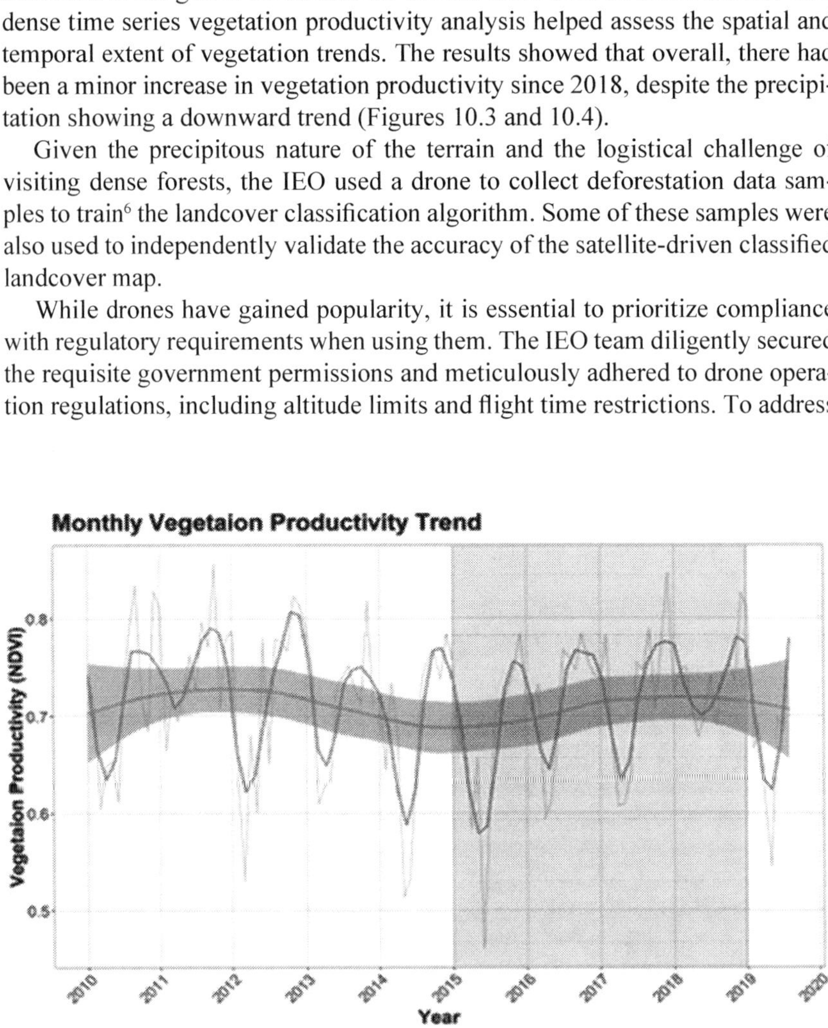

Figure 10.3 Average vegetation productivity trend.

Source: GEFIEO.

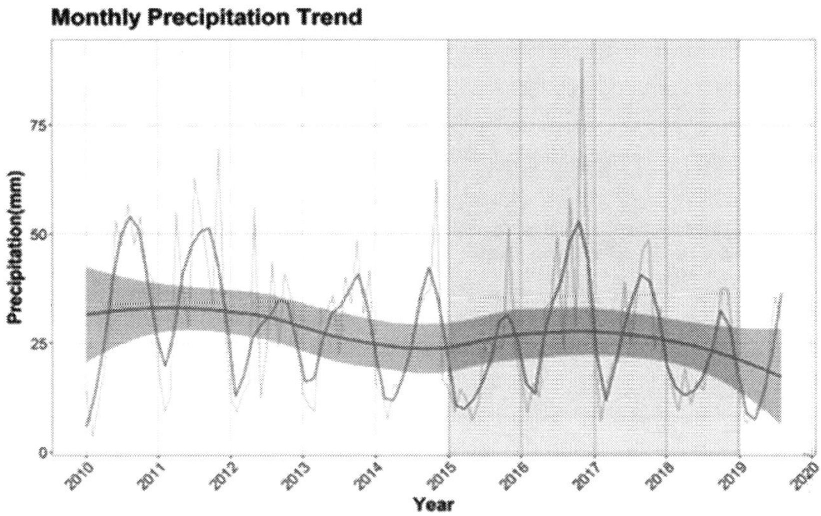

Figure 10.4 Monthly precipitation trend.

Source: GEFIEO.

local concerns, UAVs were deployed in the presence of forest department officials. Stringent measures ensured that the drones operated exclusively within designated airspace, avoiding airfields and private properties. In addition to following local regulations pertaining to UAVs, the team consulted resources such as https://www.dji.com/flysafe/geo-map for preliminary information on drone deployment (see Figure 10.5).

This use case of geospatial data in evaluation points out the need for better locally validated data, as global datasets such as WDPA might not be consistent. This is particularly important in SIDS, which are smaller in spatial extent and have highly uneven coastlines that call for a more accurate boundary delineation to carry out spatial analysis or planning.

Utilizing Geospatial Data in Challenging Evaluation Environments

The GEF has a large portfolio (33%) of environmental projects in conflict-affected countries (GEF-STAP, 2018). The map in Figure 10.6 shows the GEF-supported interventions with conflict hotspots. Development projects in fragile and conflict contexts are complex, hard to reach, isolated, and unsafe, and these areas face a wide array of challenges where a regular flow of information is difficult. Standard tools and processes have to be adapted to gather information

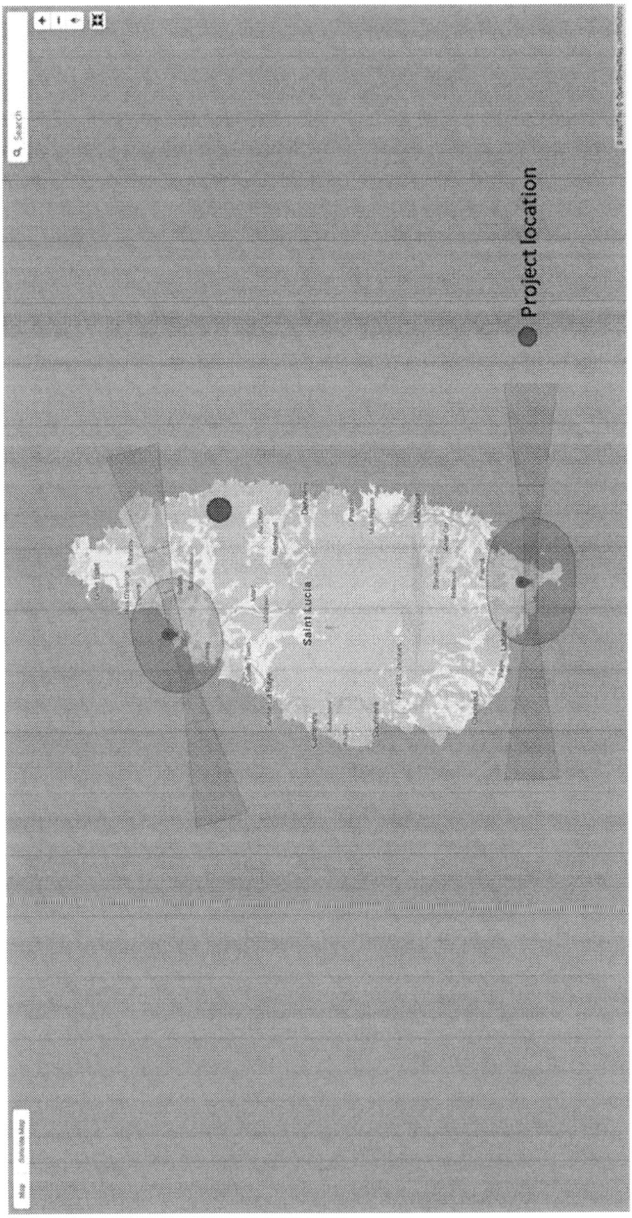

Figure 10.5 Map showing the restricted areas and project site for drone deployment.

for designing projects for implementation and conducting M&E. In this context, where constraints related to logistics, time, and resources are paramount, geospatial approaches are particularly useful. We have seen how context and baseline information derived through geospatial data and methods can help us frame evaluation questions with greater clarity. This can also be useful for mission planning. In addition, when combined with other information sources, geospatial data can help identify causal factors such as the drivers of environmental degradation and conflict and quantify the extent of habitat degradation and loss. The availability of different types of sensors enables us to see through the clouds at night or during inclement weather.

Remote sensing data collected through drones can prove to be handy for doing rapid assessments in hard-to-reach, isolated, and unsafe areas. Drones can be deployed quickly to collect more granular data within a short time frame. The exciting part of using drones is the ability to use them to capture visuals that can later be used to enhance knowledge and learning products. In an evaluation of GEF Support to Mainstreaming Biodiversity (GEF IEO, 2019), the IEO used drones to assess the extent of illegal mining and logging areas at different project sites and to collect ground truth data for validating satellite data products. During the field mission to Colombia, the evaluators did not have easy and safe access to illegal mining sites (Figure 10.7), and with permission from the authorities, had to deploy the drone from a safe distance.

Using drones for collecting geospatial data not only provided visual evidence of environmental degradation, but we also fed this data into a machine learning classification algorithm to help delineate the mining and associated environmental degradation. The Pacific region of Colombia has a high cloud cover and is opaque to optical remote sensing platforms. However, radar data that could see through clouds was used to train the classification model (Figure 10.8).

These areas overlaid with the conflict and project location data and helped answer questions on the relevance and effectiveness of the project intervention (Figure 10.9).

Further, the results were used to examine the interrelationship between the conflict and environmental outcomes more closely and were validated through field interviews and literature analysis. For example, the results indicating the increase in deforestation (Figure 10.10) during the post-conflict period were validated through data collected by interviewing key informants and project participants.[7] In a post-conflict period, many forested areas become more accessible for agriculture, large-scale cattle ranching, the spread of coca cultivation, and the speculative land market (Murillo-Sandoval et al., 2020).

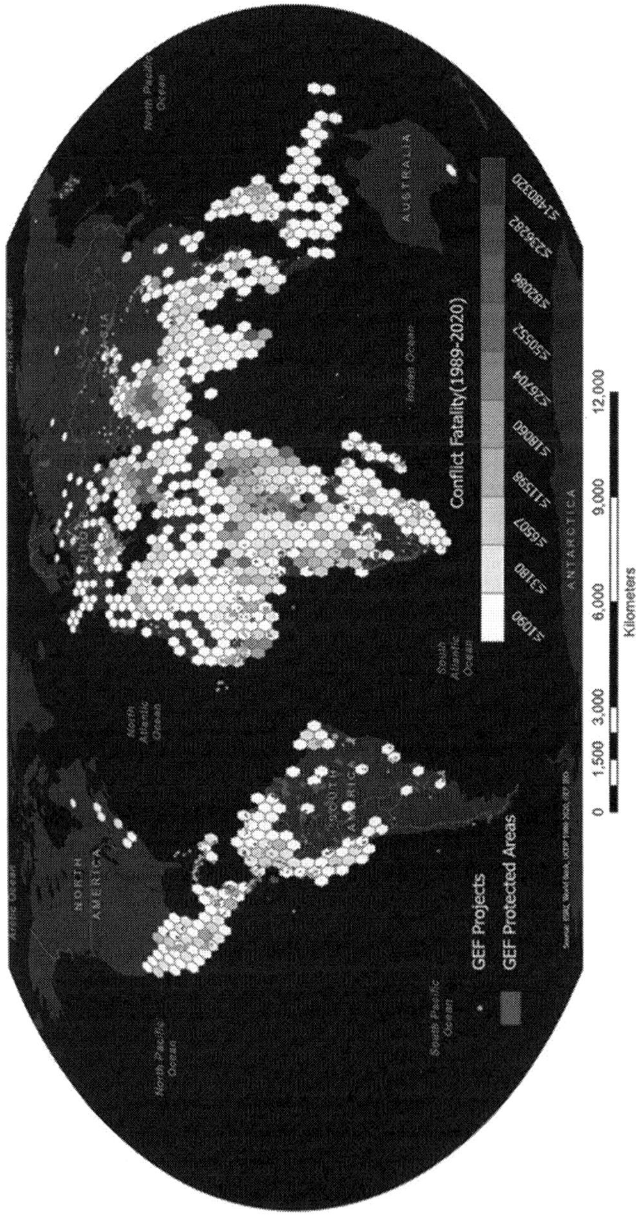

Figure 10.6 GEF-supported interventions with conflict hotspots.

Source: GEFIEO.

Figure 10.7 Drone image showing illegal mining area near a project site in Colombia; Pic: Anupam Anand.

Addressing Digital Infrastructure and Resource Requirements

Analyzing satellite data on a large scale requires access to high-performance computers, specialized software, and robust data and IT infrastructure, resources typically unavailable within evaluation offices. To address these challenges, we adopted a resourceful approach. For instance, we harnessed the capabilities of existing computing resources to mitigate some of these obstacles. One noteworthy technique involved leveraging parallel computing, a method that makes effective use of multiple processing cores found in modern desktops and laptops. This approach efficiently distributed computing tasks, resulting in a significant reduction in processing time. Implementing parallel computing generally requires only minor modifications to programming language code and is supported by widely used statistical packages like R, Python, and Stata.

To scale up our analyses, we also turned to cloud computing platforms. For instance, during the evaluation of protected areas using forest cover data (Hansen et al., 2013), we conducted our analysis on Google's cloud computing platform, which led to a substantial reduction in processing time.

Additionally, our efforts have included valuable collaborations with external organizations. Notably, we collaborated with the National Aeronautics and

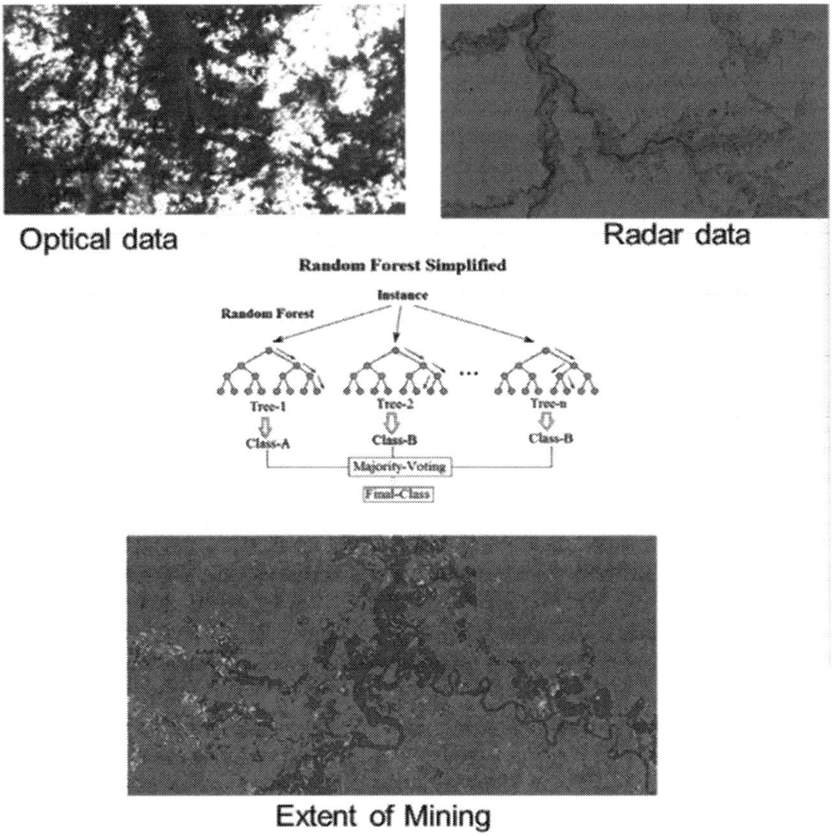

Figure 10.8 Mining area classification using machine learning based on optical and radar data.

Source: GEFIEO.

Space Administration (NASA) to gain access to high-resolution images and ecological forecasting data for ex-ante evaluation (Thieme et al., 2020), used AID data for geocoding project locations (Runfola et al., 2020), and partnered with the University of Maryland to benefit from technical guidance, data resources, and additional computing support for a mixed-methods-based impact evaluation focused on the effectiveness of protected areas (GEF IEO, 2016).

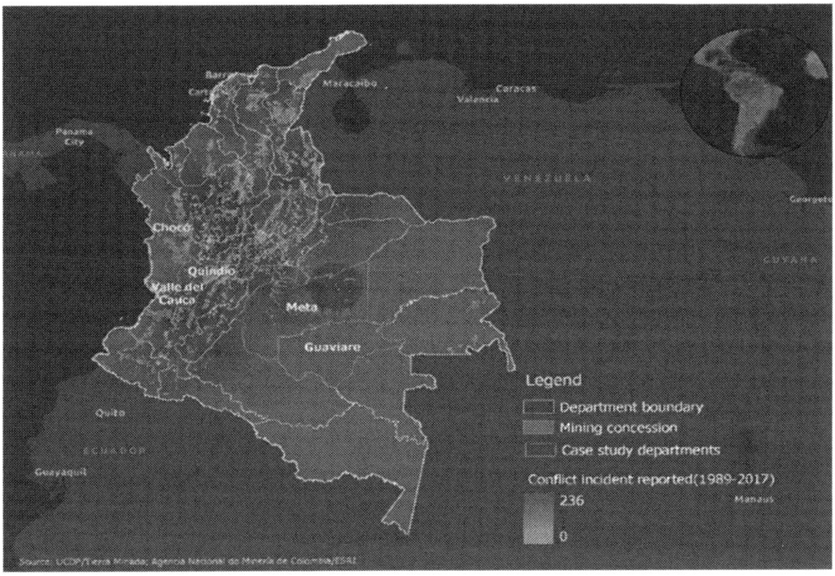

Figure 10.9 The map shows mining areas overlap with the hotspots of conflict and conservation in Colombia.

Source: GEF IEO.

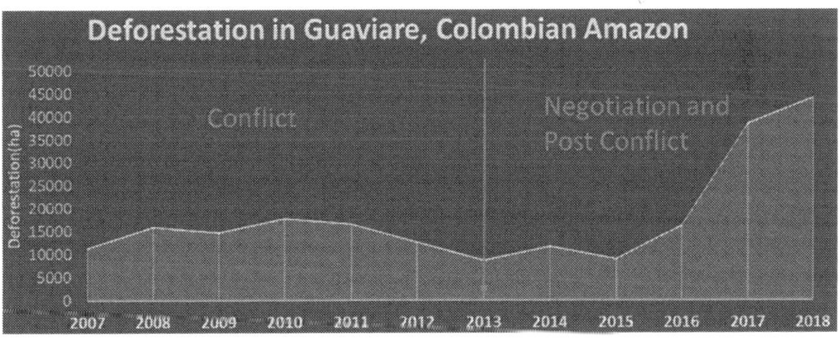

Figure 10.10 Increase in deforestation in the project area during the post-conflict period.

Discussion and Conclusions

Geospatial technology is among the rapidly evolving sectors, with the global geospatial industry ecosystem boasting approximately 339 billion U.S. dollars in market revenue in 2020. Projections indicate that this industry is set to expand further, reaching an estimated 439 billion US dollars by 2025 (Statista, 2020). The proliferation of satellites in Earth's orbit is a significant catalyst for the geospatial industry's growth, with a current count of 7,702 active satellites orbiting the Earth as of May 4, 2023, as reported by the satellite tracking website "Orbiting Now" (NanoAvionics, 2023, May 4).

This rapid growth is attributed to several factors, including the increased accessibility and affordability of location-sensing technologies, the growing demand for location-based services and solutions, the integration of machine learning and artificial intelligence with spatial data, and the emergence of new applications and innovations.

Geospatial technology finds its applications across various sectors and is applied to analyze biophysical and environmental conditions, settlement patterns, infrastructure, and socioeconomic development. In particular, digital tools and analytics have improved monitoring, evaluation, and reporting, making data collection more efficient and cost-effective. These tools have significantly enhanced our capacity to conduct evaluations efficiently and cost-effectively, spanning individual projects, portfolio assessments, and ecologically significant scales. Insights derived from these evaluations have unveiled changes and patterns related to development outcomes that were previously concealed.

This chapter illustrates that by combining the strengths of evaluation and geospatial analysis, we can gain a better understanding of the impact and effectiveness of interventions, especially in contexts where location and spatial relationships are important factors. Both geospatial science and evaluation are multidisciplinary fields that demand technical proficiency, cooperation, and a breadth of knowledge.

Throughout this chapter, it is clear that geospatial analysis introduces a crucial spatial dimension to evaluation. This dimension enhances our understanding of how interventions affect particular locations, regions, or communities. Moreover, geospatial data serves as an effective monitoring tool, offering real-time insights into the progress of interventions and enabling adjustments for improved effectiveness. Geospatial analysis facilitates comparisons between diverse locations or regions, shedding light on the factors contributing to the varying success of interventions. These tools can also be harnessed for predictive modeling of future scenarios based on different intervention strategies. Finally, geospatial analysis empowers the creation of visualizations that enhance the accessibility and clarity of evaluation findings.

The chapter also highlights how evaluations can, in turn, enhance geospatial science by providing feedback, insights, and evidence for future research. They assess the relevance, performance, and effectiveness of interventions and highlight methodological challenges and data gaps.

Evaluations play an important role in enhancing geospatial approaches by offering a comprehensive approach for the utilization of geospatial data and tools. They ensure that the data used in the evaluation is accurate and reliable by triangulating it with other sources. Evaluations also apply quality control guidelines and ethical considerations, which are particularly important when dealing with sensitive spatial data. Evaluations assess the impact and effectiveness of interventions that apply geospatial analysis, identifying areas for improvement in methodology and data quality. Through the provision of decision-making feedback and the creation of visual representations that enhance the understanding of geospatial findings, evaluations ensure that geospatial approaches are more user-friendly and comprehensible for stakeholders.

However, as these geospatial techniques gain wider acceptance among evaluators due to their utility, their challenges come into focus. These challenges include capacity limitations, data availability and access, infrastructure and resource constraints, ethical considerations, transparency, and validation, among others. These challenges may pose barriers to the use of geospatial data, especially in evaluation departments with limited resources in data-scarce regions. They could influence the willingness to embrace innovative technologies.

As evaluators, our commitment to harnessing advanced technology remains driven by the continuous expansion of data sources, open-source tools, and computational capabilities. To encourage the use of these technologies and address these challenges, evaluators and evaluation departments must actively promote awareness of these methodologies, hire experts proficient in spatial analysis, and encourage closer collaborations with institutions that have expertise in these domains. Effective partnerships are imperative among evaluation teams, academia, data providers, and research and policy institutions to establish guidelines, codes of conduct, data quality standards, transparency criteria, and ethical and legal frameworks for the responsible adoption of emerging methods, enhancing the quality of our work.

Notes

1 Spatial thinking is the guiding principle of geospatial science. It is a type of reasoning that involves organizing and integrating concepts based on space and using them to solve problems and make decisions. Understanding the characteristics of space, such as dimensionality, continuity, proximity, and separation, is essential for spatial thinking.
2 https://www.ghgsat.com/en/what-we-do/.
3 https://ieg.worldbankgroup.org/evaluations/managing-urban-spatial-growth.
4 https://ieg.worldbankgroup.org/evaluations/mexico-country-program.

5 Source: Roger Tomlinson (2019); Thinking About GIS: Geographic Information System Planning for Managers, Fifth Edition (ISBN: 9781589483484; & others (cite).
6 Training data: The data was used in a machine learning model to classify the satellite data into forest and non-forest classes.
7 The SIDS – St Lucia Case Study that used mixed-methods can be viewed here: https://www.youtube.com/watch?v=y_Al2O-PqRo.

References

Anand, A., & Batra, G. (2021). Using Big Data and Geospatial Approaches in Evaluating Environmental Interventions. In J. I. Uitto (Ed.), *Evaluating Environment in International Development* (pp. 79–92). Routledge. https://www.taylorfrancis.com/chapters/oa-edit/10.4324/9781003094821-7/using-big-data-geospatial-approaches-evaluating-environmental-interventions-anupam-anand-geeta-batra

Anand, A., & Batra, G. (2022). Application of Geospatial Methods in Evaluating Environmental Interventions and Related Socioeconomic Benefits. In J. I. Uitto & G. Batra (Eds.), *Transformational Change for People and the Planet: Evaluating Environment and Development*. Springer Nature. https://doi.org/10.1007/978-3-030-78853-7_19

Andam, K. S., Ferraro, P. J., Pfaff, A., Sanchez-Azofeifa, G. A., & Robalino, J. A. (2008). Measuring the effectiveness of protected area networks in reducing deforestation. *Proceedings of the National Academy of Sciences*, *105*(42), 16089–16094. https://doi.org/10.1073/pnas.0800437105

Awange, J. L., & Kyalo Kiema, J. B. (2013). Spatial analysis. *Environmental Geoinformatics*, 225–236. https://doi.org/10.1007/978-3-642-34085-7_17

Azzam, T. (2013). Mapping data, geographic information systems. *New Directions for Evaluation*, *2013*(140), 69–84. https://doi.org/10.1002/ev.20074

Azzam, T., & Robinson, D. (2013). GIS in evaluation. *American Journal of Evaluation*, *34*(2), 207–224. https://doi.org/10.1177/1098214012461710

Bamberger, M., Raftree, L., & Olazabal, V. (2016). The role of new information and communication technologies in equity-focused evaluation: Opportunities and challenges. *Evaluation*, *22*(2), 228–244. https://doi.org/10.1177/1356389016638598

Bruce, K., Vandelanotte, J., & Gandhi, V. (2025). Emerging Technology and Evaluation in International Development. In S. B. Nielsen, F. Mazzeo Rinaldi & G. J. Petersson (Eds.), *Artificial Intelligence and Evaluation. Emerging Technologies and Their Implications for Evaluation* (pp. 13–36). Routledge. https://doi.org/10.4324/9781003512493

Editorial: A place for everything. *Nature.* 2008;453(2), 2. Anonymous. https://doi.org/10.1038/453002a

FAO Office of Evaluation. (2017). Final evaluation of the Programme for Improvement of Irrigation Systems in Kabul, Bamyan and Kapisa Provinces. *Food and Agriculture Organization of the United Nations*. https://www.fao.org/3/bd685e/bd685e.pdf

Ferraro, P. J., & Pattanayak, S. K. (2006). Money for nothing? A call for empirical evaluation of biodiversity conservation investments. *PLoS Biology*, *4*(4), e105. https://doi.org/10.1371/journal.pbio.0040105

Global Environment Facility Independent Evaluation Office. (2016). *Impact evaluation of GEF support to protected areas and protected area systems.* Evaluation 104.

https://www.gefeo.org/sites/default/fles/ieo/evaluations/fles/BioImpactSupportPAs-2016.pdf

Global Environment Facility Independent Evaluation Office. (2017). *Land degradation focal area study.* https://www.gefeo.org/evaluations/land-degradation-focal-area-ldfa-study-2017

GEF STAP (Global Environment Facility Scientific and Technical Advisory Panel). (2018). *Environmental security: dimensions and priorities.* https://www.thegef.org/sites/www.thegef.org/files/publications/52103%20STAP%20Report_WEB.pdf

Global Environment Facility Independent Evaluation Office. (2018). *Strategic Country Cluster Evaluation (SCCE): Small Island Developing States (SIDS).* https://www.gefieo.org/evaluations/scce-sids

Global Environment Facility Independent Evaluation Office. (2019). *Evaluation of GEF support to mainstreaming biodiversity.* https://www.gefieo.org/evaluations/biodiversity-mainstreaming-2018

Goodchild, M. F., & Janelle, D. G. (2010). Toward critical spatial thinking in the social sciences and humanities. *GeoJournal*, *75*(1), 3–13. https://doi.org/10.1007/s10708-010-9340-3

Greenstein, N., & Cho, S.-W. (2025). Ethics & Equity in Data Science for Evaluators. In S. B. Nielsen, F. Mazzeo Rinaldi & G. J. Petersson (Eds.), *Artificial Intelligence and Evaluation. Emerging Technologies and Their Implications for Evaluation* (pp. 56–77). Routledge. https://doi.org/10.4324/9781003512493

Hansen, M. C., Potapov, P. V., Moore, R., Hancher, M., Turubanova, S. A., Tyukavina, A., Thau, D., Stehman, S. V., Goetz, S. J., Loveland, T. R., Kommareddy, A., Egorov, A., Chini, L., Justice, C. O., & Townshend, J. R. (2013). High-resolution global maps of 21st-century forest cover change. *Science*, *342*(6160), 850–853. https://doi.org/10.1126/science.1244693

IFAD (International Fund for Agricultural Development). (2022). *Mapping rural development: How to use GIS to monitor and evaluate projects.* https://www.ifad.org/en/web/knowledge/-/mapping-rural-development-how-to-use-gis-to-monitor-and-evaluate-projects

Lech, M., Uitto, J. I., Harten, S., Batra, G., & Anand, A. (2018). Improving international development evaluation through geospatial data and analysis. *International Journal of Geospatial and Environmental Research*, *5*(2). https://dc.uwm.edu/ijger/vol5/iss2/3

Melesse, A., Weng, Q., Thenkabail, P., & Senay, G. (2007). Remote sensing sensors and applications in environmental resources mapping and modelling. *Sensors*, *7*(12), 3209–3241. https://doi.org/10.3390/s7123209

Murillo-Sandoval, P. J., Dexter, K. V., Hoek, J. V. D., Wrathall, D., & Kennedy, R. (2020). The end of gunpoint conservation: Forest disturbance after the Colombian peace agreement. *Environmental Research Letters*, *15*(3), 034033. https://doi.org/10.1088/1748-9326/ab6ae3

NanoAvionics. (2023, May 4). How many satellites are in space? *NanoAvionics.* https://nanoavionics.com/blog/how-many-satellites-are-in-space/

Patton, M. Q. (2018). A Historical Perspective on the Evolution of Evaluative Thinking. In A. T. Vo & T. Archibald (Eds.), *Evaluative Thinking* (158th ed., pp. 11–28). New Directions for Evaluation.

Rathinam, F., Khatua, S., Siddiqui, Z., Malik, M., Duggal, P., Watson, S., & Vollenweider, X. (2021). Using big data for evaluating development outcomes: A systematic map. *Campbell Systematic Reviews*, *17*(3). https://doi.org/10.1002/cl2.1149

Renger, R., Cimetta, A., Pettygrove, S., & Rogan, S. (2002). Geographic Information Systems (GIS) as an evaluation tool. *American Journal of Evaluation*, *23*(4), 469–479. https://doi.org/10.1177/109821400202300407

Runfola, D., Batra, G., Anand, A., Way, A., & Goodman, S. (2020). Exploring the socioeconomic co-benefits of global environment facility projects in Uganda using a Quasi-experimental Geospatial Interpolation (QGI) approach. *Sustainability*, *11*(18), 3225. https://doi.org/10.3390/su12083225

Scriven, M. (1993). Hard-won lessons in program evaluation. *New Directions for Program Evaluation*, *58*, 1–103.

Statista. (2020, October 28). *Global Geospatial Industry Market Size by Technology*. Statista. https://www.statista.com/statistics/699497/worldwide-geospatial-industry-distribution-by-technology/

Thieme, A., Glennie, E., Oddo, P., McCartney, S., Ruid, M., & Anand, A. (2020). Application of remote sensing for ex ante decision support and evaluating impact. *American Journal of Evaluation*, *43*(1), 26–45. https://doi.org/10.1177/1098214020962579

Tomlinson, R. (2019). *Thinking about GIS: Geographic Information System Planning for Managers* (5th ed.). Thinking about GIS.

Uitto, J. I. (2016). The environment-poverty nexus in evaluation: Implications for the sustainable development goals. *Global Policy*, *7*(3), 441–447. https://doi.org/10.1111/1758-5899.12347

Vaessen, J., Lemire, S., & Befani, B. (2020). *Evaluation of International Development Interventions: An Overview of Approaches and Methods, Independent Evaluation Group*. World Bank.

York, P., & Bamberger, M. (2025). The Applications of Big Data to Strengthen Evaluation. In S. B. Nielsen, F. Mazzeo Rinaldi & G. J. Petersson (Eds.), *Artificial Intelligence and Evaluation. Emerging Technologies and Their Implications for Evaluation* (pp. 37–55). Routledge. https://doi.org/10.4324/9781003512493

11 The Future of Evaluation Analytics

Case Studies of Structural Causal Modeling in Action

Pete York

Introduction

The digital age, marked by the advent of big data, artificial intelligence, and machine learning, has significantly enhanced our ability to evaluate programs with improved speed, accuracy, and timeliness. This transformation is driven by the voluminous data generated by government agencies and social program administration systems, which offers a rich, complex repository of diverse behaviors and perceptions. This data reveals intricate patterns and trends, providing insights into what happened, for whom, and under what circumstances.

However, a major challenge arises because much of this data is rooted in cognitively biased human decisions and self-reported perceptions (Spitzer & Weber, 2019). Extracting meaningful, unbiased insights from this data is complex and fraught with the risk of inaccuracies. Traditional data science methodologies, while adept at making accurate correlational predictions, often fall short in rigorously evaluating causal relationships essential for understanding the impact of various treatment modalities on beneficiary outcomes (Fan et al., 2014).

The primary research question this case study chapter aims to address is: How can big data and advanced analytics enhance our understanding and evaluation of programs? This question acknowledges the increasing complexity of evaluating such programs and the need for more sophisticated methods. To address this complexity, the methodological solution underpinning this case study is the application of Structural Causal Modeling (SCM). SCM emerges as a robust framework that effectively overcomes the limitations of traditional data science, particularly in its ability to unravel and understand causal relationships within social impact evaluation.

To exemplify this, this chapter delves into two pertinent case studies: Gemma Services and the Program to Aid Citizen Enterprise (PACE). These cases are chosen for their relevance in demonstrating the application of big data and analytics in real-world scenarios. Gemma Services represents an innovative approach to individual-level case management within complex human service domains,

DOI: 10.4324/9781003512493-11

while PACE highlights the use of big data in assessing equitable resource distribution among community-based organizations.

By exploring these case studies, this chapter demonstrates the practical implications of integrating big data analytics into program evaluation, highlighting how real-time data collection and processing can lead to more informed and impactful decision-making in social programs. The Gemma case showcases the potential of data analytics in managing complex human service domains at an individual level, providing a comprehensive view of how data-driven insights can revolutionize program evaluation and decision-making processes.

Chapter Overview

This chapter presents the data science approach of Structural Causal Modeling (SCM) and its application in program evaluation, specifically using a type of SCM called Precision Analytics (PA). It is structured to provide a foundational understanding of SCM and then transitions into demonstrating its practical applications by explicating two case studies. These case studies illustrate the real-world implementation of SCM using PA, showcasing their effectiveness in diverse settings, including addressing the challenges of applying these advanced methodologies. This dual approach of theoretical exploration and practical application offers a holistic and contextualized understanding of SCM's role in enhancing program evaluation.

The chapter begins with an introductory section, *Applying Structural Causal Modeling to Advance Causal Inference in Data Science*, an in-depth exploration of the theoretical underpinnings of Structural Causal Modeling (SCM) and its role in data science. It emphasizes SCM's ability to overcome the limitations inherent in traditional associative models that primarily focus on correlations rather than causality. By explicating the fundamental principles of SCM, the chapter lays a solid foundation for understanding its significance in discerning complex causal relationships within data, a crucial aspect often overlooked in conventional data analysis methods. This initial section sets the stage for a deeper understanding of SCM's unique capabilities in providing more accurate and meaningful insights derived from program evaluation.

Following the introduction of SCM's theoretical aspects, the next section shifts to *Precision Analytics: An Applied Machine Learning Approach to Structural Causal Modeling*. This section explains how PA is a practical application of SCM in program evaluation. It details PA's methodological approach, illustrating how it refines and applies the principles of SCM to real-world data, thus enabling more targeted and effective evaluations of social programs. The chapter highlights the unique advantages of PA, such as its capacity for handling large datasets and its ability to provide nuanced, context-specific insights critical for making informed decisions in complex social environments. This detailed explanation of PA underscores its value as a powerful tool for bridging the gap between theoretical data science and practical program evaluation.

The chapter then examines two case studies using *Precision Analytics (PA) within the Structural Causal Modeling (SCM) framework*: (1) PACE's community-level evaluation for equitable resource distribution among diverse nonprofits and (2) Gemma Services' direct service evaluation, focusing on mental health interventions for youth. These cases demonstrate PA's adaptability in different contexts, showcasing its real-world effectiveness in program evaluation. They offer insights into PA's application challenges and evolving role in data-driven program evaluation, underscoring PA's potential for significant improvements and insights in diverse programmatic environments.

Finally, the closing *Discussion* and *Conclusion* sections of the chapter synthesize key insights from the case studies and theoretical exploration, reflecting on the implications and challenges of applying SCM and PA. These sections provide a comprehensive analysis of the findings, discussing how they contribute to the field of program evaluation. The conclusion encapsulates the chapter's main arguments, highlighting the transformative potential of these methodologies while acknowledging their limitations and suggesting avenues for future research in data-driven program evaluation.

Applying Structural Causal Modeling to Advance Causal Inference in Data Science

The primary challenge in applying data science, particularly machine learning, to administratively or transactionally gathered data lies in using supervised machine learning – a prevalent tool within the data science field. Supervised learning is a subcategory of machine learning that uses labeled data of measured outcomes to train algorithms to evaluate how changes in treatment (independent) variables influence a target outcome (dependent variable).

However, the knowledge derived from these algorithms is predominantly associational rather than causal. In other words, the identified correlations or associations do not necessarily indicate a causal relationship, regardless of the prediction's accuracy. This limitation becomes especially significant when data science is applied to studying social phenomena and evaluating social programs, where understanding causal relationships should take precedence over predictive accuracy.

Structural Causal Modeling (SCM) offers a solution to the causal challenges of traditional data science techniques. Structural Causal Modeling (SCM) is a theoretical framework for causal inference that unifies graphical, potential outcome, decision analytical, and structural equation approaches to causation (Pearl, 2010). It provides a mathematical foundation for the analysis of causes and counterfactuals, allowing for the inference of the effects of potential interventions, counterfactual scenarios, and the direct or indirect effect of one event on another.

SCM is built on the idea that causal relationships should be explicated in a visual diagram, called a Directed Acyclic Graph (Figure 11.1), expressing an outcome variable as a function of its direct causes, including the identification and causal assumptions about observed and unobserved contextual factors that confound or mediate the relationship between the explanatory variable and the outcome.

One of the key features of Structural Causal Modeling (SCM) is its ability to account for confounding variables. Confounding variables are those contextual factors that influence both the independent and dependent variables, creating a spurious association. SCM allows researchers to express how they should control these confounding variables when subsequently modeling, thereby isolating the true causal effect of the independent variable on the dependent variable.

With more and more available big data, and therefore the capability to add more contextual and confounding variables, the power of SCM to accommodate confounding variables in preparation for causal modeling becomes even more pronounced. With larger datasets from program administration and other sources, many of the typically 'unobserved' confounders can now be 'observed' due to the availability and addition of more metrics (variables). This allows researchers to use SCM as a framework to subsequently account for and control these confounders when conducting data modeling, rather than having unobserved variables show up in the error.

Structural Causal Modeling vs. Structural Equation Modeling

Structural Equation Modeling (SEM) and Structural Causal Modeling (SCM) differ fundamentally in their approach to data analysis. SEM is

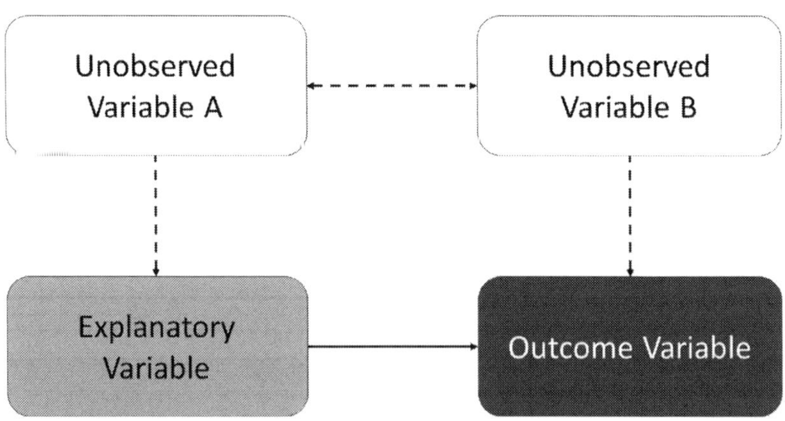

Figure 11.1 Directed acyclic graph.

primarily used for confirming hypotheses about relationships among observed and latent variables, focusing mainly on statistical correlations without asserting causality. It relies on path diagrams and covariance matrices to test and refine theoretical models (Madhanagopal, Amrhein, & McDougall Scientific Ltd., 2019).

In contrast, SCM extends beyond correlation to infer causality, integrating SEM with graphical models and adhering to stringent conditions such as ensuring directed and acyclic relationships, the conditional independence of variables, and the absence of confounding "back-door" paths. This rigorous structure, represented through Directed Acyclic Graphs (DAGs), enables SCM to not only depict correlations but also to unravel the causal mechanisms behind these relationships, making it a more robust tool for understanding the complex interplay of variables in a system.

For example, if a study was investigating the impact of a social program, variables such as socioeconomic status, educational attainment, and access to resources, which might have been unobserved or unavailable in smaller datasets, can now be included in the SCM and subsequent modeling. This increased data availability significantly enhances researchers' and evaluators' ability to accurately control confounding variables, thereby improving the validity of the causal inferences drawn.

Once a Structural Causal Model has been articulated and diagrammed, which can be explicated in the structure and format of a theory of change or logic model, as long as all elements of the DAG are diagrammed, researchers and evaluators can identify datasets that contain the variables reflecting the SCM's causal assumptions about treatment, confounders, and the target (outcome) variables. These datasets can then be used to train machine learning algorithms to build causal outcome evaluation models.

Precision Analytics: An Applied Machine Learning Approach to Structural Causal Modeling

One type of causal modeling approach that can be applied using the SCM framework is Precision Analytics (PA). PA is particularly useful for addressing complex social issues by leveraging large-scale data. It functions as a methodological version of a quasi-experimental observational data study, developed and tested by BCT Partners in consultation with various government and private funders (e.g., the National Science Foundation, the U.S. Department of Health and Human Services, the U.S. Department of Housing and Urban Development) and direct service providers - (York, 2021).

This approach involves subject matter experts (SMEs) training machine learning algorithms to identify and analyze 'natural experiments' in historical data. These experiments occur when frontline practitioners make different decisions for similar types of cases concerning combinations of program interventions, dosages, and goals for specific cases.

PA's main strength lies in its capacity to uncover nuanced insights from large datasets, providing a more accurate understanding of the causal relationships between various program elements and their outcomes. The method is particularly adept at revealing the effectiveness of different program interventions in real-world settings, which can then guide the development of more effective and targeted social programs. The steps in the PA process are as follows:

1. **Finding Matched Comparison Groups:** PA begins by training machine learning algorithms to identify matched comparison groups or subpopulations based on their predicted likelihood to engage in a program, treatment, or intervention. This process minimizes selection bias, a common issue in observational studies where the treatment group differs from the control group in ways that could affect the outcome. By matching on a range of confounder (contextual) variables, PA ensures that the comparison groups are similar in all relevant aspects except for the treatment. This effectively controls for (holds constant) the contextual confounding variables, holding them constant within each block-matched subgroup.
2. **Determining the Ideal Program Model:** Once the matched comparison groups have been identified, PA determines the optimal combination of treatment or program elements for each group. This determination is based on the ability of these historical data elements to predict the highest likelihood of achieving the desired outcome (target metric). By identifying the ideal combination of program elements that uniquely and collectively contribute to a block-matched group of cases achieving the desired outcome, PA facilitates the development of more tailored and effective interventions.
3. **Causal Evaluation of the Program Model:** The final step in the PA process is to evaluate the significance and effect size of the group-specific ideal program model. This involves assessing whether the identified program elements have a significant, albeit observationally quasi-experimental, causal impact on the desired outcome. By doing so, PA provides a robust measure of the effectiveness of the program, treatment, or intervention.

The PA approach predicts the likelihood of success for different groups based on the counterfactual analysis of different treatment variations for similarly matched cases. This process produces a ranked and weighted set of causally attributable program elements that uniquely and in aggregate improve the rate of outcome achievement. Precision Analytics includes automating the inferential analyses of treatment data to determine the "attributable" causal role it plays in achieving the desired outcome (dependent variable).

By harnessing the power of big data, focusing on causal relationships, and automating the evaluation process, these methods offer a more nuanced, accurate, and equitable approach to program evaluation. They provide the evidence needed to advocate for more individualized interventions, revolutionizing our understanding of the complexity of social phenomena and informing the development of more effective, equitable interventions.

Precision Analytics provides a powerful tool for quasi-experimentally evaluating social programs. Operationalizing the SCM framework by training machine learning algorithms to find and evaluate naturally occurring counterfactual experiences in history allows for a more nuanced understanding of the causal relationships inherent within the data. This can lead to more effective and equitable interventions and improve our understanding of social phenomena.

Structural Causal Modeling Using Precision Analytics: Two Case Studies

Precision Analytics using SCM offers an advantage to program evaluation and front-line decision support tools compared to traditional evaluation methods. It enables the development of on-demand, personalized, and equitable case-specific treatment solutions, meeting each client where they are and customizing interventions according to their unique needs. This approach promotes the equitable treatment of all clients, provides organizations with a better understanding of the nuances required to achieve attributable success for each group, and automates the production of quasi-experimental causal evaluation findings in real-time and on-demand. More specifically, PA can automate the evaluation of social programs using program administration data. By systematically analyzing this data, these methods can provide real-time evaluation insights into program effectiveness, identify areas for improvement, and guide decision-making processes.

The application of Precision Analytics (PA) will be exemplified in this chapter through two case studies: (1) Program to Aid Citizen Enterprise (PACE) and (2) Gemma Services.

- *The Program to Aid Citizen Enterprise (PACE)*, a nonprofit organization, is dedicated to fostering community economic development by providing technical assistance and capacity building to community-based organizations. PACE sought to use PA to study if and how equitable the distribution of funding, resources, and support was to nonprofits serving disadvantaged communities throughout the Pittsburgh region, particularly focusing on organizations serving communities of color. The goal was to use PA to identify local communities across a ten-county region around Pittsburgh with the most need, pinpoint nonprofits serving these communities, and assess their capacity to deliver services, informing more equitable funding distribution, capacity building, and support.

- *Gemma Services* is an organization that promotes hope and healing for vulnerable and at-risk children, teens, and families through mental health services, education, and specialized support. The organization serves over 3,000 children per year across its residential and outpatient behavioral health programs and its special needs school. Key services provided by Gemma Services include mental health services, education programs, foster care and adoption services, prevention programs, and residential treatment for children and youth who have experienced trauma or significant mental health concerns. Gemma Services planned to use PA to identify specific combinations of residential psychiatric services and treatments that would be most effective for different sub-groups of youth, thereby enabling more individualized interventions.

Both studies were designed and implemented by BCT Partners, a minority-owned mid-sized consulting firm in the United States, in collaboration with each organization. The selection of these two case studies is deliberate, as they offer unique and similar opportunities to exemplify the application of SCM in diverse settings.

On the one hand, PACE, a nonprofit organization dedicated to fostering community economic development, represents the application of PA in a broader, community-focused context. The use of SCM in this case highlights how data can be used to assess and improve the equitable distribution of resources and support among community organizations, particularly those serving communities of color. This case study illustrates the capacity of SCM to address systemic issues and inform policy-level decisions.

On the other hand, Gemma Services, an organization focused on mental health and support for vulnerable children and families, exemplifies the use of PA in a direct service delivery context. The use of PA here demonstrates how detailed data analysis can enhance the understanding and tailoring of mental health interventions to the specific needs of sub-groups, showcasing the potential of SCM in improving individualized care and treatment outcomes.

Together, these case studies demonstrate the versatility of PA using SCM in both micro (individual-focused) and macro (community-focused) applications. They showcase how SCM can be effectively employed in varied settings to improve program evaluation, enhance decision-making processes, and ultimately contribute to more effective, equitable, and personalized social interventions.

Program to Aid Citizen Enterprise: A Big Data Geospatial Study of the Equitable Distribution of Capacity Building to Nonprofits Serving Communities of Color

BCT Partners, a management consulting firm based in Newark, New Jersey, specializes in providing a range of services to government agencies, corporations, and nonprofit organizations, including research, evaluation, and analytics. The firm developed a big data architecture and platform using precision analytics to study the flow of governmental and philanthropic giving and its impact on improving community well-being across over 74,000 communities throughout the United States. This platform is called the *Equitable Impact Platform (EquIP)*, a geospatial big data platform that assesses and evaluates the nonprofit sector's contribution to equitable community improvement. EquIP combines data from IRS 990 tax forms and the U.S. Census Bureau's American Community Survey (ACS) with BCT's Precision Analytics modeling approach. This platform helps funders and donors identify communities in greatest need, prioritize marginalized communities, find the most accessible nonprofits that can serve these communities best, and receive assessment, predictive, and prescriptive insights about the types of financial and capacity-building support these organizations need to make a difference.

EquIP is a platform that has been developed with the aim of evaluating the impact of nonprofit program output on community well-being. The platform utilizes precision analytics techniques to measure the research metric of the Area Deprivation Index (ADI), as described in Knighton, Savitz, Belnap, Stephenson, and VanDerslice (2016). The precision analytics process employed by EquIP involves matching communities based on their likelihood of having access to direct services provided by 18 different types of nonprofit providers, including services like healthcare, mental health, education, employment, housing, and youth development services, among others. This matching of communities based on their well-being, community giving, socioeconomics, population density, and other factors is done to minimize selection bias and control for contextual factors affecting access to services.

Geospatial analysis was utilized to determine the distance people would travel to access nonprofit services, based on commuting patterns and levels of access to public transportation, as well as socioeconomic status and population density. Subsequently, every census tract was analyzed to determine the number of nonprofit providers accessible for each of the 18 service types and the amount of programmatic output accessible, in dollars, for each type of service. The measurement of programmatic output was derived from the nonprofit tax filing data set, IRS 990. Specifically, the aggregate amount of locally accessible service output, in dollars, was calculated using the geospatial access algorithms for every Census Tract in the U.S. This calculation was generated for every type of the 18 services, specifically:

- Animal-Related
- Arts, Culture, and Humanities
- Civil Rights, Social Action, and Advocacy
- Community Improvement and Capacity Building
- Crime and Legal-Related

- Education
- Employment
- Environment
- Food, Agriculture, and Nutrition
- Health Care
- Housing and Shelter
- Mental Health and Crisis Intervention
- Public and Societal Benefit
- Public Safety, Disaster Preparedness, and Relief
- Recreation and Sports
- Religion-Related
- Youth Development

To evaluate the impact of nonprofit program output on ADI scores, machine learning algorithms were trained to assess the level of programmatic output required for each of the eighteen types of services to significantly improve ADI scores over a three-year period. This approach facilitated the evaluation of each census tract to determine whether sufficient programmatic output was accessible to contribute to significant improvements in community well-being over multiple years. This precision causal method controlled for contextual factors, such as philanthropic and government funding, individual donations, the size of the nonprofit sector, levels of volunteerism, population density, and socioeconomic status. Furthermore, it analyzed each type of service while controlling for interactions and confounding effects with the other 17 types of services (Figure 11.2).

The Program to Aid Citizen Enterprise (PACE), which provides support and access to critical resources for nonprofit organizations to build their capacity, conducted a study on the capacity-building needs of nonprofit organizations in the Southwestern Pennsylvania region using EquIP (Program to Aid Citizen Enterprise, 2021). The study sought to evaluate capacity-building support to local nonprofits, particularly if and why organizations serving communities of color have less access to capacity-building support than organizations serving white communities.

The study used the Equitable Impact Platform (EquIP) and precision analytics to evaluate the financial health and equity of nonprofits in a region. The study analyzed factors such as revenue models, expenditure models, overhead, cash reserves, debt-to-asset ratio, and revenue streams, and found that nonprofits that invested 10–25% of their budget into capacity building were twice as likely to grow, scale, and contribute to community well-being over three years.

However, the study also found significant inequities in access to resources and support for capacity building for communities of color. After controlling for community contextual factors, nonprofits serving communities of color received statistically significantly less support than those serving white communities, with fewer individual donations, philanthropic grants, government grants,

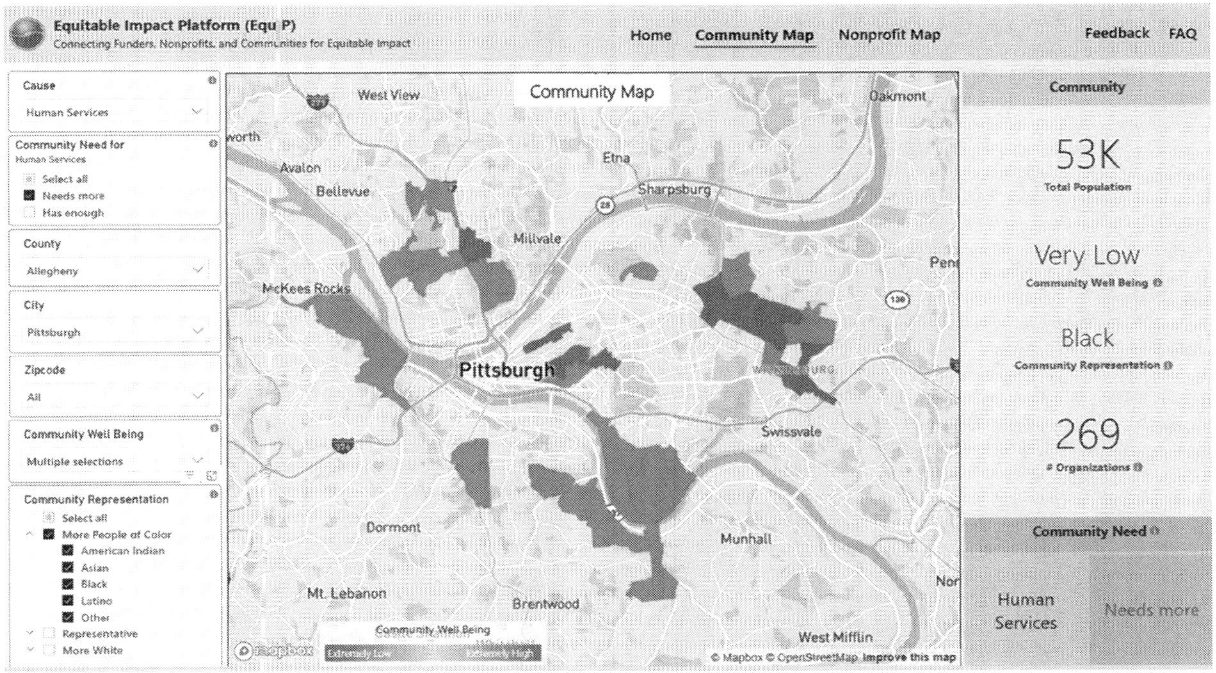

Figure 11.2 Screenshot showing how EquIP identifies deprived Pittsburgh communities (US census tracts) that need human services, determined by applying Precision Analytics.

volunteer engagement, and capacity-building support. The study found that nonprofits serving communities with more persons of color were 43% more likely to be financially unhealthy than those serving white communities.

This study was made possible through a combination of big data from the U.S. IRS 990 (tax filing data), household survey data (from the U.S. Census Bureau), and causal precision modeling, which allowed the research team to mitigate selection biases and social biases through the control of predictive community context factors via block matching, thereby identifying causal contributions to inequities.

This case study provided insightful revelations on the use of Precision Analytics (PA) within a community development context. This study highlighted how PA, underpinned by Structural Causal Modeling (SCM), could effectively assess and improve the equitable distribution of resources and support among community organizations. The key learning here was the ability of PA to leverage big data for identifying and addressing disparities in resource allocation, particularly for organizations serving communities of color. PACE's use of PA demonstrated the significant potential of these methodologies in informing policy-level decisions and fostering systemic change.

Building on these lessons, the chapter now transitions to the case study of Gemma Services, which presents a different yet complementary application of PA. While PACE focused on community-level interventions and resource distribution, Gemma Services provides a more micro-level perspective, concentrating on individualized mental health and support services for vulnerable children and families. This shift from the macro focus of community development to the micro focus of direct service delivery exemplifies the versatility of PA in varying contexts. The Gemma Services case study will further explore how PA, through SCM, can be utilized to tailor mental health interventions to the specific needs of diverse sub-groups, thus demonstrating the adaptability of PA in creating personalized, effective treatment solutions. This juxtaposition between the two case studies underscores the broad applicability and transformative potential of PA in both community-focused and individual-focused settings.

Gemma Services: Transforming Evaluation from Ex-post Accountability to a Dynamic Management Tool

Gemma Services is a multi-program social service agency that provides behavioral health, education, and prevention services to children and families experiencing emotional and behavioral difficulties (York, 2021). Their Psychiatric Residential Treatment Facility (PRTF) serves up to 72 children at a time. It aims to reduce distress and improve their lives, enabling them to return to living successfully in a home or community setting. More specifically, the residential treatment program provides 24/7 clinical treatment and trauma-focused care for children and adolescents who have experienced loss, trauma, and significant challenges. This program serves boys and girls aged 6–14 who are experiencing significant emotional and behavioral challenges that make it unsafe for them to live in the community. The program aims to provide healing treatment and teach new skills so these children can be safe and successful in their homes and communities. The residential treatment programs include clinical services provided by master's-level therapists, including individual, family, and group therapy, as well as comprehensive case management. There is a specialized residential program for adolescent girls aged 10–14 with acute behavioral health needs, with Dialectical Behavior Therapy being a central component of this program. Psychiatric services are also provided, overseeing treatment and providing medication management as needed.

Recognizing the importance of program evaluation to improve outcomes and ensure accountability, the leaders of Gemma Services invested in program administration data systems to assess, monitor, and evaluate outcomes. However, these systems did not meet their evidence-generation needs, leaving front-line staff without actionable evidence to improve program planning and engagement. The data systems collected pre- and post-test client data using scientifically validated assessment instruments, and follow-up interviews and surveys were conducted to evaluate long-term outcomes. Nevertheless, practitioners did not receive insights when needed to strengthen program planning and engagement, nor did they or the program managers or leaders receive evaluation findings across all children served to understand how they were doing with respect to achieving the programs' desired outcomes. As is all too common among organizations that have implemented program administrative data systems, Gemma Services became data-rich but was unable to extract evaluation or decision-support insights for program leaders, managers, or front-line practitioners.

To address this gap, Gemma Services explored the adoption of analytic methods and data science tools such as predictive and prescriptive analytics. They aimed to provide practitioners with actionable evidence to improve the success rate of every youth they served. Leveraging the program administration data they were already collecting, they developed predictive and prescriptive models to provide practitioners with on-demand access to actionable evidence.

Gemma Services utilized BCT Partners' causal precision analytics methodology, as previously mentioned in this chapter. This approach builds causal predictive, prescriptive, and evaluation models using program administration and case assessment data, allowing algorithms to identify naturally occurring real-world "experiments" and determine which treatment patterns achieve the greatest comparative gains. Practitioners are provided with on-demand predictive and prescriptive insights through dashboards, reports, and visualizations while generating rigorous evaluation findings for leaders and funders on how many lives are being improved. By leveraging program administration data and machine learning algorithms, this causal modeling approach provides practitioners with more contextualized, timely, and precise information than traditional summative evaluation designs. Gemma Services conducted the causal precision modeling project to identify effective treatment solutions for their residential treatment program (Gay & York, 2018). Gemma Services leveraged its administrative dataset to build what they ended up calling precision care models, reflecting the concept of precision medicine. Gemma Services' administrative dataset contained 717 cases of children, including hundreds of children and families surveyed for three years post-discharge.

The precision care modeling process involved applying machine learning algorithms to the administrative data to identify matched comparison groups of children with similar backgrounds and diagnostic histories. The algorithms then determined the treatment approaches that worked best for each sub-group, based on the naturally occurring experiments that arose from clinicians' differing intervention decisions for the same group of children. The precision care modeling process yielded several significant findings, including identifying eleven comparison sub-groups of children matched based on their diagnosis, medication protocol at intake, and caregiver situation.

The algorithms also identified specific treatment features that significantly decreased the likelihood of a child would be re-hospitalized post-discharge, such as specific dosages of behavioral modification or length of stay. The precision care modeling approach emphasizes the importance of disaggregating cases into increasingly refined groups as the data set expands (i.e., the number of cases on which to train the algorithms increases). Such a strategy allows for a more precise understanding of the unique combination of programs and treatments that work for each group. This approach acknowledges that a one-size-fits-all solution is insufficient and that tailored solutions are more likely to lead to success.

By employing a precision care model, practitioners can meet each client where they are and customize interventions according to their unique needs, including diagnosis, history, intellectual disabilities, and other relevant factors. For example, Gemma Services sought to reduce the exit acuity of children, thereby reducing the likelihood of re-hospitalization within the next year post-discharge. Using this approach, they identified the specific combination of services, treatments, and medications that psychiatrists and clinicians could use to tailor interventions for each different sub-group of youth.

The precision care model is advantageous because it reduces the problems associated with statistical modeling, which, with its reliance on measures of central tendency (i.e., averages), often ignores mild to significant outlier cases, those in the minority, especially those in the tails of the normal distribution.

As the data set grows and the number of matched groups increases, the precision and equity of the model improve, enabling practitioners to generate more accurate recommendations and increase the likelihood of success. Additionally, this approach facilitates quasi-experimental causal analysis and promotes equitable treatment for all clients.

The precision care modeling process automatically generated quasi-experimental evaluation findings for each block-matched group, providing Gemma Services with a better understanding of the nuances required to achieve attributable success for each group. This allowed them to manage expectations and make a stronger case to their payers for more tailored and customized solutions.

Although practitioners have always known that a one-size-fits-all approach does not work, the evidence from precision care modeling provided Gemma Services with the proof they needed to advocate for more individualized interventions. They used this evidence to argue for funding for more precise and customized solutions, including advocating for less standardized lengths of stay. The evaluation evidence showed that fidelity to one program model, tuned to moving the average member of the population, does not work for everyone, and that tailored interventions with different lengths of stay are needed to maximize success.

In addition to providing more group-specific findings as to what it took to achieve group-specific attributable outcomes, the precision care modeling process produced more accurate and precise evaluation findings for all levels of outcomes, allowing staff to conduct a deeper qualitative investigation into the outliers, odds beaters, etc. Machine learning algorithms evaluated and produced quasi-experimental cause-and-effect evaluation results for each child, allowing providers to count all children who received what they needed and succeeded as an "attributable success." However, there are three other categories of outcomes: (1) children who did not get what they needed but succeeded anyway (i.e., what were called the "independent success" cases); (2) those who got what they needed but did not succeed (i.e., what were called the "unknown needs" cases); and (3) those who did not get what worked for their group and did not succeed (i.e., what were called the "unmet needs" cases). The children in each of these outcome types could now be assessed by clinicians, independently and collectively, to qualitatively study what made these cases exceptions. This includes clinicians meeting with each other, with families, and with the children, to learn more qualitatively why they did or did not get what works for their group and/or succeed.

These four outcome types identified through the Precision Analytics modeling process represent the concepts of true positives, true negatives, false positives, and false negatives, which are fundamental to understanding the performance of a predictive model (refer to Figure 11.3 to view a screenshot of the generated results). True positives and true negatives represent cases where the model's predictions align with the actual outcomes. These are the instances where the existing data was sufficient to support the in-depth analysis and decision-making required in their work.

On the other hand, false positives (i.e., children who did not get what they needed but succeeded anyway) and false negatives (i.e., children who got what they needed but did not succeed) represent the error in the model, where the

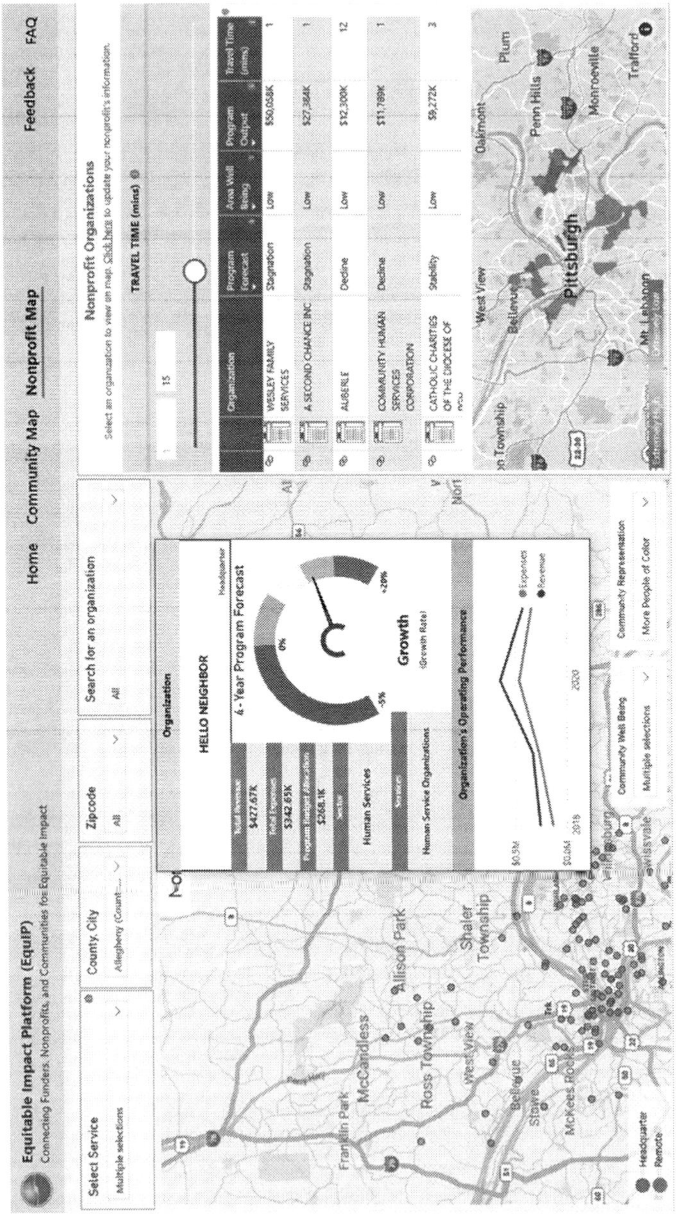

Figure 11.3 Screenshot showing how EquIP identifies nonprofit service providers most proximate to deprived communities, sharing an evaluation of their financial health, determined by applying Precision Analytics.

model's predictions do not match the actual outcomes. These are the instances where the practitioners encountered questions and hypotheses that the existing data could not adequately address, highlighting a gap in the current data collection process. The false positives and false negatives thus indicated that the data being collected was not comprehensive enough to fully support the in-depth analysis and decision-making required in their work (Figure 11.4).

The challenge, therefore, was not just in identifying these additional data variables (metrics), but also in refining the model to reduce these errors. This is where the causal modeling training of machine learning algorithms came into play, helping to identify specific cases to assess, hypothesize, and draw conclusions about what happened, why, and how. However, the presence of false positives and false negatives also underscored the inherent limitations of predictive models and the need for continuous refinement and validation against real-world outcomes.

This realization underscored the evolving nature of data requirements in precision care models. As program directors, managers, and practitioners delved deeper into their cases, they encountered questions and hypotheses that the existing data could not adequately address. This highlighted a gap in the current data collection process, suggesting that the data being collected was not always comprehensive enough to fully support the in-depth analysis and decision-making required in their work. Consequently, there was a clear need for further refinement of the data collection process to include additional data variables that could provide a more complete picture of the cases under consideration. Once new data variables (metrics) were added, this enabled more robust testing of hypotheses and ultimately led to more informed decisions and interventions.

The challenge to overcome, therefore, lay in identifying these additional data points and integrating them into the existing data collection and analysis framework in a meaningful and practical way. This would not have been possible without the causal modeling training of machine learning algorithms that helped identify these specific cases to assess, hypothesize, and draw conclusions about what happened, why, and how. Gemma Services implemented a precision care model using program administration and case assessment data to build causal, predictive, prescriptive, and evaluation models. This approach generated automated evidence-based, or causal, evaluation findings and accurate recommendations, and resultantly increased the likelihood of success for sub-groups of children. Their system has been in place for almost three years. Gemma has analyzed their length of stay and exit acuity data, before and after the implementation of the precision care model. They found that the length of stay per child was reduced by 30 days: residential care costs $10,000 per month, per child.

Additionally, Gemma found that the exit acuity of children dropped by 40%; it is important to note that Gemma tracks longitudinal post-discharge data on every child, and their analysis proved that exit acuity is highly correlated with children remaining in the community for 12 months post-discharge. It is

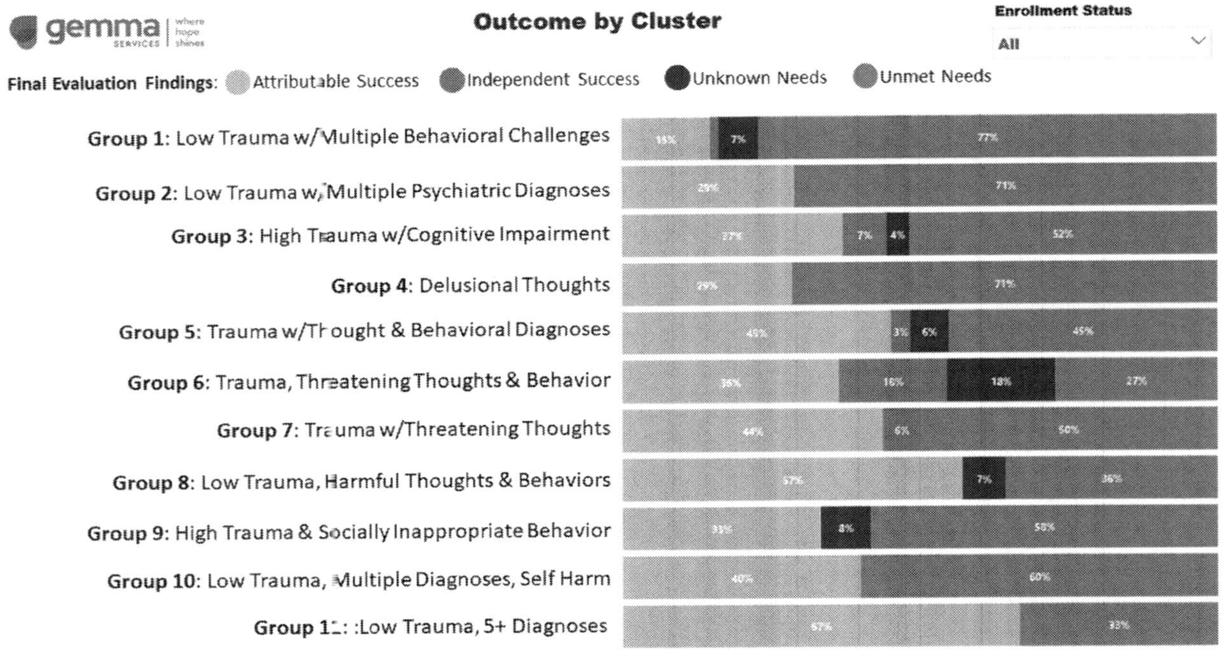

Figure 11.4 Screenshot of an evaluation dashboard, presenting precision analytics results for each block-matched comparison group of program recipients, with probabilistic outcome results.

important to note that these pre-/post-precision care model results do not reflect experimental evaluation findings, but they do serve as an early indication of the benefits of practitioners using the precision care model on a case-by-case basis (Figure 11.5).

Challenges in Applying Structural Causal Modeling Using Precision Analytics

While Gemma Services has made significant strides in implementing a precision care model, their use raised a key challenge: the need for the engagement of beneficiaries as users of the tool. Currently, precision analytics tools are primarily used by practitioners on behalf of the beneficiaries. This means that the direct beneficiaries – the children and their families – are not yet fully involved. Gemma Services has recognized this gap and expressed the goal of engaging beneficiaries directly in the use of precision learning tools. The aim is to enable beneficiaries to actively participate in interpreting outputs and planning what to prioritize to improve their likelihood of success (York, 2021).

However, achieving this goal presents its own set of challenges. It requires not only making the tools accessible and user-friendly for beneficiaries but also providing them with the necessary training and support to use these tools effectively.

Furthermore, it involves addressing potential barriers to engagement, such as lack of access to technology, low digital literacy, or reluctance to engage with digital tools. Overcoming these challenges is crucial to fully realizing the potential of the precision care model, as it is the beneficiaries themselves who stand to gain the most from the insights and recommendations generated by the model.

Discussion

The integration of Structural Causal Modeling (SCM) and Precision Analytics (PA) in evaluating social impact programs, as exemplified in the case studies of Gemma Services and the Program to Aid Citizen Enterprise (PACE), represents a significant advancement in the field of program evaluation. This integration effectively addresses the longstanding challenge of drawing valid causal inferences from big data, which is often characterized by complex, non-experimental settings.

The PACE case study shed light on the distribution of resources and support to nonprofits, especially those serving communities of color. Applying SCM and PA illuminated the disparities in resource allocation, providing a data-driven approach to understanding and addressing these inequities.

In the context of Gemma Services, the precision care model enabled a nuanced understanding of the treatment effects on different subgroups of children. The use of SCM in this context allowed for a more tailored approach to treatment, considering the unique characteristics and needs of each subgroup. This approach

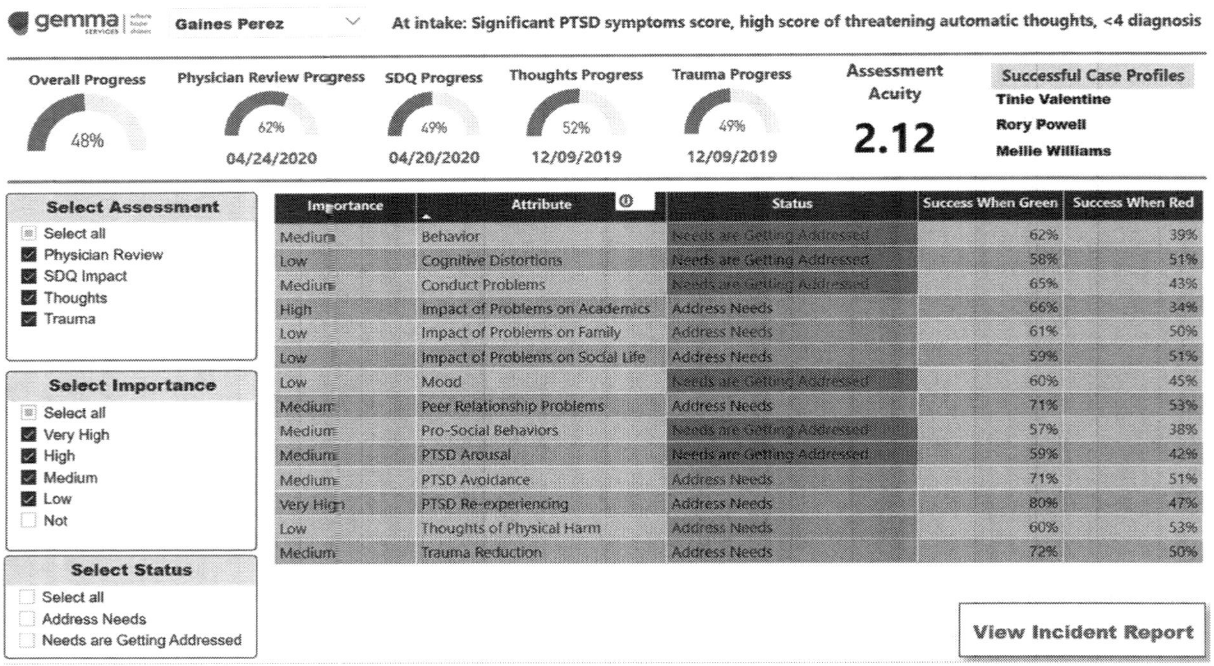

Figure 11.5 Screenshot of up-to-date case-specific recommendations, determined through precision analytics, presenting practitioners with counterfactual evidence (i.e., "Success When Green" contrasted with "Success When Red") for goals (attributes) that need to be addressed to improve the likelihood of the desired outcome (i.e., exit acuity less than 1.5). Note that names are fictional.

demonstrated a move away from one-size-fits-all solutions, acknowledging the complexity and individuality of the treatment needs.

Despite these advancements, some limitations must be acknowledged. One significant limitation is the reliance on available data, which may not always be comprehensive or representative. This limitation underscores the need for ongoing refinement of data collection methods to ensure that the data adequately captures the complexities of the social phenomena being studied. Another limitation is the potential for algorithmic bias. While SCM and PA offer powerful tools for analysis and the methods control for selection bias, they are not immune to the biases inherent in the data. This raises the need to address the potential for these models to perpetuate existing inequities, particularly when used in the context of social program evaluation.

Future research should focus on addressing these limitations, particularly through the development of more robust data collection methods and the exploration of ways to mitigate algorithmic bias. Additionally, there is a need for more studies that apply SCM and PA in different contexts to validate and refine these methods further.

That said, the integration of SCM and PA represents a significant contribution to the field of program evaluation. It offers a more rigorous and nuanced approach to understanding the causal relationships in social programs, thereby providing more accurate and actionable insights for program improvement. This approach also has implications for policy-making, as it provides a more robust evidence base for decision-making. By illuminating the causal mechanisms underlying program outcomes, SCM and PA enable policymakers and practitioners to design more effective and equitable interventions.

While the integration of SCM and PA in program evaluation marks a significant step forward, it is crucial to continue exploring and addressing the challenges identified. This will ensure that these methods are used responsibly and effectively to improve social impact programs and contribute to a more equitable society.

Conclusion

In conclusion, this paper has presented a comprehensive exploration of the application of Structural Causal Modeling (SCM) using Precision Analytics (PA), with a particular focus on two case studies: Gemma Services and the Program to Aid Citizen Enterprise (PACE). SCM using PA has demonstrated its potential to revolutionize the evaluation of social impact programs, providing real-time, on-demand, and personalized insights that can guide decision-making processes and improve outcomes.

Gemma Services' implementation of a precision care model, which used program administration and case assessment data to build causal, predictive, prescriptive, and evaluation models, demonstrated the potential of this approach to generate automated, evidence-based evaluation findings and accurate recommendations. This, in turn, increased the likelihood of success for sub-groups of children.

The Program to Aid Citizen Enterprise (PACE) study offered several key benefits. Primarily, it provided critical insights into the distribution of resources and support among non-profit organizations, especially those serving communities of color. By applying Structural Causal Modeling (SCM) and Precision Analytics (PA), the study illuminated for local funders the disparities in resource allocation, highlighting areas where equity could be improved through more targeted outreach and grantmaking. This approach enabled a more data-driven understanding of these inequities, guiding more effective and equitable policy decisions.

However, the case studies also highlighted several challenges, including the need for additional data points to test hypotheses that could not be answered by the current data, and the need for greater engagement of beneficiaries in using precision learning tools. Overcoming these challenges is crucial for the full realization of the potential of the precision care model. The findings from these case studies underscore the potential of SCM and PA to transform the evaluation of social impact programs. However, they also highlight the need for continuous refinement of these models and the importance of addressing the challenges identified. As we progress, it will be crucial to continue exploring and addressing these challenges to fully harness the power of big data-driven evaluation using SCM.

Bibliography

Fan, J., Han, F., & Liu, H. (2014). Challenges of Big Data Analysis. *National Science Review, 1*(2), 293–314. https://doi.org/10.1093/nsr/nwt03

Gay, K. E., & York, P. (2018, February). *A New Way to Use Data: Precision Care for Better Outcomes in Psychiatric Residential Treatment for Children.* Retrieved from Scattergood Foundation: https://www.scattergoodfoundation.org/wp-content/uploads/2018/02/A-New-Way-to-Use-Data.pdf

Knighton, A. J., Savitz, L., Belnap, T., Stephenson, B., & VanDerslice, J. (2016). Introduction of an Area Deprivation Index Measuring Patient Socioeconomic Status in an Integrated Health System: Implications for Population Health. *The Journal for Electronic Health Data and Methods*. https://doi.org/10.13063/2327-9214.1238

Madhanagopal, B., Amrhein, J., & McDougall Scientific Ltd. (2019). *Analyzing Structural Causal Models Using the CALIS Procedure.* Retrieved from SAS: https://support.sas.com/resources/papers/proceedings19/3240-2019.pdf

Pearl, J. (2010). Causal Inference. *Proceedings of Workshop on Causality: Objectives and Assessment at NIPS 2008*, (pp. 39–58).

Program to Aid Citizen Enterprise. (2021). *The Health of the Nonprofit Sector in Western Pennsylvania.* Retrieved from Pittsburgh Black Media Federation: https://pbmf.org/wp-content/uploads/2021/01/PACENonprofitEffectivenessstudyFINAL12.9.20-1.pdf

Spitzer, S., & Weber, D. (2019). *Reporting biases in self-assessed physical and cognitive health status of older Europeans*. PloS One, *14*(10), e0223526. https://doi.org/10.1371/journal.pone.0223526

York, P. (2021, July). *Unlocking Real-Time Evidence for Practitioners: How Evaluation and Data Analytics Are Generating On-Demand, Actionable Evidence for Front-Line Practitioners at First Place for Youth and Gemma Services.* Retrieved from Project Evident: https://projectevident.org/wp-content/uploads/2022/09/FPFY-GemmaActionableEvidenceCaseStudyJuly21.pdf

12 The Algorithmization of Policy and Society

The Need for a Realist Evaluation Approach

Frans L. Leeuw

Introduction

Across the contributions in this book, a picture is emerging that artificial intelligence (AI) has gained a foothold not only in various professional research fields but also in organizations, where decision-making, developing and implementing policies, programs, interventions, regulations, therapies, drugs, and legal arrangements take place.

Misuraca, G. and van Noordt (2020: 430) present a database of 250 cases to illustrate the use of artificial intelligence by (European) governments. AI appears to have functions in policy making (to detect social issues more quickly, to improve policy decisions, and to monitor implementation), in public services (to improve service delivery and to develop innovative services), and in internal management (developing management innovations, human resources, procurement and finances, to name a few). Sharma et al. (2020) focus on how AI is applied in different government sectors ranging from health care, environmental sustainability, and education to transportation and economic applications. The Netherlands Scientific Council for Government Policy frames the development of AI as follows:

> In recent years, AI has left the confines of the lab and proliferated through-out society. AI is now being used throughout the economy and society at large, affecting the daily lives of citizens in manifold ways. Like what have been called general-purpose technologies, AI is characterized by pervasiveness, continual improvement and innovational complementarities. [We] argue that AI is a system technology, comparable to the steam engine, electricity, the combustion engine and the computer.
>
> (2021:5)

Burrell and Fourcade (2021) refer to the "Society of Algorithms," while Meijer et al. use "algorithmization" (2022: 837 ff)[1]: the process in which an organization rearranges its working routines around the use of algorithms. In this chapter,

DOI: 10.4324/9781003512493-12

the focus is on the algorithmization of policies, programs, interventions, and related activities.

The recent introduction of Large Language Models (LLMs), such as ChatGPT, trained to act in a conversational way and able to be prompted to generate new content, is a powerful illustration of this technological development that is relevant for policy making and implementation.

The Netherlands Scientific Council added that it is crucial "to ensure strong feedback loops between the developers of AI, its users, and the people that experience its consequences" (2021: 35). However, that is far from being a standing practice.

> Evaluation of the outcomes of AI systems and their feedback to developers, and other stakeholders sound like a logical requirement for AI systems, but do not happen enough in practice. One emergent trend is the use of real-life experiments which, after an initial test phase, are deployed without further evaluation.
>
> (p.36; 16)[2]

It held a plea for (more) real-world evaluations of AI use. Sharma et al. make a similar point: "[there] is a dire need for analyzing the implementation of [AI] technologies (pre-adoption) and their evaluation (post-adoption)" (2020, p. 5). There is a need to start considering how evaluators respond to this new need and apply evaluative thinking and methods and to address it. Added to this is the point that AI has been often characterized as a *black box* (that needs to be turned into a white box) and even as *magic.*[3] In this chapter, I posit that theory-based evaluation writ large and particularly realist evaluation hold promise to address this pertinent need.

I focus on realist evaluations because unpacking black boxes is one of its central themes. This not only applies to conventional programs and interventions but probably even more to the world of algorithms. Pasquale (2015) writes about the "Black Box Society" in the finance and information world. Others analyze the need for "demystifying the AI black box" in health (care) and other fields (Price, 2018). In more than two decades, realist evaluation has evolved its approaches to unpacking black boxes. This includes addressing CMO configurations (Context/Mechanisms/Outcomes). Therefore, I explore this perspective in this chapter.

Questions Asked and Structure of the Chapter

The first question is how widespread the attention for AI (use) is in evaluation journals. The answer I provide leads to a second question: why is there such limited attention to this topic in evaluation journals? The third question is what impact evaluations focusing on AI would look like? The fourth question is what the contribution of realist evaluations of algorithmization to the knowledge base may be.

One definitional issue needs to be mentioned. I focus on AI (tools/systems) that are used as building blocks or components, of interventions (programs and policies, regulation, legislation, therapies, etc.). I also refer to algorithmization; I treat these terms as interchangeable.

The structure of the paper is as follows: first, attention paid to evaluations of AI (use) in ten evaluation journals over the last 10 years, followed by a short discussion on possible explanations for these findings.

Next, components of evaluations of the impact of AI (use) are mentioned. Then the focus shifts in the core part of the chapter to what realist evaluations can contribute to help realize more (and better) evaluations of the algorithmization of policies, programs, and interventions. Some fifteen mechanisms were found in the research literature I studied that can be characterized under five realist constructs of mechanisms. I also mention a realist evaluation case study of robot-assisted surgery to share an example of how this approach works in practice. Finally, in the discussion and conclusion part of the chapter, a few points regarding the future of AI are suggested.

Question 1: Is Attention Paid to Evaluations of AI (Use) in Ten Evaluation Journals?

This question is relevant as evidence from (half) a decade ago on the interests of evaluators in picking up and evaluating digital developments revealed that the evaluation practice was hesitant in doing so. See Box 1 for more information.

Box 1: Indicators of evaluators' interests in digital developments (2012–)

As of 2012

Leeuw and Leeuw (2012) counted the frequency with which the words like "internet," "digital," "web," "cyber," and "digital policy" were used in evaluation journals since their inception and till 2012. They concluded that there "appears to be a gap between the pace at which internet and digital policies are penetrating society and the attention professional evaluators are paying to these policies" (p. 111).

As of 2017

Høljund, Olejniczak, Petersson and Rok (2017) used an e-survey that was focused on big data and was disseminated via an online survey link to LinkedIn groups that are active and characterize themselves as (professional) evaluators (data collected in 2015). They found that only 0.5

percent of the sample (fifteen respondents) said that they have used Big Data in their evaluations.
Forss and Norén (2017) studied a sample of twenty-five evaluation Terms of Reference from selected international development agencies and found that Big Data is not much used, partly because Terms of Reference sometimes close the door for Big Data.

Since 2012–2017, in the digital world things have changed and probably dramatically. AI, big data, and digitization are more and more talk of the day, including developments like ChatGPT. Therefore, one could expect for the years between the first word count (Leeuw & Leeuw, 2012) and 2023 a much greater focus on the algorithmization of programs and policies. I therefore searched again in the ten evaluation journals but now using two terms: artificial intelligence and digital. It is fair to conclude from this table that the number of times these two terms have been used is marginal, which is an indicator of evaluators' still very limRited interest in this development (Table 12.1).

Evidence presented by Nielsen, Mazzeo Rinaldi and Petersson (2025) confirms this point. They searched in nine major evaluation journals over a ten-year period (going back from 2023) for evaluators' work in the AI field and found only eighteen distinct articles with "Big Data," "Artificial intelligence," "Machine learning," "Text analytics," or "Internet of Things" as title, keyword, or in the abstract (see also Nielsen, 2023).

Kates and Wilson also present evidence as well as a debatable explanation:

Table 12.1 Appearance of the words "digital" and "artificial intelligence" in ten evaluation journals' (since their inception)

Journal	*Search term: Artificial intelligence*	*Search term: Digital*
American Journal of Evaluation	21	102
Evaluation	9	54
Evaluation Review	24	105
Evaluation and Program Planning	31	252
Assessment and Evaluation in Higher Education	58	292
Educational Research and Evaluation	15	96
Journal of Multidisciplinary Evaluation	1	2
Educational Evaluation and Policy Analysis	17	57
Evaluation and the Health Professions	4	36
New Directions for Evaluation	2	42

> Because it [AI] has emerged so recently,[4] there is little writing about the application of AI in evaluation. Currently there are no scholarly articles in the evaluation literature (e.g., American Journal of Evaluation, New Directions for Evaluation, Journal of Multidisciplinary Evaluation) that directly address the topic. The only piece we could find is a blog post.
>
> (2023: 101[5])

Question 2: Why Still So Little Attention for AI? Four Suggestions?

One suggestion is that evaluators are not very familiar with these developments and stay away from them. Kates and Wilson (2023) give an example of limited familiarity when they refer to AI as a "recent" issue. A second suggestion is that evaluators might fear complexity and methodological difficulties in designing and conducting evaluations that try to sort out the consequences of the algorithmization of interventions. Stick to what you know and have always been doing, could be the adage.

Third, what is also possible is that a number of evaluators could be categorized as the Luddites of the Knowledge Profession in the twenty-first century.[6] The Luddites from the eighteenth/nineteenth century "refers to British weavers and textile workers who objected to the introduction of mechanized looms and knitting frames. The new machinery posed a threat to their livelihood and after receiving no support from government, they took matters into their own hands"[7] and "protested against manufacturers who used machines in what they called "a fraudulent and deceitful manner" to get around standard labor practices.[8]

A fourth suggestion is the belief that evaluation is not needed, as it is a priori evident how "bad," "unacceptable," "biased," or "dangerous" algorithmization is. Simultaneously, but a probably smaller number of evaluators firmly believes in the "good and the great" of AI and do therefore not consider it necessary to do impact evaluations.

This chapter is not the place to investigate these and other suggestions, so take them for what they are worth. I turn my perspective to a way forward, discussing what important characteristics would be of (impact) evaluations of algorithmization and – next – what realist evaluations have to offer to those interested in evaluating AI use.

Question 3: What Are Basic Characteristics of Impact Evaluations?

I start by pointing out what cannot be considered as (impact) evaluations of AI.

Often, when news media or consultants refer to the impact of AI and mention measurably successful, long-lasting, and significant deployments (of AI), such labeling is useless, unless these terms are operationalized and measured through systematic evaluations (Press, 2022).[9] References are often missing.

Nor is there any question of an impact evaluation, when criteria are assessed and/or audited like explainability (aka interpretability) of the algorithms, their accuracy, transparency, safety, privacy, and the ways in which different biases of AI(-use) are mitigated.

These criteria are important, but impact in the real world concerns the question of to which extent the solution of problems and challenges within society and for people is improved (without unintended negative side-effects) when AI is used, compared to (similar) situations where that has not happened.

Despite their relevance, testing AI's data models, including test/training data, the performance of the model, metrics, and examples of (incorrect) predictions, is also still far away from evaluating the impact on behavior in society.[10]

However, work done by Price (2018), Park & Han (2018), Park et al. (2020), Havrda & Klocek (2023), and Leeuw (2023) discuss research designs and data collection methods, ranging from data quality control and algorithm testing, ethnographic research to identify needs, and understanding the workflows to cultural contexts, usability testing including A/B tests[11] and expert reviews. Also, clinical trials and evaluations in real-world settings are discussed, using RCTs and observational (cohort) studies. User feedback and "continuous monitoring and surveillance for unexpected adverse effects" are also mentioned, while sometimes attention is paid to assumptions underlying AI applications (the AI black box problem) and to procedures for ground truthing.[12]

Hernandez-Orallo (2017: 399) distinguishes between

> AI systems and AI components. Systems (such as AI agents, cognitive architectures or robots) can be evaluated as they are, since they take some sort of problem (by a specification or by rewards) and can be evaluated in terms of a utility function. Components (such as particular techniques, algorithms, methods or tools) cannot be evaluated if there is no specification for the component.

An impact evaluation of algorithmization (of interventions, policies, programs, etc.) would therefore have to pay attention to at least these issues:

- Describing AI systems, tools, models, data, and their goals;
- Addressing assumptions underlying the use of AI, opening the AI black box(es) and ground truthing;
- Specifying people and organizations for which AI is used, in which contexts and under which restrictions (time, money);
- Measuring AI's implementation (processes) in society/organizations/institutions;
- Applying research designs/approaches in order to measure the outcomes or consequences (intended, unintended) of AI use for the addressees (and

others), like contribution analysis, [theory-driven] counterfactual analysis, and comparative case studies.

- As AI is used often with "humans in the loop" (HI, human intelligence), the evaluation design needs to be able to focus on HI/AI interactions and their consequences (Crootof, Kaminski and Price, 2023).

Question 4: What Can Realist Evaluations Contribute to Impact Evaluations of Algorithmization of Programs, Policies, and Interventions?

Why a Realist Evaluations Approach?

The point made by Pawson (2006, p. 26 ff) that "interventions are theories incarnate" has proved to be one of the cornerstones of the realist tradition. These theories usually have to be reconstructed from what politicians, officials, and stakeholders say, write (and sometimes do); therefore evaluators refer to the need to unpack a policy or intervention's black box.

This applies to conventional interventions where AI does not play a role, but probably even more when hybrid or fully digital interventions are involved (Pasquale, 2015).[13] This is related to the opacity of AI, its plasticity, and the need for "demystifying the AI black box."[14] Black boxes consist, among others, of assumptions on the role of contexts, mechanisms, and outcomes, aptly referred to by realist evaluators as CMO-configurations (Pawson and Tilley, 1994; Astbury & Leeuw, 2010; Pawson, 2013; Lemire et al., 2020).

Why would this approach be relevant for evaluating algorithmization? Regarding mechanisms, a signal of the relevance of this approach can be found in Pedersen & Johansen (2020) concept of "BAI": behavioral artificial intelligence, implying that attention should be paid to "the artificial inferences inherent in, and the manifested behavior of, artificial intelligence systems in the same way as the social sciences have studied human cognition, inference and behavior."

Related to this is the "humans in the loop" issue (Crootof, Kaminski & Price, 2023: 10–11). A "human in the loop" is defined as "an individual who is involved in a single, particular decision made in conjunction with an algorithm." The "human in the loop" is contrasted with the "human on the loop" – the human overseeing an algorithmic decision-making process – and the "human off the loop" – algorithmic decision-making processes without human involvement or oversight." How these roles work in practice is at least partly dependent on behavioral and cognitive mechanisms driving the relevant behavior, including their "environments" (or ecosystems, i.e., contexts). Christen et al. (2023: 4) make the point that "effective human control of AI relies on the ability of humans to predict AI 'behavior', e.g., recognize when AI is likely going to commit an error." Results from several of their studies show findings that stimulate the authors to suggest a different approach, namely "AI controlling a human operator." A Swedish study analyzed behavioral reactions of civil servants when they are confronted with the introduction of "robotic process automation," described the contexts in which the introduction took place and "unpacked the actors" [behavior] (Ranerup & Henriksen, 2020: 6, 13).

Another indicator of its relevance is given by Price (2019: 8), who studied contexts when working with and evaluating AI in health care. He distinguishes top-level clinical practices from rural practices and investigates the consequences these differences in contexts have for the success of algorithmization.

These are considerations why realist evaluations can and will contribute to a better understanding of how algorithmization of programs and interventions works: they can address mechanisms behind AI-driven interventions (that combine AI and humans in the loop); they can add to the explainability of AI, and they can stimulate confronting the knowledge funds available from evaluations of 'conventional' interventions to the more hybrid ones.

Searching for (Assumptions About) CMO's in the Literature: Methods, Findings, and a Case Study

Methods

Given that evaluation journals hardly report on evaluations (of the impact) of AI use, I broadened the search for studies to other sources in other fields of research. I applied two strategies. First, I used the Maastricht University library databases to locate articles, books, and chapters for which the abstract and/or title suggested that relevant information on mechanisms, contexts, and outcomes of algorithmization was presented. I searched the period from 1998 to 2023 and used several keywords (Table 12.2).

It turned out that a large part of the references dealt with (completely) technical and IT issues. The second strategy was to snowball search for documents in which mechanisms and contexts of human–machine interaction and interaction between human intelligence (HI) and AI were mentioned, and, preferably, evaluated or assessed. Again, I used the Maastricht University database. The result of the two strategies was a collection of some fifteen papers that I used in my search for mechanisms, contexts, and outcomes.

Table 12.2 Frequency of key terms found in the Maastricht University Library database (1998–2023)

Search term	*Frequency*
algorithmization of governments	93
algorithmization of society	162
evaluating algorithmization	295
evaluating the impact of algorithmization on society	40
evaluating the impact of algorithmization of governments	15

Findings: Mechanisms

In the selected papers, the search for mechanisms resulted in the following list, including the source(s) and background information on assumed (functions of) mechanisms. The approach I followed is linked to Westhorp's (2018: 49–50) work on constructs (categories) of mechanisms. She distinguished between powers and liabilities, forces, interactions, feedback/feedforward, reasoning, and resources. In my search, I stumbled upon some of these, like "reasoning," "interactions," and (cognitive/attitudinal) "resources." However, I also found other constructs like emotion handling and anthropomorphism.

Figure 12.1 summarizes the findings of the search, which I will discuss more in depth thereafter.

REASONING: RATIONAL CHOICE DECISION-MAKING MECHANISMS

Source: Godin (2015); Rogers and Agarwala-Rogers (1976); Pinto et al (2023).

Background: Pinto et al. (2023) analyzed filter bubbles or echo chambers active in the digital world from a rational choice theory (RCT) perspective. Also in line with rational choice theory on reasoning is that persons confronted with an innovation like algorithmization get involved in decision-making, adopting (or rejecting) the innovation, and weighing attributes like the ones mentioned below. RCT does not assume that people have full or complete knowledge of these attributes. Nevertheless, RCT tries to take opportunity costs involved in

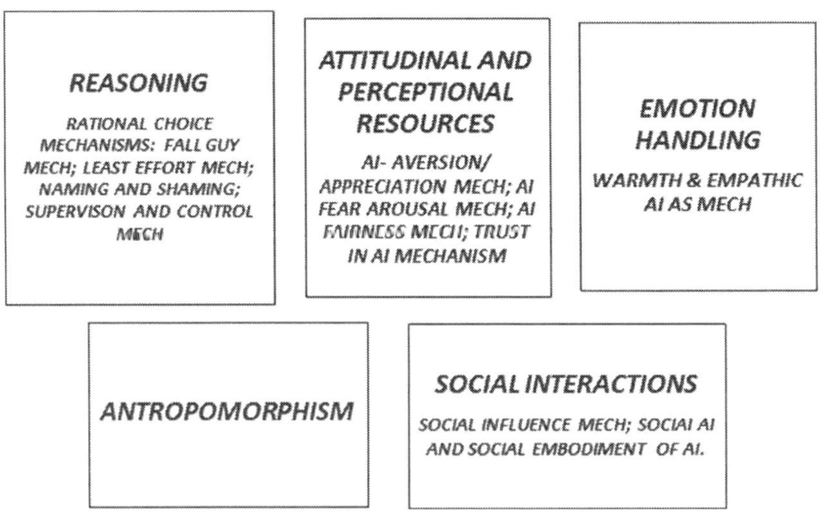

Figure 12.1 Constructs and mechanisms found in selected papers.

the decision on board. The more people are positive about the attributes, the larger the likelihood that they will engage in AI.

These are relevant attributes from the research literature:

- demonstrability of the (AI) innovation;
- relative advantage (over other possibilities);
- compatibility (consistency of the innovation with the values, experiences, and needs of the potential adopters);
- triability (the extent to which the innovation can be tested or experimented with before a commitment to adopt is made); and
- observability (the extent to which the innovation provides tangible results).

Research on AI has also found a few other mechanisms related to reasoning and rational decision-making. Croothof, Kaminski and Price (2022: 57) refer to the fall guy mechanism: a human is made "legally liable, morally responsible, or otherwise accountable for the system's decisions," enabling others to duck responsibility for difficulties or unintended side-effects, when AI&BD is used in policy-making and implementation. This mechanism is linked to the "principle of least effort," which comes from organizational sociology and economics, and to the naming and shaming mechanism: "it is not me (or my unit) that may have made mistakes, but 'the other(s).'' An opposite mechanism can also be at work: if an "ineffective human" is added to the loop, it can serve as "ethics washing," distract from other, more effective forms of regulation" (p. 9).

SUPERVISION AND CONTROL BY HUMANS AS MECHANISMS OF AI GOVERNANCE

Source: Christen et al. (2023); EU Parliament (2020); Koulu (2020); and Green, (2022).

Background: Koulu (2020) presents evidence that within the EU

> human oversight is advocated by a range of actors as a focal ethical principle for AI development and deployment. For example, the EU Commission's Communication in 2019 portrayed human agency and oversight as the first of seven key requirements, which AI applications must follow to be considered trustworthy. The risks and challenges hoped to be addressed by human oversight include dangers to human autonomy, lack of transparency and opaque algorithmic models, privacy and data protection issues, as well as discrimination.
> (p. 721)

However, Koulu (2020) is critical toward this approach:

> For law, human oversight provides an attractive, easily implementable and observable procedural safeguard. However, without awareness of its inherent

> limitations, human oversight is in danger of becoming a value in itself, an empty procedural shell used as a stand-in justification for algorithmization but failing to provide protection for fundamental rights.
>
> (p. 720)

A similar point is made by Green (2022), studying AI policies and legislation in 41 countries worldwide. He found that

> the mechanism that has become a centerpiece of global efforts to regulate government algorithms is to require human oversight of algorithmic decisions. [However,] despite the widespread turn to human oversight, these policies rest on an uninterrogated assumption: that people are able to effectively oversee algorithmic decision-making.

He found that

> these policies suffer from two significant flaws. First, the vast majority of evidence suggests that people cannot adequately provide the envisioned forms of oversight. Second, the incorrect assumption of effective human oversight legitimizes the use of flawed and unaccountable algorithms in government.
>
> (2022:16ff)

ATTITUDINAL AND PERCEPTUAL MECHANISMS

AI-aversion mechanism, which means that people are negative about AI and therefore not inclined to make use of AI;

AI-appreciation mechanism, which means that people adhere more to suggestions/advice when they think it comes from an algorithm than from a person.[15]

> Source: Logg et al. (2019); Lanz et al. (2022); Kim et al., 2022; Siemon (2023). Background: Logg et al. (2019) and Lanz et al. (2022) study

> why humans are averse to receiving instructions from algorithms, and especially in the moral domain or – the opposite – prefer and rely on AI rather than on human instructions in important life domains (as they assume that AI is fair, fast and unbiased.
>
> (p. 2)

Lanz et al. (2022) identified four theoretically meaningful mediators from the AI aversion/appreciation and the leadership literature and from qualitative data from interviews.

Kim et al. (2022: 2) present the AI device use acceptance theory. The theory addresses "the influence of the perceived characteristics of AI chatbots (like personalization, anthropomorphism) on the willingness to use AI chatbots (outcome stage) by sequentially mediating the perceptions of humanness (competence, warmth) and emotion (empathy)."

Siemon (2023: 459) tested the hypothesis "whether individuals feel concerned about evaluation when a computer evaluates their idea (instead of another human)." This hypothesis is linked to the evaluation apprehension theory, which states that participants, who are working in groups or teams, are not presenting their more original ideas because of the fear of negative evaluations from other members. The results of Siemon's experiment show that people do not feel evaluation apprehension when they present their idea to an AI-based system, but, in contrast, feel concerned when they present their idea to a human.

AI-FEAR-AROUSAL MECHANISM

Source: Cugurollo and Acheampong (2022).

Background: Previous studies identified fear as a behavioral determinant capable of influencing people's attitudes towards AIs (robots and autonomous vehicles) (Acheampong and Cugurullo 2019, Hinks 2020). This study answered the question "to what extent people's fears and concerns in relation to AI impact their intention to adopt AI as part of their daily life?" The subject matter is on AI-driven cars and the answer is "yes they do, and quite substantially. We have illustrated the plethora of fears and concerns that our participants feel in relation to AI-driven cars" (Cugurullo and Acheampong, 2022:1; 14). However, they also report that there is no Frankenstein Complex (the assumption that "most people were going to be afraid of AI as a potential source of harm and that, consequently, AI technologies were bound to be rejected by society").

AI FAIRNESS MECHANISM

Source: Narayanan et al (2023).

Background: The authors found (for different dimensions of fairness) that often human decision making is believed to be (more) fair than AI-augmented decision-making, except "when explanations in AI-driven decision-making are recognized as reasonable, understandable, and responsive to the users' needs and concerns. Then they are more likely to be perceived as informationally fair" (2023: 9).

Connected to this is what Kieslich et al. (2022: 4 ff) found out about persons' perceptions of AI: "perceiving AI as unethical [and unfair] has detrimental implications for an organization in terms of a lower reputation as well as a higher likelihood for protests and for pursuing litigation."

TRUST IN AI-MECHANISM

Source: Chen et al. (2019); Okamura & Yamada (2020) and Tschobb (2020).

Background: Studies suggest that human-automation trust shares some important features with human-to-human trust. Okamura & Yamada (2020: 22035) define trust in human–AI cooperation as “an attitudinal judgment of the degree to which a user can rely on an agent to achieve their goals under conditions of uncertainty.” One key aspect of human–AI cooperation is that human users should trust AI systems, just as humans normally do with other human partners.” However, this implies that users of AI would have to be able to “adjust their level of trust to the actual reliability of AI systems (‘trust calibration’’). In their research, they studied cognitive cues called Trust Calibration Cues to stimulate users to adjust their trust levels (Okamura & Yamada, 2020: 220335).

EMOTION HANDLING AND EMPATHY AS MECHANISMS

Source: Bagozzi, Brady & Huang (2022); Kim and Hur (2023).

Background: Emotions have long been considered a unique human trait, but recent advances in AI have led to increasing interest in the incorporation of emotional intelligence into machines. Natural language processing, computer vision, and affective computing techniques, among others, are becoming involved in the detection and analysis of human emotions. Bagozzi, Brady & Huang (2022: 499) discuss a theory on AI-emotions. One of their points is that

> interacting with AI can generate three categories of emotion: basic, self-conscious, and moral emotions. Basic emotions are how persons feel as a result of interacting with the AI robot; self-conscious emotions are pride, shame, guilt, embarrassment, envy, and jealousy persons may feel when interacting with AI and the third category are the moral emotions: they arise when people observe or become aware of another person (or organization [or machine]) doing good or bad things towards other people, technologies, or organizations.

In service interactions, AI is capable of generating all three categories of emotions – basic, self-conscious, and moral emotions – in customers and in frontline employees (Assunção et al., 2022; Bagozzi, Brady & Huang, 2022:501). Kim and Hur’s (2023) work can be located under the “AI device use acceptance (AIDUA) theory” and addresses aspects of AI like its warmth (“the degree to which individuals perceive caring and sociability in non-human entities. Several previous studies revealed that the characteristics of AI artifacts could evoke users’ inference of the warmth of AI artifacts”), empathy, and anthropomorphism.

ANTHROPOMORPHISM MECHANISM

Source: Li & Suh (2022).

Background: Anthropomorphism is the attribution of human characteristics to nonhuman beings or entities like AI systems. Li and Suh (2022) carried out a systematic literature review, categorizing six definitions of anthropomorphism and summarizing some twenty theoretical foundations of this mechanism. "Our review showed an increasing attention on understanding how anthropomorphism leads to the development of the human-AIET relationship,[16] such as rapport building, intimacy, emotional closeness and a parasocial relationship" (p. 2263). These developments have an influence on the behavior of (future) users of algorithms.

SOCIAL INTERACTIONS AND SOCIAL AI MECHANISMS

The social influence mechanism (through peers, networks).

Source: Cabiddu et al (2022) and Caplan & Boyd (2018).

Background: This mechanism can influence attitudes and behavior regarding AI through imitation by individuals and their peers, and through homogenization of organizations, leading to isomorphism. Caplan and Boyd (2018: 1) examined "how algorithms and data-driven technologies, enacted by an organization like Facebook, can induce similarity across an industry." Using theories from organizational sociology and neo-institutionalism, this paper traces the bureaucratic roots of Big Data and algorithms to examine the institutional dependencies that emerge and are mediated through data-driven and algorithmic logics. This type of analysis sheds light on how organizational contexts are embedded into algorithms, which can then become embedded within other organizational and individual practices." It considered the possibilities that algorithmization is acting as "an extension of the concept of bureaucratic mechanisms" rather than adding to innovation and thinking outside the box (2018: 4).

THE SOCIAL AI MECHANISM, INCLUDING THE SOCIAL EMBODIMENT OF AI

Source: Bolotta and Duman (2022); Dafoe et al (2021).

Background: Bolotta and Duman make the point that while

> the complex human cognitive architecture owes a large portion of its expressive power to its ability to engage in social and cultural learning, the field of AI has mostly embraced a solipsistic perspective on intelligence.... Social interactions not only are largely unexplored in this field, but also are an essential element of advanced cognitive ability, and therefore constitute metaphorically the 'dark matter' of AI.
>
> (2022: 1)

Dafoe et al. refer to

> the canonical AI problem, that of a solitary machine confronting a non-social environment. Historically, this was a sensible starting point. An AI agent – much like an infant – must first master a basic understanding of its environment and how to interact with it. Even in work involving multiple AI agents, the field [of AI research] has not yet tackled the hard problems of cooperation. (2021:34 ff)

The question then is how AI can (inter)act cooperatively with people, which can increase AI's acceptance and impact. One of their suggestions is the social embodiment of AI.

As such, it is related to the anthropomorphism mechanism. It refers to "states of the body, such as postures, arm movements, and facial expressions that arise during social interaction and play central roles in social information processing" (Bolotta & Dumas, 2022: 1). The idea of social embodiment in artificial agents is "supported by evidence of improvements in the interactions between embodied agents and humans" (2022: 6). Studies have shown positive effects of physical embodiment on the feeling of an agent's social presence, the evaluation of the agent, the assessment of public evaluation of the agent, and the evaluation of the interaction with the agent. In robots, social presence is a key component in the success of social interactions.

Dafoe et al. (2021: 34) mention four elements to realize cooperative AI: (1) understanding (the ability to take into account the consequences of actions/predict another's behavior), (2) communication (the ability to explicitly and credibly share information with others), (3) commitment (the ability to make credible promises when needed for cooperation), and (4) norms and institutions as the social infrastructure or context that reinforces the three other elements. The first three can be considered as mechanisms, while the fourth concerns contexts.

Findings: Contexts

Moving to the C (Contexts), Pawson and Tilley (1997) consider context as a set of factors influencing when and how an intervention is delivered and how mechanisms are triggered. Thus, recognizing such contextual conditions that enable or impede mechanisms is crucial to realist evaluation. Pawson et al. (2004) suggest that contextual factors can be identified at four different levels:

1. The individual capacities of the key actors and stakeholders, such as interests, attitudes, knowledge, and skills.
2. The interpersonal relationships required to support the intervention, such as lines of communication, management, and administrative support, as well as professional relations and contracts.

3. The institutional setting in which the intervention is implemented, such as the culture and norms, leadership, and governance of the implementing body.
4. The wider (infra-)structural system, such as political support, the availability of funding resources, as well as competing policy priorities and influences (Pawson et al., 2004: 7–8; Nielsen, Lemire and Tangsig, 2022; Solaiman et al., 2023)

evaluated the impact of generative AI systems in systems and society[17] and their first sentence draws attention to contexts:

> Understanding an AI system requires insight into aspects such as training data, the model itself, material infrastructure, and the context in which the system is deployed. It also requires understanding people, society, and how societal processes, institutions, and power are changed and shifted by the AI system.

Only a few lines further they further stress contexts again: "the social impact aspects of an AI system are often largely dependent on context, from the sector in which they are developed to the use-cases and contexts in which they are deployed."

Makarius et al. (2020) studied the introduction of AI in organizations from the perspective of STSt: socio-technical [systems] theory. STSt has at its core the idea that the design and performance of an organizational system can only be understood and improved if both the "social" and "technical" aspects are brought together, and they are treated as interdependent parts of a complex system. They point to the importance of contextual awareness in AI and mention the possibility that when AI is "involved in decision-making [this will] enable flattened organizational structures and empower employee decision-making at lower levels" (p. 264). If contexts contribute to the realization of the integration of AI technology and humans (i.e., employees), this may create "sociotechnical capital," where both entities act as a tightly coupled system exhibiting increased responsiveness" (Makarius et al., 2020: 265).

Price (2019: 8) discussed the relevance of contexts when working with AI in healthcare. Meijer et al. (2021) studied the role of contexts in police organizations when experiencing "algorithmization" and Van Noordt & Masagurka (2022) studied the environments around the use of AI&BD in their case studies. Using a quasi-experimental design, Henkel et al. (2020) introduced in the traditional context of call **centers**. Other examples are studies looking into how supply chains in different contexts work with AI (Helo & Hao 2022), LLM's are used for legal analysis (van Dijck, 2024) and AI functions in the development/SDG's world (York & Bamberger, 2020; Leeuw & Bamberger, in press).

Findings: Outcomes, Results, Consequences

Finally, the O stands for Outcomes/results/consequences that can be almost everything: results, performance, failures, impact (health and/or welfare), gains, efficiency, etc.

Babic et al. (2021: 285) analyzed the broadly accepted drive to make medical AI "explainable," in addition to accurate and (medically) effective. "As a result, lawmakers [in the US] have been moving in the direction of requiring the availability of explanations for black-box algorithmic decisions. Indeed, a near-consensus is emerging in favor of explainable AI/ML among academics, governments, and civil society groups" (2021: 286). They questioned this development, as it

> both overstates the benefits and undercounts the drawbacks of requiring black-box algorithms to be explainable (in medicine). For health AI/ML-based medical devices at least, it may be preferable not to treat explainability as a hard and fast requirement but to focus on their safety and effectiveness.
> (2021: 284)

The next box summarizes the approach of a realist evaluation of the role AI is playing in surgery (i.e., robot-assisted surgery), pointing to the relevance of addressing "theories incarnate" and context-mechanism-outcomes configurations.

Box: A realist evaluation of robot-assisted surgery (Randell et al., 2014; 2017).

The question is how and under what circumstances robotic surgery (aka RAS, robot-assisted surgery) is effectively introduced into routine practice and how and under what circumstances robotic surgery impacts teamwork, communication, decision-making, and subsequent patient outcomes. The evaluation started with a literature review to identify theories concerning how RAS becomes embedded into practice and impacts teamwork and decision-making. These were refined through interviews across nine NHS trusts with theater teams, following the "teacher–learner cycle" approach, more often used by realist evaluators. This stage is referred to as "theory elicitation and refinement" and produced an "initial theory of robot-assisted surgery." This theory and the ones to follow over the years were presented as CMO-configurations. Next, a multisite case study was conducted; data were collected using observation, video recording, interviews, and questionnaires. The "initial theory" was then refined. A third phase focused on interviews in other surgical disciplines to assess the generalizability of the findings. This phase also included a design workshop which allowed the evaluators "to explore the significance of our findings for a wide group of stakeholders and to work with them to explore the potential practical implications of the study."

Discussion and Conclusions

The roles and functions of AI when designing, developing, and implementing policies, programs, and interventions are in need of more impact evaluations. While the diffusion and uptake of AI innovations in society is rapid,[18] the number of papers in (evaluation) journals on AI evaluations remains very limited. This chapter makes the case for (more) realist evaluations.

One argument is that as black boxes are involved in both conventional programs and interventions and in AI-driven ones and realist evaluators are trained to unpack them, transform them into white boxes and review their content, it is almost their duty to contribute to this endeavor.

A second argument is that when doing this work, understanding the role of CMOs is important. This chapter has found, albeit not from articles in evaluation journals, some fifteen mechanisms that (can) play a role in making AI-driven interventions work and understanding why and how AI-driven interventions (don't) work in particular contexts. Some refer to reasoning and rational choice decision-making, some to attitudinal and perceptual components, and some to social interactions. We also found mechanisms of a more bio-social nature, like empathy, anthropomorphism, and embodied AI. As repositories of studies of these and related mechanisms in the non-digital world already exist, trying to apply them to the world of algorithms would be worthwhile.

A third argument is future-focused. Given the possibilities that sentient AI, including "feeling AI" may in a not-too-distant future reach our shores, knowledge about the functioning of mechanisms and contexts will probably become even more relevant than is the case now. Sentient AI refers to an artificial intelligence system[19] that has the ability to think, feel, and perceive the physical world around it, just like humans do. Insights from studies by primatologists on "sentient animals"[20] are brought into the perspective of AI researchers and developers.[21] There is a simple reason for that: most animals are not capable of communicating like humans do, but biologists and psychologists work hard to try to understand and create other communication possibilities with animals.

The realist approach is also well suited to play an important role in understanding the algorithmization of interventions and programs with respect to the "beneficiaries" or end-users, and also for intermediate persons in organizations like governments, health care, education, and safety and security institutions.

Notes

1 The term is older than might be thought. In 1976, L. Landa & F. Kopstein referred to algorithmization in a book on education and instruction (Englewood Cliffs, NJ, Educational Technology Publications, 1976).
2 The references to pages 35 and 36 regard the English summary of the full report in Dutch; the reference to page 16 is a translation of the text.
3 See Pasquale (2015); Rai (2019); and Schinckus, Gasparin, and Green (2020).
4 One should ask: what is "recent? AI was already a topic decades ago.
5 This claim is not entirely true; see Nielsen (2023) for a few articles.
6 Thanks to Steffen Bohni Nielsen for this suggestion.
7 https://www.historic-uk.com/HistoryUK/HistoryofBritain/The-Luddites/ (visited October 25, 2023).
8 http://www.history.com/news/ask-history/who-were-the-luddites.
9 https://www.forbes.com/sites/gilpress/2022/05/29/what-is-ai-understanding-the-real-world-impact-of-artificial-intelligence/). See also https://www.mckinsey.com/industries/public-sector/our-insights/the-potential-value-of-ai-and-how-governments-could-look-to-capture-it#/ which presents very elementary suggestions on how to capture impact, based on debatable assumptions and approach. Visited in September 2023.
10 How to Test Machine Learning Models (deepchecks.com) (visited in September 2023).
11 A/B tests refer to (randomized) experimentation processes wherein two or more versions of a variable (web page, page element, color, etc.) are shown to different segments of website visitors at the same time. They are often applied in digital marketing.
12 Ground truth is considered to be the "correct" answer to the prediction problem that the AI tool is learning to solve. That data (set) then becomes the standard against which developers measure the accuracy of the AI system's predictions. See Lebovitz et al. (2023) for results from a study on how managers apply "ground truthing" and refer to Kang (2023) for a more epistemological discussion.
13 See Schinckus, Gasparin, and Green (2020) for an example of opening up AI black boxes in the financial world.
14 https://www.scientificamerican.com/article/demystifying-the-black-box-that-is-ai/; https://www.scientificamerican.com/article/why-we-need-to-see-inside-ais-black-box/.
15 See Logg et al. (2019: 90-103). They also found in their experiments that "experienced professionals, who make forecasts on a regular basis, relied less on algorithmic advice than lay people did."
16 AIET = AI-enabled technology.
17 It is remarkable that in the hundreds of references, not one time an Evaluation journal is mentioned.
18 Mitchell (2021) is more doubtful. She asked the question "Why AI is Harder Than We Think" and one of her arguments is this one: "Even with today's seemingly fast pace of AI breakthroughs, the development of long-promised technologies such as self-driving cars, housekeeping robots, and conversational companions has turned out to be much harder than many people expected."
19 https://emeritus.org/blog/ai-and-ml-what-is-sentient-ai/.
20 https://experiencemachines.substack.com/p/dangers-on-both-sides-risks-from.
21 https://asteriskmag.com/issues/03/are-we-smart-enough-to-know-how-smart-ais-are.

References

Acheampong, R., & Cugurullo, F. (2019). Capturing the behavioural determinants behind the adoption of autonomous vehicles: Conceptual frameworks and measurement models to predict public transport, sharing and ownership trends of self-driving cars. *Transportation Research Part F: Traffic Psychology and Behaviour, 62,* 349–375. https://doi.org/10.1016/j.trf.2019.01.009

Astbury, B., & Leeuw, F. L. (2010). Unpacking black boxes: Mechanisms and theory-building in evaluation. *American Journal of Evaluation, 31*(3), 363–381. https://doi.org/10.1177/1098214010371972

Assunção, A., Patrão, B., Castelo-Branco, M., & Menezes, P. (2022). An overview of emotion in artificial intelligence. *IEEE Transactions on Artificial Intelligence, 3*(6). 867–886. https://doi.org/10.1109/TAI.2022.3159614

Babic, B., et al (2021). Beware explanations from AI in health care. The benefits of explainable artificial intelligence are not what they appear. *Science, 373*(6552). https://doi.org/10.1126/science.abg1834

Bagozzi, R., Brady, M., & Huang, M.-H. (2022). AI service and emotion. *Journal of Service Research, 25*(4), 499–504. https://doi.org/10.1177/10946705221118579

Bolotta, S., & Dumas, G. (2022). Social neuro AI: Social interaction as the "dark matter" of AI. Preprint arXiv:2112.15459v3.

Burrel, J., & Fourcade, M. (2021). The society of algorithms. *Annual Review of Sociology, 47*, 3.1–23.25. https://doi.org/10.1146/annurev-soc-090820-020800

Cabiddu, F., et al (2022). Why do users trust algorithms? A review and conceptualization of initial trust and trust over time. *European Management Journal, 40*, 685–706. https://doi.org/10.1016/j.emj.2022.06.001

Caplan, R., & Boyd, D. (2018). Isomorphism through algorithms: Institutional dependencies in the case of Facebook. *Big Data & Society, 9*, 1–12. https://doi.org/10.1177/2053951718757

Chen, J., et al (2019). Trust architecture and reputation evaluation for internet of things. *Journal of Ambient Intelligence and Humanized Computing, 10*, 3099–3107. https://doi.org/10.1007/s12652-018-0887-z

Christen, M., et al (2023). Who is controlling whom? Reframing "meaningful human control" of AI systems in security. *Ethics and Information Technology, 25*(10). https://doi.org/10.1007/s10676-023-09686-x

Crootof, R., Kaminski, M., & Price, W. Humans in the Loop. (2022). Vanderbilt Law Review, Forthcoming 2023/ U of Colorado Law Legal Studies Research Paper No. 22-10.

Cugurullo, F., & Acheampong, R. A. (2023). Fear of AI: An inquiry into the adoption of autonomous cars in spite of fear, and a theoretical framework for the study of artificial intelligence technology acceptance. *AI & Society*. https://doi.org/10.1007/s00146-022-01598-6

Dafoe, A., Bachrach, Y., Hadfield, G. K., Horvitz, E., Larson, K., & Graepel, T. (2021). Cooperative AI: Machines must learn to find common ground. *Nature, 593*, 33–36. https://doi.org/10.1038/d41586-021-01170-0

EU Parliament. (2020). Artificial intelligence: How does it work, why does it matter, and what can we do about it? *European Parliamentary Research Service*, 2020.

Forss, K., & Norén, J. (2017). Using "Big Data" for Equity-Focused Evaluation - Understanding and Utilizing the Dynamics of Data Ecosystems. In G. Petersson & J. Breul (Eds.), *Cyber Society, Big Data and Evaluation* (pp. 171–191). New Brunswick, NJ: Transaction Publishers.

Godin, B. (2015). Models of innovation: Why models of innovation are models, or what work is being done in calling them models? *Social Studies of Science*, *45*(4), 570–596. https://doi.org/10.1177/0306312715596852

Green, B. (2022). The flaws of policies requiring human oversight of government algorithms. *Computer Law & Security Review*, *45*, 10568. https://doi.org/10.1016/j.clsr.2022.105681

Havrda, M., & Klocek, A. (2023). Well-being impact assessment of AI: A search for causality and proposal for an open platform for well-being impact assessment of AI systems. *Evaluation and Program Planning*, *99*, 102294. https://doi.org/10.1016/j.evalprogplan.2023.102294

Helo, P., & Hao, Y. (2022). Artificial intelligence in operations management and supply chain management: An exploratory case study. *Production Planning & Control*, *33*(16), 1573–1590. https://doi.org/10.1080/09537287.2021.1882690

Henkel, A., Bromuri, S., Iren, D., & Urovi, V. ((2020). Half human, half machine-augmenting service employees with AI for interpersonal emotion regulation. *Journal of Service Management*, *31*(2), 247–265. https://doi.org/10.1108/josm-05-2019-016

Høljund, S., Olejniczak, K., Petersson, G., & Rok, J. (2017). The Current Use of Big Data in Evaluation. In G. Petersson & J. Breul (eds.), *Cyber Society, Big Data and Evaluation* (pp. 35–61). New Brunswick, NJ: Transaction Publishers.

Hernández-Orallo, J. (2017). Evaluation inartificial intelligence: From task-oriented to ability-oriented measurement. *Artificial Intelligence Review*, *48*, 397–447. https://doi.org/10.1007/s10462-016-9505-7

Hinks, T. S. (2020). Fear of robots and life satisfaction. *International Journal of Social Robotics*, *13*, 327–340.

Kang, E. (2023). Ground truth tracings (GTT): On the epistemic limits of machine learning. *Big Data & Society*, *10*, 1–12. https://doi.org/10.1177/20539517221146122

Kates, A., & Wilson, K. (2023). AI for evaluators: Opportunities and risks. *Journal of Multidisciplinary Evaluation*, *19*(45), 99–104.

Kieslich, K., Keller, B., & Starke, C. (2022). Artificial intelligence ethics by design. Evaluating public perception on the importance of ethical design principles of artificial intelligence. Artificial intelligence ethics by design. *Big Data & Society*, *8*, 1–15. https://doi.org/10.1177/20539517221092956

Kim, W., & Hu, H. (2023). What makes people feel empathy for AI chatbots? Assessing the role of competence and warmth. *International Journal of Human–Computer Interaction*. https://doi.org/10.1080/10447318.2023.2219961

Koulu, R. (2020). Proceduralizing control and discretion: Human oversight in artificial intelligence policy. *Maastricht Journal of European and Comparative Law*, *27*(6), 720–735. https://doi.org/10.1177/1023263X20978649

Lanz, L., Briker, R., & Gerpott, F. H. (2022). Employees adhere more to unethical instructions from human than AI supervisors: Complementing experimental evidence with machine learning. *Journal of Business Ethics*. https://doi.org/10.1007/s10551-023-05393-1

Lebovitz, S., Lifshitz-Assaf, H., & Levina, N. (2023). Determining whether an AI solution is worth implementing requires looking past performance reports and finding the ground truth on which the AI has been trained and validated. *MIT Sloan Management Review*, March 7. https://papers.ssrn.com/sol3/papers.cfm?abstract_id=3839601

Leeuw, F. L. (2020). Program Evaluation B: Evaluation, Big Data, and Artificial Intelligence: Two Sides of One Coin. In R. V. Dana., et al (eds.), *The Handbook of Research Methods in Public Administration, Management and Policy: Breaking New Frontiers*. Cheltenham: EE Publishers, Cheltenham, chapter 16.

Leeuw, F. L., & Leeuw, B. (2012). Cyber society and digital policies: Challenges to evaluation? *Evaluation*, *18*(1), 111–127. https://doi.org/10.1177/1356389011431777

Lemire, S., et al (2020). What is this thing called a mechanism? Findings from a review of realist evaluations. *New Directions for Evaluation, 167*, 73–86. https://doi.org/10.1002/ev.20428

Li, M., & Suh, A. (2022). Anthropomorphism in AI-enabled technology: a literature review. *Electronic Markets, 32*, 2245–2275. https://doi.org/10.1007/s12525-022-00591-7

Logg, J., et al (2019). Algorithm appreciation: People prefer algorithmic to human judgment. *Organizational Behavior and Human Decision Processes, 151*, 90–103. https://-doi.org/10.1016/j.obhdp.2018.12.005

Makarius, E. et al (2020). Rising with the machines: A sociotechnical framework for bringing artificial intelligence into the organization. *Journal of Business Research*, 120 (November): 262–273.

Meijer, W., Lorenz, L., & Wessels, M. (2021). Algorithmization of bureaucratic organizations: Using a practice lens to study how context shapes predictive policing systems. *Public Administration Review*, *81*(5), 837–846. https://doi.org/10.1111/puar.13391

Mitchell, M. (2021). Why AI is harder than we think Melanie Mitchell. *arXiv*:2104.12871v2. https://doi.org/10.48550/arXiv.2104.12871

Misuraca, G. & van Noordt, C. (2020). *Overview of the Use and Impact of AI in Public Services in the EU, EUR 30255 EN*. Luxembourg: Publications Office of the European Union. ISBN 978-92-76-19540-5. https://doi.org/10.2760/039619, JRC120399

Narayanan, D. (2023). Fairness perceptions of artificial intelligence: A review and path forward. *International Journal of Human–Computer Interaction.* https://doi.org/10.1080/10447318.2023.2210890

Netherlands Scientific Council for Government Policy. (2021). *Mission AI: The New System Technology*. Den Haag.

Nielsen, S. B. (2023). Disrupting evaluation? Emerging technologies and their implications for the evaluation industry. *New Directions for Evaluation*. https://doi.org/10.1002/ev.20558

Nielsen, S. B., Lemire, S., & Tangsig, S. (2022). Unpacking context in realist evaluations: Findings from a comprehensive review. *Evaluation*, *28*(1), 91–112. https://doi.org/10.1177/13563890211053032

Nielsen, S. B., Mazzeo Rinaldi, F., & Petersson, G. J. (2025). Evaluation in the Era of Artificial Intelligence. In S.B. Nielsen, F. Mazzeo Rinaldi & G.J. Petersson (eds.), *Artificial Intelligence and Evaluation. Emerging Technologies and Their Implications for Evaluation* (pp. 1–12). London: Routledge. https://doi.org/10.4324/9781003512493

Okamura, K., & Yamada, S. (2020). Empirical evaluations of framework for adaptive trust calibration in human-AI cooperation. *IEEE Access*, *8*, 220335–220351. https://doi.org/10.1109/ACCESS.2020.3042556

Park, S., & Han, K. (2018). Methodologic guide for evaluating clinical performance and effect of artificial intelligence technology for medical diagnosis and prediction. *Radiology, 286*(3), 800–809. https://doi.org/10.1148/radiol.2017171920

Park, Y., et al (2020). Evaluating artificial intelligence in medicine: Phases of clinical research. *JAMIA Open, 3*(3), 326–331. https://doi.org/10.1093/jamiaopen/ooaa033

Pasquale, F. (2015). *The Black Box Society : The Secret Algorithms That Control Money and Information*. Cambridge: Harvard University Press.

Pawson, R. (2006). *Evidence-based Policy: A Realist Perspective*. London: Sage.

Pawson, R. (2013). *The Science of Evaluation*. London: Sage.

Pawson, R., & Tilley, N. (1997). *Realistic Evaluation*. London: Sage.

Pawson, R., et al (2004). *Realist Synthesis: An Introduction*. ESRC Research Methods Program, University of Manchester Methods Paper # 2. https://www.researchgate.net/publication/228855827_Realist_Synthesis_An_Introduction

Pedersen, T., & Johansson, C. (2020). Behavioral artificial intelligence: An agenda for systematic empirical studies of artificial inference. *AI & Society, 35*, 519–532. https://doi.org/10.1007/s00146-019-00928-5

Pinto, B-.-J., et al (2023). Bioethics and AI: Between deliberation on values and rational choice theory. *Frontiers in Robotics and AI*, 1140901. https://doi.org/10.3389/frobt.2023.1140901

Press, G. (2022). What is AI; understanding the real world impact of AI. https://www.forbes.com/sites/gilpress/2022/05/29/what-is-ai-understanding-the-real-world-impact-of-artificial-intelligence/?sh=5d8cd59c4507

Price, W. (2018). Big data and black-box medical algorithms. *Science Translational Medicine, 10*(471) https://doi.org/10.1126/scitranslmed.aao5333

Price, W. (2019). Artificial intelligence in the medical system: Four roles for potential transformation. 21 *Yale Journal of Law and Technology, 31*, 122–132.

Rai, A. (2020). Explainable AI: from black box to glass box. *Journal of the Academy of Marketing Science*48:137–141 https://doi.org/10.1007/s11747-019-00710-5

Randell, R., et al (2017). A realist process evaluation of robot-assisted surgery: Integration into routine practice and impacts on communication, collaboration and decision-making. *Health Service Delivery Research, 20*(5), 1–140.

Randell, R., et al (2014). Integration of robotic surgery into routine practice and impacts on communication, collaboration, and decision making: A realist process evaluation protocol. *Implementation Science, 9*, 52. https://doi.org/10.1186/1748-5908-9-52

Ranerup, A., & Hendrikson, H. (2020). Digital discretion: Unpacking human and technological agency in automated decision making in Sweden's social services. *Social Science Computer Review, 20*, 1–17. https://doi.org/10.1177/0894439320980434

Rogers, E. M., & Agarwala-Rogers, R. (1976). *Communication in Organizations*. New York: Free Press.

Schinckus, C., Gasparin, M., & Green, W. (2020). Opening the black boxes: Financial algorithms and multi-paradigmatic research in information technology. *Journal of Systems Information Technology, 24*(3), 284–303. https://doi.org/10.1108/JSIT-01-2020-0006

Sharma, G., et al (2020). Artificial intelligence and effective governance: A review, critique and research agenda. *Sustainable Futures, 2*, 10004. https://doi.org/10.1016/j.sftr.2019.100004

Siemon, D. (2023). Let the computer evaluate your idea: Evaluation apprehension in human-computer collaboration. *Behaviour & Information Technology*, *42*(5), 459–477. https://doi.org/10.1080/0144929X.2021.2023638

Solaiman, et al. (2023). Evaluating the Social Impact of Generative AI Systems in Systems and Society. *arXiv*:2306.05949v2.

Tschopp, M. (2020). *Pas, The Perfect Automation Schema.* BLOG, SCIP, ZURICH, May 7.Retrieved from https://www.researchgate.net/publication/341458775_PAS_-The_Perfect_Automation_Schema_Influencing_Trust

van Dijck, G. (2024). Legal Complexity and Evaluating Legal Phenomena: The Relevance of Linking Data and Legal Data Analytics. In F. Leeuw & M. Bamberger (eds.), *Big Data,Rule of law and Development: The Evaluators Perspective*. Cheltenham: EE Publishers, in press.

Westhorp, G. (2018). Understanding mechanisms in realist evaluation and research. In N. Emmel (eds.), *Doing Realist Research* (pp. 41–59). London: Sage.

York, P. & Bamberger, M. (2020). Measuring results and impact in the age of big data: The nexus of evaluation, analytics, and digital technology. The Rockefeller Foundation. Retrieved from: https://www.rockefellerfoundation.org/report/measuring-results-and-impact-in-the-age-of-big-data-the-nexus-of-evaluation-analytics-and-digital-technology/

13 The Evaluation Industry and Emerging Technologies

Steffen Bohni Nielsen

Introduction

Across the globe, digital technologies have radically changed social life in the twenty-first century. Digital devices have become all-pervasive in how private, public, and not-for-profit institutions operate. Additionally, in private lives, digital devices are endemic. According to recent reports, 5.2 billion of the globe's 8 billion inhabitants are online (Datareportal, 2023).

In other words, we are currently witnessing an exponential growth in globally generated data, and accordingly new ways that such data are put to use prevail in all sectors (Nielsen, Ejler & Schretzmann, 2017).

In the private sphere, digital users leave footprints about their interests, preferences, consumer habits, and physical whereabouts. Behind user interfaces, powerful computers capture, store, and process data about our online behavior. For Big Tech companies such as Amazon, Google, Meta, Apple, and Microsoft, data is the new gold. The ability to exploit these data is based on digitally driven emerging technologies (ET).

Based on a review of the literature, Rotolo and colleagues define

> emerging technology as a radically novel and relatively fast growing technology characterised by … the potential to exert a considerable impact on the socio-economic domain(s) which is observed in terms of the composition of actors, institutions and patterns of interactions among those, along with the associated knowledge production processes. Its most prominent impact, however, lies in the future and so in the emergence phase is still somewhat uncertain and ambiguous.
>
> (Rotolo, Hick & Martin, 2015: p. 1828)

Indications are that we have only seen the tip of the iceberg. In a comprehensive analysis offered by the management consulting firm McKinsey, they concluded:

> Our analysis of more than 2000 work activities across more than 800 occupations shows that certain categories of activities are more easily automatable than others. They include physical activities in highly predictable and

DOI: 10.4324/9781003512493-13

> structured environments, as well as *data collection and data processing*. These account for roughly half of the activities that people do across all sectors. The least susceptible categories include managing others, *providing expertise, and interfacing with stakeholders*.
>
> (2018, p. 2, author's italics)

Also, a more recent study focusing solely on the impact of large language modeling (LLM) provided remarkable insights. The study focused on more than 1,016 occupations, 19,265 job tasks, and 2,087 daily work activities and applied both expert assessment and LLM approaches to assess the consequences of LLM for each occupation. The study concluded that knowledge-intensive industries are among those most affected by the emergence of generative artificial intelligence (AI), especially those with routine cognitive tasks such as quantitative data collection and analyses, and technical and scientific services (Eloundou, Manning, Mishkin & Rock, 2023). In other words, it remains clear that social scientific research, foundational and applied, conducted by academics or practitioners is equally likely to be affected by ET. This is corroborated by other analyses (Cotton, Cotton & Shipway, 2023). Evaluation as a form of knowledge production applying social science methods is no exception.

Writ large, no profession, or industry, is left unaffected by digital ETs. It is not a question of *whether* an industry and its professions will be affected. It is a question of *how*. According to the aforementioned McKinsey study, the public and social sectors are among those that will be affected the most (2018). These are the sectors within which evaluation activities prevail. It was documented not too long ago that organizations within these sectors are also vastly dominant procurers of evaluation services in the United States of America (U.S.) and elsewhere (Kinarsky, 2018; Lemire, Fierro, Kinarsky, Fujita-Conrads & Christie, 2018).

It remains debatable to what extent evaluation can be considered a profession in its own right (Picciotto, 2011) and its practice demarcates an industry (Nielsen, Lemire & Christie, 2018a). However, it is clear that it is a form of expert knowledge, which is mostly procured, contracted, and delivered in the context of market conditions (Nielsen, Lemire & Christie, 2018a, 2018b). Here, we understand the market in broad terms, often involving public service organizations as a procurer, partner, or provider of evaluation services.

Procurers demand evaluation services in competition with adjacent forms of knowledge production (Nielsen & Hunter, 2013). Nielsen et al. (2018c) demarcated this form of expert knowledge as "evaluation services [that] share an evaluative purpose and are offered through a set of activities informed by specialist evaluative knowledge" (pp. 21–22).

When gauging the interface between ETs on the one hand and evaluation practice on the other hand, one cannot ignore the market forces shaping this interface. Effectively, market dynamics shape *whether* we are evaluating (or demanding other services), *how* we evaluate (ex-ante, ongoing, ex-post – and sometimes with what methods), and *what* we evaluate (the characteristics of the evaluand). The contributions to this anthology are suggestive that ETs already are, or are likely to become, part and parcel of *what* and *how* evaluators evaluate.

However, we must look at its implications from a market perspective. What are the consequences of emerging technology for evaluation as an industry?

In this chapter, I seek to answer this overarching question. More specifically, the research questions are: (1) will ETs affect evaluation as an industry; and (2) if so, how will ETs affect evaluation practice?

To do so, I review the relevant literature on the evaluation market and relate this to the emerging technologies as presented previously by Bruce and her colleagues (Chapter 2, this volume) and York and Bamberger (Chapter 3, this volume). I will then relate the findings to a wider discussion on the practice of evaluation.

Review of Literature

In this section of the chapter, I review the literature on ET and the evaluation market and its industry. This literature review provides the empirical and conceptual grounding for the analysis at hand. This chapter draws from and expands upon my previous analyses of the topic (Nielsen, 2023).

Emerging Technologies and Evaluation

In this volume, Bruce, Gandhi and Vandelanotte (2025) and York and Bamberger (2025) provide an overview of the different emerging technologies relevant for evaluation practice. They describe what they are, their potential uses, and pitfalls. I refer to these chapters for further analysis of the extant technologies.

The premise of the anthology, and this article, is that these technologies for data capture, data storage, and data processing will have profound consequences for evaluation practice.

Characteristics of the Evaluation Market and Its Industry

Research on the commercial aspects of evaluation is scarce. This is paradoxical given that most evaluation practice is carried out under contracted agreements and thus is part of the evaluation market. Nielsen, Lemire and Christie (2018c) mapped the available literature as part of a special issue of New Directions for Evaluation (NDE), which was entirely dedicated to the topic (Nielsen, Lemire & Christie, 2018a). They observed a dearth of empirical research on the evaluation market in general. This included major markets such as the United States (U.S.), Canada, United Kingdom (U.K.), and elsewhere.

In their NDE volume, various contributors offered analyses of the largest segments of evaluation commissioners in the U.S., federal government (Lemire, Fierro, Kinarsky, Fujita-Conrads & Christie, 2018), and philanthropies (Kinarsky,

2018). Peck offered an in-depth analysis of the large evaluation providers in the U.S. market and how they operate (2018). Hwalek and Straub surveyed the segment of smaller evaluation providers and described how they operate (2018). Lahey, Elliott and Heath offered an analysis of the Canadian evaluation marketplace (2018). They focused mainly on the Federal segment. Davies, Morris and Fox also provided a similar analysis of the evaluation market in the U.K. (2018). Elsewhere, Nielsen and his colleagues analyzed the Danish evaluation market (Nielsen, Lemire & Christie, 2018d; Nielsen & Winther, 2014). Based on these empirical analyses, Lemire, Nielsen and Christie collated findings and offered a number of conclusions as to the nature of the evaluation market and its actors in general (2018). I shall return to these in short order.

Since the NDE issue was published, a number of new publications have appeared that offer, at least tangential new data to our knowledge (i.e., Askim, Døving & Johnsen, 2021; Jarosewich, Feldmann, Martínez-Rubin & Clark, 2019); Martínez-Rubin, Germuth & Feldmann, 2019). These contributions add important further empirical findings to our knowledge about the evaluation market and its industry. Overall, they support the conclusions offered by Lemire and colleagues (2018). Let me therefore summarize the key findings here. Here I also draw on my earlier attempts to summarize the findings (Nielsen, 2023).

Globally, indications are that evaluation continues to be a growth industry. That being said, there are notable national or regional differences. Shifting priorities in procurement strategies by major evaluation procurers may radically change market size and composition. One example is the national government's cutback on management consulting services in Denmark. This has effectively obliterated the market for external evaluations. Similarly, the Canadian Federal government changed its sourcing strategy in the 2000s and built significant internal evaluation functions. The market for external evaluation shrank as a consequence (Lahey, Elliott & Health, 2018: Lahey & Nielsen, 2013). Market dynamics may rapidly change market size and composition. This is an important observation in the face of fast-growing ETs.

When talking about the evaluation market, it is difficult to assert that one coherent market exists. In some industries, professional associations track market trends, sizes, and shares. None of the Voluntary Organizations for Professional Evaluation (*VOPEs*) do this. Writ large, the evaluation market is segmented and multi-layered. It is divided into segments. Segmentation is based on national (U.S., Canada, U.K., etc.), regional (i.e., Pacific Northwest, type of client (i.e., philanthropies, federal, state, local government), domain (i.e., international development, public health, social services, education), or methodological (i.e., experimental impact evaluations, participatory evaluation models) differentiators. Different clients and different providers may dominate in each segment. Notably, there are no globally dominant market actors across national markets, which is seen in adjacent professional service fields such as auditing and management consulting.

On the demand side, the public sector, particularly national government agencies, is a dominant procurer of evaluation services. Over the years, the procurement, management, and the practice of evaluation services have been increasingly institutionalized (Jacob, Speer & Furubo, 2015). The dominance of public institutions as major procurer of evaluation services also implies that they are subject to public procurement regulations in most countries. Such procurement regulations are often translated into framework contracts that provide access to a portfolio of potential contracts and Requests for Proposals (RfP) for individual contracts. The nature of the contracting schemes, that is, framework contracts as gatekeepers and individual contract awards (win or nothing), and a limited number of contracts effectively limits market access. Framework contracts tend to favor larger evaluation providers (see Peck, 2018).

As noted above, the dominance of public government implies that shifting government priorities and sourcing strategies may effectively, and quickly, alter market size and composition. The above examples of government policies to establish internal evaluation functions in Denmark and Canada dramatically, and rapidly, reduced the demand for evaluation services by external providers. Such policies contribute to the longer-term ebbs and flows in evaluation demand. In countries such as the U.S., large philanthropies are a significant source of funding for evaluation, but it is often funneled through grant recipients (Kinarsky, 2018). As such, large philanthropy funders do not yield the same direct influence on market composition.

On the supply side, a range of management and research consulting firms (semi-public), research institutes, and individual evaluation consultants offer evaluation services. As mentioned, market position is dependent on which segment they cater to. For the largest contracts, there is a limited number of larger consultancies that are dominant (see Lemire, Fierro, Kinarsky, Fujita-Conrads & Christie, 2018; Peck, 2018). Peck argues that this has to do with the potent composition of consortia that combine subject matter and methodological expertise for individual contracts in the U.S. For contracts, there are strategic partnerships between firms, research institutes, universities, and experts. Larger evaluation contracts are often carried by consortia consisting of a number of different firms and experts that deliver different bit parts into a bigger project (Peck, 2018). Nielsen and colleagues found a similar pattern in Denmark (2018d). Thus, evaluation providers differentiate themselves in terms of methodological expertise, domain expertise, and utility focus. As posited by Peck, larger firms tend to have a broader offering and offer full service capacity that goes beyond evaluation services. Their economic acumen, breadth of services, and expertise provide a distinct competitive advantage.

Let us move on to the final points. For one, market access barriers to evaluation services are low. There are no formal entry barriers such as credentialing programs (not even in Canada, one of few countries with a professional designation program for evaluators). One can easily self-ascribe as an evaluator and offer evaluation services. Informally, access barriers for larger contracts are typically in the form of resources towards responding to RfPs, which may be extremely time-consuming for small businesses, and access to framework contracts.

Second, there are no clear boundaries between evaluation services and adjacent services such as performance auditing, policy analysis, monitoring, and business intelligence. Nielsen and Hunter (2013) noted that in the eyes of the commissioner, there is no sharp distinction between evaluation and other forms of knowledge production.

Methodology

The research is based on an environmental scan of existing data from published reports and articles (see Nielsen, 2023). There exist relatively few published peer-reviewed articles on ETs and their implications on evaluation practice. A search of research articles in nine major evaluation journals from 2013 to 2023 identified eighteen distinct articles with "Big Data," "Artificial intelligence," "Machine learning," "Text analytics," or "Internet of Things" in the title, keyword, or in the abstract. The journals included were: *American Journal of Evaluation, Canadian Journal of Program Evaluation, Educational Evaluation and Policy Analysis, Evaluation, Evaluation and the Health Professions, Evaluation and Program Planning, Evaluation Journal of Australasia, Evaluation Review, Journal of Multidisciplinary Evaluation, and New Directions for Evaluation.*

In contrast, a Google Scholar search conducted in May 2023 with the same search terms yielded between 5-1,480 results when combining the search term with "evaluation." When scanning the documents, it remains clear that the majority of the documents refer to the evaluation of the predictive performance of AI or ML models. They do not focus on integration with the evaluation of policies or programs. Nevertheless, the relative scarcity of peer-reviewed articles suggests that gray literature is the most likely source of data on the topic.

Other than online searches, hand searching and citation chasing were applied as search strategies. Herein, I identified relevant reports, books, and articles. The criteria for inclusion were that the text included (1) a discussion of, (2) empirical findings of, or (3) demonstrations of ET for evaluation practice, and preferably provided empirical evidence to that effect. Initial scans indicated that most documents could be categorized as use cases or reflective case studies on the application of ET in evaluation. Only a few manuscripts, if any, provided data pertaining to broader market coverage. No data extraction form was developed as the disparity of the content made such an approach unwieldy.

Instead, I applied an inductive thematic analysis. Thematic analysis is a qualitative research method that is used across a range of epistemologies and research questions. Thematic analysis can be used for a range of purposes, such as identifying, analyzing, organizing, describing, and reporting themes found within a data set, such as existing literature in the present case (Nowell, Norris, White

& Moules, 2017). As noted, empirical data on the implications of ET for the evaluation industry is limited. Therefore, most analyses are based on available knowledge of industry dynamics and practical ET case applications identified in the literature and in this volume. As such, the analysis expands upon previous analyses conducted by Nielsen elsewhere (2023).

Findings

In this section, I present the findings for each of the two research questions: (1) will emerging technologies affect evaluation as an industry? and (2) how will emerging technologies affect evaluation practice? The section is structured accordingly.

Will ET Affect Evaluation as an Industry?

In their analysis of the dynamics of the evaluation market, Nielsen, Lemire and Christie (2018c) offer an analytical framework wherein they note that salient contextual drivers affecting the market are societal trends such as the increased digitization of social lives, and drivers in technological development, particularly Big Data (see also Nielsen, Lemire & Christie, 2018d). Yet, some professions and related industries have faced digital disruption for some time and more rapidly embraced (or been forced to adapt to) increased digitization and digital technologies. Evaluation has been slower in adopting ET (Petersson & Breul, 2017; Picciotto, 2020; Raftree & Bamberger, 2014). Among the first books to focus on the interlacing between ET, specifically Big Data, and evaluation was the anthology edited by Petersson and Breul (2017). Herein, a survey among self-reported evaluators in the mid-2010s documented that about 10 percent had experience with Big Data (Højlund, Olejniczak, Petersson & Rok, 2017). The survey was based on convenience sampling and yielded a remarkably low response rate (324 responses from up to 85,000 population). No other survey of the demand or supply side was identified. Since, York and Bamberger have echoed these findings (2020).

Of late, the evaluation community appears to have shown increasing interest in the application of ET. Examples of applications of ET in evaluation are beginning to appear in the peer-reviewed literature (Bonfiglio, Camaioni, Carta & Cristiano, 2023; Cintron & Montrosse-Moorhead, 2022; Roy & Rambo-Hernandez, 2021). Professional development programs have started to include modules on AI and ML, topical groups such as the MERL Tech group have emerged, and protagonists make concerted calls for further cooperation and integration with data science (Bruce, Gandhi & Vandelonotte, 2020; Hejnowicz & Chaplowe, 2021; Raftree, 2020; York & Bamberger, 2020).

As also put forward in the introductory chapter to this volume (Nielsen, Mazzeo Rinaldi & Petersson, 2025), Linda Raftree, writing in the context of international development evaluation, has suggested that ETs have already begun to proliferate in (international development) evaluation in three distinct waves (2020). The proliferation is to some extent corroborated by Rathniman and colleagues' review (2021).

According to Raftree, *the first wave* essentially allowed evaluators to keep doing what they did, but their approach was augmented by new sources for data capture (such as geo-spatial data, large administrative registries, and mobile phones). *The second wave* focused on new forms of data capture such as the internet of things, satellites, and drones, and a burgeoning focus on data analytics techniques such as AI and ML. I refer to the chapter by Anand, Batra and Uitto (this volume) for an analysis of the application of geo-spatial data in evaluation.

The third wave came in close with the second wave and explored new technologies for data capture, storage, and data processing. Importantly, Raftree observes, "new disciplines (such as software development and data science) are entering the MERL field, bringing new ideas and ways of working" (2020: p. 15). I refer to Næss and colleagues' analysis of the application of text mining (Chapter 6, this volume) as an exemplar of this wave.

While confined to international development evaluation, Raftree's identification of waves may be an appropriate metaphor for the adoption of ET in the evaluation industry at large. Currently, only tangential empirical evidence exists about how ET has spread across domain segments in the industry, and to what extent practitioners today have more competencies and experience with ET. The recent proliferation of peer-reviewed articles suggests that new ways of data processing, such as texts and photographic images, are part of the third wave (Cintron & Montrosse-Moorhead, 2022; York & Bamberger, 2020). An entire issue of New Directions for Evaluation dedicated to AI and evaluation was recently published (Montrosse Moorhead & Mason, 2023).

These developments, particularly in the international development segment of the evaluation industry, suggest that evaluation practice is already affected by ETs. However, one must bear in mind that the evaluation market is demand-driven (Lemire, Nielsen, and Christie, 2018). Commissioners of evaluation services have a large say in framing what is in demand in terms of scope, budget, timeframe, and competencies (and often methodology). If commissioners' demand for ET is explicit, it is more likely to spread throughout the industry. Only one, somewhat dated, study has noted that RfPs in international development evaluation did not request the application of ETs (Forss & Norén, 2017). Such observations may help to explain the relatively slow proliferation of ETs in evaluation. However, there are several indications of emerging use in this domain (i.e., Franzen et al., 2022).

When considering these developments, evidence suggests that ET are already affecting evaluation practice. Yet, *how*, it affects evaluation practice remains under-analyzed. Let us therefore consider this question in further detail in the next section.

How Will ET Affect Evaluation Practice?

In the introduction chapter to this volume, Nielsen, Mazzeo Rinaldi and Petersson argued that ETs could affect tasks in evaluation practice by way of displacing, augmenting, facilitating, or generating new tasks (this volume). However disruptive today's proliferation of generative AI is considered, technological innovation has been part of the industry for years. Two decades ago, Rebecca Maynard observed that the growth in the evaluation industry was driven by application into new policy domains and innovations in methodologies that gave a competitive advantage in the market (2000). This observation still rings true today. In the current digital era, these technological innovations (ETs) are digital. Yet a more granular analysis must take several factors into consideration. These factors include:

- Evaluation of providers' competitive strategies
- Size and duration of evaluation contracts
- Nature of the evaluation service
- Breadth and depth of the evaluation provider's capability
- Appropriateness of the technology

In what follows, an analysis of each of these factors is presented.

Competitive strategies. Most evaluation practice is contracted, and the financial terms are fixed. Evaluation commissioners set the terms through requests for proposals (RfPs). Evaluation of providers' competitive strategies compete on price, quality, and timeframe. Put differently, these pertain efficiency (at what cost the evaluation tasks can be delivered), effectiveness (at what quality the evaluation tasks can be delivered), and expedience (how quickly the evaluation tasks can be delivered).

Empirical analyses suggest that quality differentiators are methodology, subject matter expertise, and utility (Lemire, Nielsen & Christie, 2018). Price differentiators are the overall sum (and sometimes for each staffing category). The timeframe differentiator is how expediently the work can be delivered.

A first differentiator appears when evaluators compete on price. Here, technology enables the provider to lower the price or deliver more for a fixed price. One example is applying drones or satellite imagery to collect data on indicators for household income rather than field visits in difficult-to-reach areas (York & Bamberger, 2020). The latter at a far higher cost. Another example is provided in this volume by Næss and his colleagues (this volume). Here, the evaluation team applied a machine learning algorithm to classify the emergence of cybercrime in the *entire* dataset rather than in a *sample* using manual coding of police crime registries (Næss et al., 2025).

A second differentiation strategy focuses on providing a higher quality evaluation service through the application of new technology. One driver may be *empirical,* wherein new sources, new data collection methods, or new kinds of analyses are introduced. An example is the use of algorithms for building predictive and prescriptive models for child abuse or neglect, combining machine learning and quasi-experimental design (Schwartz et al., 2017, see also York & Bamberger, 2020).

A third differentiator may be *utility* (variability, immediacy, frequency, and granularity of reporting and process management). One case exemplar is the building of a comprehensive evaluation system in New York City used for predicting and preventing evictions in a homelessness prevention program, where real-time reporting delivered granular data at the street/block level for immediate action (Nielsen, Ejler & Schretzmann, 2017). Technology may also be used to include difficult-to-reach populations or intensify contact and channels for process use.

Size and duration of contracts. The size and duration of the contract, set by the commissioners of evaluation studies, are likely to be of importance for how ET affects the evaluation market. The human capability, technology, and time invested in conducting algorithm-based text analytics are significant (Franzen et al., 2022; Næss et al., this volume). It may therefore prove too costly, time-consuming, and cumbersome for small-budget evaluations. As long as sophisticated ET are not commodified, application by smaller evaluation providers may be too cost-intensive. Others will have to shoulder developmental investments. Examples of access through commodified solutions such as ChatGPT, Rayyan, and QDAminer are starting to appear (Head et al., 2023).

Nature of the evaluation service. Using Nielsen and colleagues' definition of evaluation service (2018c), such services encompass building monitoring and evaluation systems, conducting evaluation studies, and building evaluation capacity. While rooted in expert evaluative knowledge, the tasks to be carried out are quite different. The compatibility of ET to these different services varies.

The aforementioned McKinsey study identified human tasks such as managing others, interfacing with stakeholders, and providing expertise as least replaceable by technology (2018).

ET is more likely to be applied in services wherein more tasks concern large, recurrent, potentially automated, and scalable tasks are evident. Most likely, building monitoring and evaluation (M&E) systems will be most directly affected (Mazzeo Rinaldi, Giuffrida & Negrete, 2017). Such M&E systems provide recurrent streams of data as opposed to episodic data collection from singular evaluation studies (Nielsen & Ejler, 2008). The streams of data imply a recurrent flow of data collection, analytic and reporting activity that holds the potential for automation. In some instances, M&E system data streams may be more or less automated by ET and displace some human tasks. Consider automated customer engagement surveys in the private sector or employee engagement surveys. Once questionnaire items are conceptualized (by experts), the activities in collecting, managing, analyzing, and reporting data are more or less fully front-end and back-end automated.

In episodic evaluation, where design, data collection, analysis, and reporting are one-off, the potential for automation is lower. Yet, certain tasks may be augmented by AI, such as succinctly summarizing bodies of text and drafting parts of the report. Such use comes with a number of implications, but there are examples of reporting done, in part, by generative AI solutions (Cotton, Cotton & Shipway, 2023).

Overall, this is less likely in short-term, episodic, human interaction-intensive engagements that rely more heavily on specialist evaluator knowledge. Relations management will be affected by ETs. Evaluation capacity-building services tend to be such kinds of service. Here such expert knowledge and relational skills are important (but may be augmented by AI tools in some activities such as training).

Table 13.1 presents an annotated general assessment (high, medium, low) of displacing human tasks with ET automation. Given the segmentation of the industry into different subject matter domains, the emergence may differ.

Capability of the evaluator. The application of ET in evaluation practice will rely on the combined skills of the evaluation team. Evaluation teams often combine evaluation methodology and subject matter expertise (Hwalek & Straub, 2018; Peck, 2018). A number of observers have argued that if evaluators are to make full use of ET, evaluation teams must add competencies from data science to their team composition (Petersson, Leeuw & Olejniczak, 2017; Raftree, 2020; York & Basmberger, 2020; this volume). As of yet, training in ETs is only beginning to find its way into evaluation training (see Nielsen, Mazzeo Rinaldi, & Petersson, this volume).

Appropriateness of the technology. Much has been written about the potential ethical problems and inequities of AI (Greenstein & Cho, 2025; Head et al., 2023; Reid, 2023). Undoubtedly, further issues will be raised as ETs become more widespread. Evaluators' critical thinking pertaining to equity issues, potential biases, and design is crucial in assessing its appropriateness.

Across these different factors that will affect how ET will be applied in the evaluation industry beckons the question of whether ETs such as generative AI will replace humans. An in-depth analysis of the implications would require a more thorough analysis at the task level for each service comprising the evaluation industry than permitted here (see Eloundou, Manning, Mishkin & Rock, 2023). However, the evaluation industry is composed of a workforce with uneven levels of evaluation expertise. As in other professional services, there are finders (i.e., partners), minders (i.e., project managers), and grinders (i.e., analysts) (Maister, 1997). A composite analysis at the task and activity level would be needed to estimate implications.

As mentioned above, we assume four overarching ways ETs may affect human tasks: (1) displacing, (2) augmenting, (3) facilitating, or (4) generating new human tasks.

Table 13.1 Potential application of ET in different evaluation service lines across categories of tasks (adapted and expanded from Nielsen, 2023)

Categories of tasks	*Monitoring and evaluation systems*	*Evaluation studies*	*Evaluation capacity building*
	Potential ET application		
Design	*Low*	*Low*	*Low*
	• Need for expertise to design the system, identify appropriate indicators, sources, and technology. • Need for expertise to manage stakeholders.	• Need for expertise to design the evaluation study, sources, means, and sources for data collection. • Need for expertise to establish evaluation criteria and standards. • Need for expertise to manage stakeholders.	• Need for expertise to design ECB actions and determine organizational success criteria. • Need for expertise to manage stakeholders.
Data capture	*High*	*Medium*	*Low*
	• ET may provide new sources and methods for data collection. • ET has the potential to collect more and diverse data. • Recurrent data collection enables the automation of tasks.	• ET may provide new sources and methods for data collection. • ET has the potential to collect more and diverse data. • Episodic data collection is restrictive for harvesting economies of scale.	• ECB tasks are based mostly on existing capability and some limited data collection for readiness assessment. • Limited and episodic data collection is restrictive for harvesting economies of scale.
Data storage and management	*High*	*Medium*	*Low*
	• Recurrent and automated data collection enable structured data management and storage. • ET enables the automated collation of large datasets. • ET enables automated data management and data cleansing.	• Collation of small to large data sets may be augmented by ET. • Episodic data collection is restrictive for harvesting economies of scale. • Larger data sets are more amenable for displacing human tasks.	• Limited need for data collection in ECB. • Episodic data collection limits the potential for ET to displace human tasks.

(*Continued*)

Table 13.1 Continued

Categories of tasks	*Monitoring and evaluation systems*	*Evaluation studies*	*Evaluation capacity building*
	Potential ET application		
Data processing	*High*	*Medium*	*Low*
	• Machine Learning enables a broad range of quantitative analyses, both descriptive and inferential. • Recurrent data flow enables setting predefined forms of reporting tailored to different users. • Automated reporting will lower costs. • Text analytics (LLM) enables summarization of data across different rubrics. • Recurrence of reporting enables scale in in prompts. • Expediency of automated processing is significant.	• Off the shelf AI solutions for transcription and translation have great potential for efficiency and expediency and will partially displace human tasks. • Machine Learning enables a broad range of quantitative analyses, both descriptive and inferential. • Machine Learning and NLP enable text mining, screening, and coding of large volumes of texts. • Text analytics (LLM) enables summarization of data across different rubrics. • Episodic nature of reports limits automation.	• Limited need for data analysis. • Episodic reporting limits the use of ET, but summarization may augment reporting.

Most likely, we anticipate a future where evaluators will work alongside digital ETs and apply them to a much larger extent than today (Leeuw, 2020). Arguably, some more menial tasks will be fully, or partially, displaced by technology, while others will be augmented by access to and processing of new and larger data sets. More or less automatable tasks, such as interview transcriptions (Da Silva, 2021), translations, screening of administrative documents, abstracts in literature reviews, and high-level coding of texts, are likely to be delivered by AI-powered solutions (Leeuw, 2020). The consequences for entry-level evaluators (grinders) remain to be seen.

Overall, some tasks and activities may be more at risk of displacement than others. For example, sources of data, types of data collection, and tools for data management, processing, and reporting are likely to change. The revolution in technologies such as IoT and AI-driven tools for data capture and instantaneous processing implies that *what* data is being collected, *how* it is processed, and *who* analyzes and reports data will undergo significant changes (Head et al., 2023). Following this vein, McKinsey Global Institute wrote: "… the transitions that will accompany automation and AI adoption will be significant. The mix of occupations will change, as will skill and educational requirements. Work will need to be redesigned to ensure that humans work alongside machines most effectively" (2018, p. 3).

At a more general level, Wilson and Daugherty have introduced the notion of collaborative intelligence, where professionals (and non-professionals) play a critical role in training, explaining, and sustaining AI to make full use of its potential (2018). Frans Leeuw pointedly argues that we foresee a future wherein artificial and human intelligence will need to collaborate in evaluation also (Leeuw, 2020; this volume). One must assume that the use of LLMs may be stretched as far as possible.

Some tasks are most likely still left to humans, such as those associated with *evaluative thinking*. Tasks such as defining overall research design, establishing relevant evaluation criteria and standards, selecting appropriate sources, critical questioning, assessment of data credibility, evaluative synthesis, and judgment are most likely still going to be handled by expert evaluators. Yet, there remains a potential that such tasks can be augmented by AI. In other fields, such as medicine, AI solutions function as decision support and thus augment professional reasoning, diagnosing, and judgment

These considerations all assume a continuous demand for evaluation services in the field of knowledge production. However, the market size may be affected by shifts in demand toward competing forms of knowledge. Currently, the promise of generative AI appears to be right at the summit of the proverbial hype curve. This implies that demand for evaluation services is at risk of being, at least partially, substituted by demand for other services. Previously, ebbs and tides in demand have come and gone in the evaluation industry.

Depending on the evaluations provers' market segment and position, they are likely to be affected differently.

As argued above, larger evaluation providers tend to have more business lines (and requisite competencies in adjacent services such as auditing, business intelligence, and computer and data science) (Peck, 2018). Given their larger financial acumen, they may be more able to cope with shifts in demand, both within and beyond evaluation services. Highly specialized boutique evaluation firms may be more exposed because of a narrower field of expertise and less financial acumen to invest in ETs.

In sum, there are several indications that evaluation at large is in the *emerging approaches* phase as suggested by Linda Raftree. Several factors are likely to influence how ET is used by different evaluation providers. In the next section, I discuss the wider implications and challenges for the evaluation industry as a whole.

Discussion

Many observers consider the appearance of digital ETs a profound challenge to the application of social science methods at large (Alvarez, 2016). Evaluation as a practice that applies social scientific methods is no exception. Yet some market dynamics may imply that ETs' proliferation is different than in academia.

As noted, the evaluation market is characterized by considerable buyer power from, particularly, the public sector. Currently, the potential use of LLM is hotly debated in governments across Europe. The outcome will affect further proliferation in the public sector of AI. Therefore, much will depend on the response of evaluation commissioners. If they demand certain sources or technologies, then they are likely to get them.

McKinsey Global Institute posited that the public sector is expected to embrace ET (particularly AI) to a much larger extent than hitherto seen. This is the case in the U.S. (DeSouza, 2018), Europe (Misuraca & van Noordt, 2020), and elsewhere. I observed elsewhere: "When *what* is being evaluated is bound to change, *how* it will be evaluated, and by *whom* will likely change too" (2023: 55)

As noted above, AI and other emerging technologies will most likely challenge evaluation practice by way of displacing some human tasks, facilitating and augmenting others, while also generating new tasks such as drone operations, and installing and retrieving data from sensors, etc.

Emerging technologies will most likely not entirely disrupt the industry and its core need for evaluation expertise as suggested in Table 13.1. As presciently observed by Petersson, Leeuw and Olejniczak (2017), the competency challenge is significant and very real in a rapidly changing technological environment. It is therefore important to ponder what different actors in the evaluation landscape should do (see also Leeuw, Willemsen & Leeuw, 2017).

Individual evaluators may want to develop basic AI literacy to critically understand and appraise ETs, identify ethical issues, and potentially apply ETs (Greenstein & Cho, this volume; Ng, Leung, Chu & Qiao, 2021). This kind of upskilling will be necessary as AI evolves rapidly. Leeuw discusses the importance of (critically) embracing the potential of ET rather than dismissing them (this volume).

Evaluation of providers, as actors in the wider professional service industry, must choose adaptation strategies when changes in market dynamics emerge. They must *grow* requisite competencies from within their organization, *hire* the talent, or *collaborate* with data science providers. Obviously, economic acumen, market position, and strategy may prompt evaluation providers to opt for different strategies. Larger providers will likely invest in competency development programs and hire talent, whereas smaller providers are more likely to seek collaboration, or await the commodification of AI-powered solutions to be exploited at lower investment costs.

Voluntary Organizations for Professional Evaluation (VOPEs) potentially play a crucial role in positioning evaluation as an indispensable form of knowledge for decision-makers. Yet, the boundaries between evaluation and other adjacent knowledge-producing services are permeable. Evaluation is one of several forms of knowledge competing to inform decision-making. Evaluation does not hold a privileged position.

VOPEs must also face the challenge of ETs by strategically appraising future evaluator competencies. AI literacy should be incorporated in existing evaluator competency frameworks. This should lead to targeted upskilling and reskilling programs focused on building AI literacy widely in the evaluation community. VOPEs need to advocate what skills and competencies evaluators can bring into the development and exploitation of ET, such as its understanding of theory, causality, validity, ethics, equity, valuing, and judgment (Leeuw, 2020). As argued throughout this volume, evaluators need to learn for collaborating with data science.

Educational institutions offering evaluation programs and training must incorporate ET in the curriculum and as learning outcomes. To this day, this has been neglected and must be rectified swiftly.

Conclusion

In this chapter, I have argued that ETs will significantly affect the evaluation industry. They will do so in different ways: by displacing, augmenting, and facilitating existing human tasks and generating new ones. The actual consequences will depend on a number of factors such as the evaluation providers' competitive strategies, size and duration of contracts, nature of the evaluation service, capability of the evaluator, and the appropriateness of the technology. Emerging technologies will not entirely disrupt the evaluation industry and its reliance on specialist evaluation knowledge, yet evaluators and their institutional structures must embrace ETs' potential to remain relevant to buyers of knowledge-producing services.

References

Alvarez, R. M. (ed) (2016). *Computational Social Science: Discovery and Prediction.* New York, NY: Cambridge University Press.

Anand, A., Batra, G., & Uitto, J. I. (2025). Harnessing Geospatial Approaches to Strengthen Evaluative Evidence. In S. B. Nielsen, F. Mazzeo Rinaldi and G. J. Petersson (eds.). *Artificial Intelligence and Evaluation. Emerging Technologies and Their Implications for Evaluation* (pp. 196–218). London: Routledge. https://doi.org/10.4324/9781003512493

Bonfiglio, A., Camaioni, B., Carta, V., & Cristiano, S. (2023). Estimating the common agricultural policy milestones and targets by neural networks. *Evaluation and Program Planning*. https://doi.org/10.1016/j.evalprogplan.2023.102296

Bonfiglio, A., Camaioni, B., Carta, V., & Cristiano, S. (2023). Estimating the common agricultural policy milestones and targets by neural networks. *Evaluation and Program Planning*. https://doi.org/10.1016/j.evalprogplan.2023.102296

Bruce, K., Vandelanotte, J., & Gandhi, V. (2025). Emerging Technology and Evaluation in International Development. In S. B. Nielsen, F. Mazzeo Rinaldi and G. J. Petersson (eds.). *Artificial Intelligence and Evaluation. Emerging Technologies and Their Implications for Evaluation* (pp. 13–36). London: Routledge. https://doi.org/10.4324/9781003512493

Bruce, K., Gandhi, V. J., & Vandelanotte, J. (2020). *Emerging Technologies and Approaches in Monitoring, Evaluation, Research, and Learning for International Development Programs*. MERL Tech Report # 4. Retrieved from https://merltech.org/wp-content/uploads/2020/07/4_MERL_Emerging-Tech_FINAL_7.19.2020.pdf

Cintron, D. W., & Montrosse-Moorhead, B. (2022). Integrating big data into evaluation: R code for topic identification and modeling. *American Journal of Evaluation*, *43*(3), 412–436. https://doi.org/10.1177/10982140211031640

Cotton, D., Cotton, P. A., & Shipway, J. R. (2023). Chatting and cheating. ensuring academic integrity in the era of ChatGPT. *EdArXiv*. https://doi.org/10.1080/14703297.2023.2190148

Datareportal. (2023). *Digital around the World*. Retrieved from https://datareportal.com/global-digital-overview

Da Silva, J. (2021). Producing 'good enough' automated transcripts securely: Extending Bokhove and Downey (2018) to address security concerns. *Methodological Innovations*, *14*(1), 2059799120987766.

Davies, P., Morris, S., & Fox, C. (2018). The evaluation market and its industry in England. *New Directions for Evaluation*, *160*, 29–43. https://doi.org/10.1002/ev.20347

Desouza, K. C. (2018). *Delivering Artificial Intelligence in Government: Challenges and Opportunities*. IBM Center for The Business of Government, 48. Retrieved from https://www.businessofgovernment.org/report/delivering-artificial-intelligence-government-challenges-and-opportunities

Eloundou, T., Manning, S. Mishkin, P., & Rock, D. (2023). GPTs are GPTs: An early look at the labor market impact potential of large language models. *arXiv:2303.10130* [econ.GN]. https://doi.org/10.48550/arXiv.2303.10130

Forss, K., & Norén, J. (2017). Using "Big Data" for Equity-Focused Evaluation – Understanding and Utilizing the Dynamics of Data Ecosystems. In G. J. Petersson and J. Breul (eds.). *Cyber Society, Big Data and Evaluation* (pp. 171–190). London: Routledge.

Franzen, S., Quang, C., Schweizer, L., Budzier, A., Gold, J., Vellez, M., Ramirez, S., & Raimondo, E. (2022). *Advanced Content Analysis: Can Artificial Intelligence Accelerate Theory-Driven Complex Program Evaluation?* IEG Methods and Evaluation Capacity Development Working Paper Series. Independent Evaluation Group. Washington, DC: World Bank. https://doi.org/10.1596/IEG167127

Greenstein, N., & Cho, S.-W. (2025). Ethics & Equity in Data Science for Evaluators. In S. B. Nielsen, F. Mazzeo Rinaldi and G. J. Petersson (eds.). *Artificial Intelligence and Evaluation. Emerging Technologies and Their Implications for Evaluation* (pp. 56–77). London: Routledge. https://doi.org/10.4324/9781003512493

Head, C. B., Jasper, P., McConnachie, M., Raftree, L., & Higdon, G. (2023). Large language model applications for evaluation: Opportunities and ethical implications. *New Directions for Evaluation, 2023*, 33–46. https://doi.org/10.1002/ev.20556

Hejnowicz, A., & Chaplowe, S. (2021). Catching the wave: Harnessing data science to support evaluation's capacity for making a transformational contribution to sustainable development. *Canadian Journal of Program Evaluation, 36*(2), 162–186. https://doi.org/10.3138/cjpe.71527

Hwalek, M. A., & Straub, V. L. (2018). The small sellers of program evaluation services in the United States. *New Directions for Evaluation, 160*, 125–143. https://doi.org/10.1002/ev.20340

Højlund, S., Olejniczak, K., Petersson, G. J. & Rok, J. (2017). The use of big data in evaluation. In G. J. Petersson and J. Breul (eds.). *Cyber Society, Big Data and Evaluation* (pp. 35–60). London: Routledge.

Jacob, S., Speer, S., & Furubo, J.-E. (2015). The institutionalization of evaluation matters: Updating the International Atlas of Evaluation 10 years later. *Evaluation, 21*(1), 6–31. https://doi.org/10.1177/1356389014564248

Jarosewich,T., Feldmann, M.L., Martínez-Rubin, N., & Clark, N. (2019). Who we are: Findings from the American Evaluation Association's Independent Consulting Topical Interest Group 2015 Decennial Survey. *New Directions for Evaluation*, 164, 27– 41. https://doi.org/10.1002/ev.20387

Kinarsky, A. R. (2018). The evaluation landscape: U.S. Foundation spending on evaluation. *New Directions for Evaluation, 160*, 81–96. https://doi.org/10.1002/ev.20342

Lahey, R., Elliott, C., & Heath, S. (2018). The evolving market for systematic evaluation in Canada. *New Directions for Evaluation, 160*, 45–62. https://doi.org/10.1002/ev.20346

Lahey, R., & Nielsen, S. B. (2013). Rethinking the relationship among monitoring, evaluation, and results-based management: Observations from Canada. *New Directions for Evaluation, 137*, 45–56. https://doi.org/10.1002/ev.20045

Leeuw, H. B. M., Willemsen, F., & Leeuw, F. L. (2017). Big data and evaluation: Background, benefits and challenges. *RIV Rassegna Italiana di Valutazione, 68*, 24–47. https://doi.org/10.3280/RIV2017-068003

Leeuw, F. L. (2020). Program evaluation B: Evaluation, big data, and artificial intelligence: Two sides of one coin. In E. Vigoda-Gadot and D. R. Vashdi (eds.). *Handbook of Research Methods in Public Administration, Management and Policy* (pp. 277–297). https://doi.org/10.4337/9781789903485.00024

Leeuw, F. L. (2025). The algorithmization of policy and society: The need for a realist evaluation approach. In S. B. Nielsen, F. Mazzeo Rinaldi and G. J. Petersson (eds.). *Artificial Intelligence and Evaluation. Emerging Technologies and Their Implications for Evaluation* (pp. 242–265). London: Routledge. https://doi.org/10.4324/9781003512493

Lemire, S., Nielsen, S. B., & Christie, C. A. (2018). Toward understanding the evaluation market and its industry — Advancing a research agenda. *New Directions for Evaluation, 160*, 145–163. https://doi.org/10.1002/ev.20339

Lemire, S., Fierro, L. A., Kinarsky, A. R., Fujita-Conrads, E., & Christie, C. A. (2018). The U.S. federal evaluation market. *New Directions for Evaluation, 160*, 63–80. https://doi.org/10.1002/ev.20343

Maister, D. H. (1997). *Managing the Professional Service Firm*. New York, NY: Free Press.

Maynard, R. A. (2000). Whether a sociologist, economist, psychologist or simply a skilled evaluator: Lessons from evaluation practice in the United States. *Evaluation, 6*(4), 471–480. https://doi.org/10.1177/13563890022209433

Martínez-Rubin, N., Germuth, A. A., & Feldmann, M. L. (eds.). (2019). Independent evaluation consulting. approaches and practices from a growing field. *New Directions for Evaluation*, 164. https://doi.org/10.1002/ev.20338

Mazzeo Rinaldi, F., Giuffrida, G., & Negrete, T. (2017). Real-time monitoring – Emerging news as predictive process using big data-based approach. In G. J. Petersson and J. D. Breul (eds.). *Cyber Society, Big Data and Evaluation* (pp. 191–214). New York, NY: Routledge.

McKinsey Global Institute. (2018). *Ai, Automation, and the Future of Work: Ten Things to Solve for*. Briefing Note Prepared for the Tech4good Summit. Retrieved from https://www.mckinsey.com/featured-insights/future-of-work/ai-automation-and-the-future-of-work-ten-things-to-solve-for

Misuraca, G., & van Noordt, C. (2020). *Overview of the Use and Impact of AI in Public Services in the EU*. Luxembourg: Publications Office of the European Union. Retrieved from: https://joinup.ec.europa.eu/sites/default/files/document/2020-07/jrc120399_Misuraca-AI-Watch_Public-Services_30062020_DEF_0.pdf

Montrosse-Moorhead, B., & Mason, S. (eds.). (2023). Editors' note. *New Directions for Evaluation, 178–179*, 1–6. https://doi.org/10.1002/ev.20551

Ng, D. T. K., Leung, J. K. L., Chu, S. K. W., & Qiao, M. S. (2021). Conceptualizing AI literacy: An exploratory review. *Computers and Education: Artificial Intelligence, 2*, 100041. https://doi.org/10.1016/j.caeai.2021.100041

Nielsen, S. B. (2023). Disrupting evaluation? Emerging technologies and their implications for the evaluation industry. *New Directions for Evaluation, 178–179*, 47–57. https://doi.org/10.1002/ev.20558

Nielsen, S. B., Mazzeo Rinaldi, F., & Petersson, G. J. (2025). Evaluation in the Era of Artificial Intelligence. In S. B. Nielsen, F. Mazzeo Rinaldi and G. J. Petersson (eds.). *Artificial Intelligence and Evaluation. Emerging Technologies and Their Implications for Evaluation* (pp. 1–12). London: Routledge. https://doi.org/10.4324/9781003512493

Nielsen, S. B., Lemire, S., & Christie, C. A. (eds.). (2018a). The evaluation marketplace. Exploring the evaluation industry. *New Directions for Evaluation, 160*. https://doi.org/10.1002/ev.20282

Nielsen, S. B., Lemire, S., & Christie, C. A. (eds.). (2018b). Editors' notes. *New Directions for Evaluation, 160*, 1–6. https://doi.org/10.1002/ev.20345

Nielsen, S. B., Lemire, S., & Christie, C. A. (2018c). The evaluation marketplace and its industry. *New Directions for Evaluation, 160*, 13–28. https://doi.org/10.1002/ev.20344

Nielsen, S. B., Lemire S., & Christie, C. A. (2018d). The Commercial Side of Evaluation: Evaluation as an Industry and as a Professional Service. In J. E. Furubo and N. Stame (eds.). *The Evaluation Enterprise: Evaluating Evaluation*. New York: Routledge.

Nielsen S. B., Ejler, N., & Schretzmann, M. (2017). Exploring Big (Data) Opportunities: The Case of the Center for Innovation through Data Intelligence, New York City. In G. J. Petersson and J. D. Breul (eds). *Cyber Society, Big Data and Evaluation* (pp. 147–170). New York, NY: Routledge.

Nielsen, S. B., & Winther, D. M. (2014). A Nordic evaluation tradition? A look at the peer-reviewed evaluation literature. *Evaluation, 20*(3), 311–331. https://doi.org/10.1177/1356389014540703

Nielsen, S. B., & Hunter, D. E. K. (2013). Challenges to and forms of complementarity between performance management and evaluation. *New Directions for Evaluation, 137*, 115–123. https://doi.org/10.1002/ev.20051

Nielsen, S. B., & Ejler, N. (2008). Improving performance? Exploring the complementarities between evaluation and performance management. *Evaluation, 14*(2), 171–192. https://doi.org/10.1177/1356389007087538

Nowell, L. S., Norris, J. M., White, D. E., & Moules, N. J. (2017). Thematic analysis: Striving to meet the trustworthiness criteria. *International Journal of Qualitative Methods, 16*(1). https://doi.org/10.1177/1609406917733847

Næss, T., Prabhu, C., Mjaaland, M., Holtermann, H., & Engebretsen, L. S. (2025). Text Mining and Machine Learning in an Evaluation of Police Handling of Cybercrime in Norway. In S. B. Nielsen, F. Mazzeo Rinaldi and G. J. Petersson (eds.). *Artificial Intelligence and Evaluation. Emerging Technologies and Their Implications for Evaluation* (pp. 103–119). London: Routledge. https://doi.org/10.4324/9781003512493

Peck, L. R. (2018). The big evaluation enterprises in the United States. *New Directions for Evaluation, 160*, 97–124. https://doi.org/10.1002/ev.20341

Petersson, G. J., & Breul, J. (eds.). (2017). *Cyber Society, Big Data, and Evaluation*. London: Routledge.

Petersson, G. J. Leeuw, F. L., & Olejniczak, K. (2017). Cyber Society, Big Data and Evalaution: A Future Perspective. In G. J. Petersson and J. Breul (eds.). *Cyber Society, Big Data, and Evaluation* (pp. 237–254). New York, NY: Routledge.

Picciotto, R. (2020). Evaluation and the big data challenge. *American Journal of Evaluation, 41*(2), 166–181. https://doi.org/10.1177/1098214019850334

Picciotto, R. (2011). The logic of evaluation professionalism. *Evaluation, 17*(2), 165–180. https://doi.org/10.1177/1356389011403362

Raftree, L. (2020). *MERL Tech State of the Field. The Evolution of MERL Tech*. MERL Tech Report # 1. Retrieved from https://merltech.org/resources/merl-tech-state-of-the-field-the-evolution-of-merl-tech/

Raftree, L., & Bamberger, M. (2014). *Emerging Opportunities: Monitoring and Evaluation in a Tech-Enabled World.* Retrieved from https://www.rockefellerfoundation.org/report/emerging-opportunities-monitoring/

Rathinam, F., Khatua, S., Siddiqui, Z., Malik, M., Duggal, P., Watson, S., & Vollenweider, X. (2021). Using big data for evaluating development outcomes: A systematic map. *Campbell Systematic Review, 17*(3), e1149. https://doi.org/10.1002/cl2.1149

Rotolo, D., Hicks, D., & Martin, B. R. (2015). What is an emerging technology? *Research Policy, 44*(10), 1827–1843. https://doi.org/10.1016/j.respol.2015.06.006

Reid, A. M. (2023). Vision for an equitable AI world: The role of evaluation and evaluators to incite change. *New Directions for Evaluation*, (178–179), 111–121.

Roy, A., & Rambo-Hernandez, K. E. (2021). There's so much to do and not enough time to do it! A case for sentiment analysis to derive meaning from open text using student reflections of engineering activities. *American Journal of Evaluation, 42*(4), 559–576. https://doi.org/10.1177/1098214020962576

Schwartz, M., York, P., Nowakowski-Sims, E., & Ramos-Hernandez, A. (2017). Predictive and prescriptive analytics, machine learning and child welfare risk assessment: The Broward County experience. *Children and Youth Services Review, 81*, 309–320. https://doi.org/10.1016/j.childyouth.2017.08.020

Wilson, J., & Daugherty, P. R. (2018). Collaborative intelligence: Humans and AI are joining forces. *Harvard Business Review*, July-August, 2018. Retrieved from https://hbr.org/2018/07/collaborative-intelligence-humans-and-ai-are-joining-forces

York, P., & Bamberger, M. (2025). The Applications of Big Data to Strengthen Evaluation. In S. B. Nielsen, F. Mazzeo Rinaldi and G. J. Petersson (eds.). *Artificial Intelligence and Evaluation. Emerging Technologies and Their Implications for Evaluation* (pp. 37–55). London: Routledge. https://doi.org/10.4324/9781003512493

York, P., & Bamberger, M. (2020). *Measuring Results and Impact in the Age of Big Data: The Nexus of Evaluation, Analytics, and Digital Technology*. New York, NY: The Rockefeller Foundation. Retrieved from https://www.rockefellerfoundation.org/report/measuring-results-and-impact-in-the-age-of-big-data-the-nexus-of-evaluation-analytics-and-digital-technology/

14 Artificial Intelligence

Challenges for Evaluators

Francesco Mazzeo Rinaldi and Steffen Bohni Nielsen

Introduction

The key premise of this book is that evaluators must pay vivid attention to AI. The real question is not *if* AI is relevant for evaluators, nor *when*. It's already happening: AI has entered offices, factories, and businesses and works among us. Sometimes alongside humans, increasing productivity. Sometimes – a growing concern – in their place.

Many observers argue that we are in the midst of the fourth industrial revolution. A key driver in this revolution is the rapid evolution of different forms of AI. Whenever a new technology emerges with the potential to disrupt an industry, concerns inevitably arise about its potential to replace human workers (Felten et al., 2023; Vedantam, 2023).

A recent study by Goldman Sachs suggested that up to 300 million full-time jobs worldwide, including approximately two-thirds of employment in the United States and Europe, are susceptible to some level of replacement by generative AI alone. The economists at the investment bank emphasized in their report that if generative AI fulfills its potential capabilities, it could significantly disrupt the labor market (Hatzius et al., 2023). These estimations are echoed by other reports (e.g., Eloundou et al., 2023; McKinsey & Co, 2023). We are facing a future with significant disruption to many industries, displacement of entire jobs and tasks, and augmentation of others. Historically, technological progress that initially displaces human workers has also led to long-term job creation and economic growth. AI has the potential to serve as a job creator. However, it will require workers to adapt and oversee the technology under various industry-wide regulatory frameworks.

Evaluation as an industry is no exception to these changes (Nielsen 2023; 2024, Chapter 13). As observed by Nielsen: "When *what* is being evaluated is bound to change, *how* it will be evaluated, and by *whom* will likely change too" (Nielsen, 2023, 55). Our operable approach to address these pertinent issues was to pose three overarching research questions to the contributors of this volume.

DOI: 10.4324/9781003512493-14

1. What are the emerging digital technologies?
2. What requisite skills do evaluators need?
3. What contribution can evaluation make to AI and vice versa?

The purpose of this chapter is twofold. First, we want to extract some of the key findings concerning these questions from the volume and place these in a larger context of the evaluation field and AI. Second, based on the main findings, we discuss what we consider critical challenges that the global evaluation community must address in the era of AI.

Key Findings

In Table 14.1, we provide an overview of how each chapter addressed the main research questions of the volume. Given the specific scope of each chapter, they focused directly on at least one of the main research questions. For readers who want to explore these questions in-depth, we refer to the separate chapters of the volume.

Therefore, let us move on to the key findings pertaining to each question.

Research Question 1. What Are the Emerging Digital Technologies?

Several observers have noted that evaluation has been slow in adopting emerging digital technologies (see Nielsen et al., 2025, Chapter 1; Nielsen, 2025, Chapter 13) and thus harvesting the potential of the exponential growth of data generation (Nielsen et al., 2017; Nielsen, 2023; York & Bamberger, 2020). However, there are indications that this is changing (Nielsen et al., 2025, Chapter 1).

In this volume, the emerging digital technologies, mostly powered by AI, are described. Bruce and her colleagues (2025, Chapter 2) and York and Bamberger (2025, Chapter 3) provided a *broad* overview of different emerging technologies (ETs) and their potential applications (and caveats) in evaluation. We refer to these chapters for an overview. These technologies cover new ways of capturing, storing, and processing data. Other chapters provide an *in-depth* analysis of applying a particular technique in evaluation.

Anand and colleagues (2025, Chapter 4) provided an analysis of the application of geospatial analysis using several ETs, such as imagery from drones, satellites, and sensors. York provided insights into how particular quantitative analytical techniques, such as precision analytics and structural causal modeling, enabled evaluation use and program improvement (2025, Chapter 11).

However, the preponderance of chapters focused on different forms of text analytics. We posit that in recent years, the most significant advancements have been in digitally driven, sometimes commodified, tools to conduct text analytics. Such tools enable the analysis of much larger datasets, and sometimes, in greater detail than otherwise possible. Text analytics covers several different

Table 14.1 Overview of chapters and their relation to the main research questions

Chapter	*Title*	*Research question*		
		1. What are the emerging digital technologies?	*2. What requisite skills do evaluators need?*	*3. What contribution can evaluation make to AI and vice versa?*
2	Emerging Technology and Evaluation in International Development	Yes*	Yes	To some extent
3	The Applications of Big Data to Strengthen Evaluation	Yes*	No	Yes
4	Ethics & Equity in Data Science for Evaluators	No	Yes	To some extent
5	Extracting Meaning from Textual Data for Evaluation. Lessons from Recent Practice at the Independent Evaluation Group of the World Bank	Yes	To some extent	To some extent
6	Text Mining and Machine Learning in an Evaluation of Police Handling of Cybercrime in Norway	Yes	To some extent	To some extent
7	Big data for big investments: making responsible and effective use of data science and AI in research councils	Yes	To some extent	To some extent
8	The Use of Quantitative Text Analysis in Evaluations	Yes	To some extent	To some extent
9	Artificial Intelligence and Text Analysis in Evaluating Complex Social Phenomena. The Russia-Ukraine Conflict	Yes	To some extent	Yes

(*Continued*)

Table 14.1 Continued

Chapter	*Title*	*Research question*		
		1. What are the emerging digital technologies?	*2. What requisite skills do evaluators need?*	*3. What contribution can evaluation make to AI and vice versa?*
10	Harnessing Geospatial Approaches to Strengthen Evaluative Evidence	Yes	To some extent	Yes
11	The Future of Evaluation Analytics: Case Studies of Structural Causal Modeling in Action	To some extent	No	Yes
12	The Algorithmization of Policy and Society: The Need for a Realist Evaluation Approach	No	No	Yes
13	The Evaluation Industry and Emerging Technologies	No	Yes	To some extent

*Provides an overview of emerging digital technologies.

combinations of technologies applying machine learning (Holm et al., 2025, Chapter 7; Næss et al., 2025, Chapter 6; Ziulu et al., 2025, Chapter 5), natural language processing (Ziulu et al., 2025, Chapter 5), quantitative textual content analysis (Gatto & Bundi, 2025, Chapter 8), sentiment analysis, and emotion detection (Mazzeo Rinaldi et al., 2025, Chapter 9).

These chapters are essentially use cases documenting the opportunities (and limitations) offered by ETs to evaluation. Outside this book, recent publications highlight the potential of AI-driven technologies for coding and thematic analysis of qualitative data, including a literature review (Sabarre et al., 2023).

Since the fall of 2022, the enormous potential of Large Language Models (generative AI) solutions for the summarization of large bodies of quantitative and qualitative data has become apparent in evaluation and beyond (Mason & Montrosse-Moorhead, 2023; Mazzeo Rinaldi et al., 2024). The potential of this nascent technology for professionals is expected to be significant (Eloundou et al., 2023). However, recent experimental findings from the management consulting industry indicate that the application must be judicious and carefully measured (Dell'Acqua et al., 2023). Further analysis of the implications for evaluation is needed.

Therefore, let us move on to the second research question of this book; what skills do evaluators need to use ETs?

Research Question 2. What Requisite Skills Do Evaluators Need?

A recurring theme across the chapters is the pressing need for evaluators to adapt to the digital era. The requisite skills for evaluators encompass competencies in managing and analyzing large data sets and collaborating with data scientists. The digital era demands evaluators to bridge the gap between traditional evaluation expertise and innovative data science methodologies.

Most of the chapters emphasize the importance of evaluators acquiring skills related to data sharing, rapid feedback on data, and proficiency in visualization technologies. Machine-readable, API-accessible formats are becoming integral, underlining the need for evaluators to stay technologically fluent. Integrating traditional evaluation expertise with innovative data science methodologies becomes crucial in navigating the evolving technological landscape.

Evaluators who want to meet the demands of the digital revolution should acquire a range of competencies and skills that enable them to effectively assess and adapt to the changing landscape of AI technologies and their impact on various domains. The list is undoubtedly numerous. Below, we highlight the main competency requirements that emerge from the chapters in this volume.

First of all, *Digital Literacy*. Evaluators need to be well versed in AI technologies, including software, data analytics, and emerging technologies. They should understand the basics of data capture, storage, and analysis, as well as the capabilities and limitations of various digital tools and platforms (Nielsen, Chapter 13). Digital literacy is not a static skill but an evolving one. Digital literacy also evolves as technology advances, requiring ongoing learning and adaptation. It is an essential skill for individuals who work in any capacity with digital information and technology. Developing and maintaining digital literacy are critical for evaluators to stay competitive in the modern workforce (Mazzeo Rinaldi et al., 2025, Chapter 9). Evaluators should be able to quickly adjust their evaluation methods to keep up with technological advancements.

Second, *Data Analysis*. Data is at the core of the digital revolution. Evaluators should be skilled in collecting, analyzing, and interpreting data from new sources. This includes not only understanding statistical analysis, data visualization, and working with large datasets. It may involve concrete skills to use AI tools such as machine learning (ML) algorithms, natural language processing, and predictive analytics to make predictions or classify data into categories (Ziulu et al., 2025, Chapter 5; Næss et al., 2025, Chapter 6; Gatto and Bundi, Chapter 8); text and sentiment analysis to analyze text, extract insights, and determine sentiment and emotion (Mazzeo Rinaldi et al., 2025, Chapter 9); geospatial analysis to uncover spatial patterns and relationships (Anand et al., 2025, Chapter 10); BD analytics to work with technologies and distributed computing frameworks to process and analyze large datasets efficiently (York & Bamberger, 2025, Chapter 3; Holm et al., 2025, Chapter 7). The ability to extract meaningful insights from large datasets, such as social media posts and online content, is crucial today.

Third, *Programming and Coding Basics*. While not every evaluator needs to be a software developer, having a basic understanding of programming concepts is important. Automation and technology will become increasingly prevalent in evaluation processes. Coding skills can help evaluators automate specific tasks, analyze data more efficiently, and streamline evaluation processes, thus increasing productivity (Gatto & Bundi, 2025, Chapter 8). A basic understanding of programming languages (especially R and Python) allows evaluators to adapt to changing technologies and stay relevant in their field. Programming skills are beneficial for handling and processing data programmatically. This knowledge can help evaluators communicate effectively with data analysts and understand the capabilities and limitations of digital solutions (Mazzeo Rinaldi et al., 2025, Chapter 9).

Fourth, *Data Ethics and Privacy*. Ethical concerns, such as data privacy, security, and bias, have become paramount with the increased use of AI. Evaluators need to understand these ethical issues and ensure that their evaluations adhere to ethical standards. Evaluators must be aware of ethical considerations related to data analysis, including privacy, consent, and responsible data handling (Greenstein & Cho, 2025, Chapter 4).

Fifth, *Collaborative Skills*. It may help to bridge the gap between the technical aspects of data analysis and the evaluation goals. Such collaboration must encourage interdisciplinary learning (Næss et al., 2025, Chapter 6). By aligning objectives, fostering open communication, and jointly managing projects, evaluators and data scientists create a synergistic partnership that enhances the quality and relevance of evaluations, leading to actionable insights for program improvement (Leeuw, 2025, Chapter 12).

Research Question 3. What Contribution Can Evaluation Make to AI and Vice Versa?

This question covers two aspects: the contribution AI (and other ETs) can make to evaluation, and what contribution evaluation can make to AI. Let us consider these in turn.

This volume is full of examples of how evaluators make use of ETs to collect new forms of data, such as geospatial data (Anand et al., 2025, Chapter 10), social media data (Mazzeo Rinaldi et al., 2025, Chapter 9), or cover much larger datasets through the application of text mining and textual analysis (Gatto and Bundi, 2025, Chapter 8; Holm et al., 2025, Chapter 7; Næss et al., 2025, Chapter 6; Ziulu et al., 2025, Chapter 5), or apply AI/ML techniques to carry out more accurate and quotidian analyses to inform decision-making (York, 2025, Chapter 11). The examples are many and attest to the vast potential of AI for evaluators. For instance, York (2025, Chapter 11) emphasizes how integrating structural causal modeling with precision analytics represents an advancement in the field of program evaluation. This integration harnesses the power of big data, machine learning, and causal modeling to provide real-time, personalized, and nuanced insights. York highlights how this approach offers a more rigorous and data-driven understanding of social programs, contributing to the evolving landscape of data science applications. Precision care modeling involves identifying patterns and specific treatment features through machine learning algorithms. Additionally, the process facilitates the identification of outliers and exceptions, encouraging a deeper qualitative investigation.

Another example is provided in this volume by Næss and his colleagues. Here, the evaluation team applied a machine learning algorithm to classify the emergence of cybercrime in the *entire* dataset rather than in a *sample* using manual coding of police crime registries (Næss et al., 2025, Chapter 6). A similar point is made by Ziulu and colleagues (2025, Chapter 5). They discuss the increasing use of text as data in the evaluations conducted by the Independent Evaluation Group (IEG) of the World Bank Group (WBG), illustrating, through different applications, the substantial benefits of employing textual data in evaluations. Efficiency improves with automated and semi-automated techniques like portfolio identification and streamlining labor-intensive tasks. Quality enhancements arise from innovative methods like NLP-based data classification, offering higher accuracy rates than traditional coding. The use of textual data broadens evaluative inquiry, leveraging new forms of text analytics, such as social media data for network or sentiment analyses. Overall, these approaches enhance efficiency, accuracy, and the scope of evaluative work under favorable conditions, demonstrating the transformative potential of leveraging textual data in evaluations.

However, throughout the volume, there is also a vocal warning that application must be done conscientiously and judiciously with keen attention to ethics and equity issues (Bruce et al., 2025, Chapter 2; York & Bamberger, 2025, Chapter 3). Greenstein and Cho propose concrete steps for evaluators to address such equity and ethics concerns before applying said techniques. This ties into a broader debate on what evaluation criteria should be used for applying AI in evaluation.

Montrosse-Moorhead (2023) recently proposed a number of evaluation criteria that must be considered by evaluators when deciding on using AI in their evaluation practice.

1. *Design and implementation.* That is, is AI appropriate for the evaluation purpose?

2. *Efficiency of process*. That is, is AI technology more efficient than other alternatives in performing a task?
3. *Equity of process*. That is, does the AI technology attend to equity issues?
4. *Effectiveness*. That is, does the AI technology provide better solutions (analyses, narratives, etc.) than other alternatives?
5. *Trust*. That is, are the results from the AI technology considered trustworthy by stakeholders?
6. *Methodological validity and trustworthiness*. That is, are the results from the AI technology considered valid when subject to scrutiny?
7. *Understandability*. That is, do the AI technology results provide sufficient transparency so that they can be understood?
8. *Equity of resulting information and evidence*. That is, does AI-produced information and evidence attend to equity?

These evaluation criteria bring us closer to an actionable framework for deciding when, if, and under what conditions AI can contribute to specific evaluation work.

Throughout the book, we find a number of arguments that evaluation has much to offer AI. First, while we are facing a highly potent technology, many solutions are still immature, and we have not yet fully understood their positive and negative implications, intended and unintended. Therefore, evaluation must play a crucial role in assessing the consequences of interventions with (components of) AI. Leeuw (2025, Chapter 12) argues that evaluation must develop frameworks that can capture what is at play and that realist evaluation is particularly adept for this purpose. A similar notion is entertained by Mazzeo Rinaldi and his colleagues (2025, Chapter 9). York and Bamberger (2025, Chapter 3) argue that essential tools from the evaluation toolbox, such as construing program theories, research designs, triangulation, and establishing sound evaluation criteria, have much to offer the field of data science and its incumbent technologies. Some of the key contributions of evaluation to AI can be summarized as:

- *Contextual understanding*: Evaluation, particularly realist evaluation, is highlighted as essential for understanding the contextual conditions in which complex social phenomena, events, and interventions occur. This understanding is crucial for accurately interpreting the results obtained through data science tools.
- *Validation and interpretation*: Evaluation can contribute to validating the results obtained through AI applications and interpreting them using other sources. Evaluation can, for example, help validate the performance of AI algorithms, particularly in real-world scenarios where ground truth data is essential for training and testing these algorithms.
- *Human-centered analysis*: While AI tools can process and analyze large datasets, the importance of understanding the human context in interventions

remains critical. Evaluation, with its focus on human factors, can contribute to a more comprehensive and sensitive analysis of the impact of events and interventions on individuals and communities.

- *Identifying biases and limitations*: Evaluation can play a role in identifying biases, limitations, and ethical considerations in AI applications, ensuring a more robust and accurate use of technology.
- *Adaptability and long-term impacts*: Evaluation can consider the dynamic nature of political decisions and societal contexts over time, contrasting with the static nature of machine learning systems. Evaluation methodologies emphasizing adaptability and understanding long-term impacts can complement data science by providing insights into the evolving reality of political interventions.

In the introduction chapter, Nielsen and colleagues argue that AI must be considered against evaluation criteria such as expedience, efficiency, effectiveness, equity, and ethics (Nielsen et al., 2025, Chapter 1). To be sure, the evaluation toolkit must further evolve and adapt as AI technologies become part and parcel of what we evaluate – and offer its pertinence to AI.

So far, we have extracted some of the key messages from the current volume. We will now move on to place these in a broader context for evaluation.

Discussion. Future Perspectives for Evaluation?

In this section, we intend to discuss four key issues that we consider essential in shaping evaluation practice in the digital era. First, we discuss impending legislative frameworks designed to framework AI's application. Herein, we address *why* evaluation should be pivotal in such legislative frameworks. Then, we move on to discuss *what* evaluator competencies are needed in the future. Subsequently, we discuss *how* this should be done by composing future evaluation teams and professional identities and *which* pertinent considerations ought to shape future approaches when evaluating AI's consequences.

Legislative Frameworks and the Role of Evaluation

Some technologies – such as AI – highlight epistemic, normative, ethical, and political issues. The dynamics of power and surveillance increasingly leverage digitization to integrate "expert knowledge" into decision-making processes automatically. A crucial aspect comes into play: how to manage this transformation and regulate AI. Additionally, how could evaluation contribute to overseeing AI regulation?

The agreement reached on December 8, 2023, between European Union (EU) member states and the European Parliament regarding the AI Act marks the beginning of a new era in government approaches to managing AI, moving beyond mere codes of conduct and voluntary guidelines (European Parliament, 2023). The legislation prohibits certain AI applications that violate human rights and civil liberties, such as the use of subliminal techniques to manipulate behavior – and subjecting "high-risk" applications, from critical infrastructure to education, healthcare, and justice, to a certification and labeling process that ensures transparency, human oversight, security, and sustainability, and prevents the discrimination algorithms "learn" from studying the real world.

The goals outlined in the AI Act align with those articulated in the recent executive order on AI and the Biden administration's Blueprint for an AI Bill of Rights (White House, 2022). However, the European Commission has translated these aspirations into tangible regulations backed by substantial potential fines.

The text introduces two key innovations in the EU's approach to regulating AI. Firstly, it expands the list of prohibited technologies, banning using a "social credit system" akin to China's, emotional reading algorithms in sensitive areas, and predictive policing algorithms measuring criminal propensity. Real-time identification using biometric data is restricted to severe crimes with judicial authorization. The second innovation responds to advancements like ChatGPT, requiring registration (not certification) for certain AI systems, with additional conditions like human oversight. AI-generated content must transparently indicate its nature to counter disinformation, and companies using general AI systems receive protection. While this crackdown on intrusive technologies is comprehensive, clashes with the Council are anticipated due to varying government stances on providing such tools to law enforcement. The measures aim to balance technological advancements with safeguards, addressing concerns of copyright infringement lawsuits against AI developers. Therefore, the AI Act not only represents a significant endeavor to regulate the rapidly evolving field of AI but also serves as a guiding framework for democracies dealing with AI's multifaceted challenges and opportunities.

However, there are critical voices on the AI Act. Critics argue that the EU AI Act's emphasis on high-risk applications may lead to an under-regulation of lower-risk AI systems. Some contend that risks associated with specific lower-risk AI applications may still warrant regulatory attention to ensure consumer protection and ethical use. Also, the compliance-based approach, particularly for high-risk AI systems, could impose significant burdens on businesses. Critics expressed concerns that the compliance requirements may be complex and resource-intensive, affecting innovation and competitiveness, especially for smaller enterprises.

In a recent paper, for instance, Laux and colleagues critique the EU AI Act's risk-based framework, arguing that it may be inadequate for fostering trust in AI, especially in government use (Laux et al., 2024). The Act categorizes AI risks into tiers, seeking to increase trust by making certain risks more acceptable. However, the authors contend that additional factors, such as government efficiency and transparency, are crucial for building public trust in AI technology, mainly when used by the public sector. They call for greater clarity from government actors regarding the use of AI to promote trust, transparency, and legitimacy, warning against the oversimplification of trust-building through risk assessment and emphasizing the multidimensional nature of trust that varies across communities, sectors, and settings. The authors suggest a more participatory approach to AI risk assessment involving laypeople on boards responsible for assessment tasks. Establishing trust in the use of AI within the public sector requires the European Parliament to craft a framework that goes beyond the basic stipulations of the EU's AI Act. They advocate for a more nuanced and participatory approach to address the intricate nature of trust in AI systems (Laux et al., 2024).

In this regard, transparency and explainability are essential principles in managing and regulating AI systems. These principles are closely related and focus on making AI algorithms more understandable and accountable. Transparency refers to the openness and clarity of AI systems and their operations. It makes the AI system's inner workings, data, algorithms, and decision-making processes accessible and understandable to relevant stakeholders. It is crucial because it helps users, developers, regulators, and affected individuals understand how an AI system makes decisions and predictions. It promotes trust, accountability, and the ability to identify and address issues like bias or discrimination. Explainability goes beyond transparency by explicitly focusing on the ability to provide understandable and coherent explanations for the AI system's actions and outputs. It aims to answer *why* a particular decision or prediction was made. It supports users and stakeholders in grasping the rationale behind AI decisions, which is especially important when those decisions have significant consequences, such as in healthcare, finance, and criminal justice. It supports user trust, accountability, and the ability to challenge or correct unjust or incorrect decisions.

A challenge arises from the fact that within these regulations, the concept of "man in the loop" is often, though at times implicitly, suggested as a potential solution. Many regulations advocate human oversight of algorithmic decisions, assuming people can effectively monitor and explain. However, a survey of 41 policies reveals two significant flaws (Green, 2022): people struggle with oversight and policies legitimize flawed algorithms without addressing core issues. Green argues for a shift to institutional oversight, where agencies must justify algorithm use with empirical evidence and undergo democratic review before adoption, aiming to enhance accountability and mitigate algorithmic harms in government decision-making. There are also some specific strategies for fostering transparency and explainability in AI, such as using model types that are inherently more interpretable (e.g., decision trees, linear models, rule-based systems) or employing specific techniques like Local Interpretable Model-Agnostic Explanations (LIME) and SHapley Additive exPlanations (SHAP), which provide tools for interpreting and explaining model predictions, even for complex models. These strategies might facilitate the accessibility and comprehension of AI system operations and decision-making processes, fostering trust, accountability, and the ability to challenge or correct AI decisions (Nohara et al., 2022; Wanner et al., 2022). Balancing model performance with interpretability remains a challenge, but transparent and explainable AI is vital for the development and diffusion of AI itself.

Evaluation could be crucial in fostering transparent and explainable AI by contributing expertise, developing frameworks, and promoting best practices. For instance, developing standardized metrics for evaluating the explainability of AI models by creating benchmarks that assess the clarity and comprehensibility of model outputs makes it easier to compare different approaches. Standardization can provide a common framework for evaluation and implementation, making it easier for practitioners to adopt transparent AI practices. Establishing benchmarks for evaluating the transparency of AI models could also be of great benefit. This would include assessing how well models disclose their decision-making processes and whether they provide understandable insights into their internal workings. Integrating user feedback to understand how individuals perceive and interact with transparent AI systems might also help assess areas for improvement and ensure that transparency measures align with user expectations.

The EU AI Act would benefit from a comprehensive evaluation framework. Clear and measurable metrics aligned with social betterment goals should be defined to operationalize the legislation's values. The evaluation community might play a pivotal role in overseeing AI regulation by developing assessment methodologies, benchmarks, and frameworks to gauge compliance with regulatory standards. Evaluators might monitor ethical and societal impacts, provide feedback to regulatory bodies, and conduct independent audits to ensure transparency and accountability. By identifying regulatory gaps and challenges, evaluators could contribute to regulatory impact assessments by providing data and analyses related to the effects of AI regulations on industry, innovation, and societal well-being. This includes recognizing areas where regulations may be insufficient, outdated, or unclear, updating evaluation criteria to address new challenges and opportunities, and providing recommendations for improvement. Evaluators can also be valuable in ensuring the reliability and accuracy of AI algorithms by contributing to the ground-truthing process, which involves on-site data collection to confirm or enhance the accuracy of information obtained from remote sources or algorithms. Their role could be beneficial in scenarios where ground truth data is essential for training and testing algorithms, a crucial concern often overlooked in the knowledge validation process using AI.

This proactive contribution might ensure that regulations remain relevant in the face of rapidly evolving AI technologies, supporting policymakers in making informed decisions about the necessity and effectiveness of regulations (see Nielsen, 2025, Chapter 13). Through these efforts, evaluators can help maintain the relevance and effectiveness of AI regulations, fostering a balance between innovation, ethical considerations, and societal well-being.

Managing the rapid development of AI techniques alongside the slower pace of evaluations and regulatory adjustments is challenging. One strategy to tackle this issue might be establishing regulatory sandboxes or pilot programs. These initiatives would enable the agile assessment of new AI applications within controlled environments, allowing regulators to gather insights on potential risks and benefits before implementing widespread regulations.

Despite the critical role evaluation could play in this context, the involvement of evaluators in AI regulation is practically nil. This may be due to several reasons. Regulatory bodies and policymakers may be unaware of the expertise and contributions the evaluation community can provide in AI. Additionally, the complex interdisciplinary nature of AI, resistance to changing established regulatory processes, and the intricate workings of AI systems present obstacles.

Efforts to bridge these gaps could involve raising awareness among regulatory bodies about the value of evaluation, fostering interdisciplinary collaboration, updating regulatory frameworks, allocating resources for external evaluation, and developing standardized evaluation criteria for AI systems. Encouraging dialogue between evaluators, policymakers, and industry stakeholders is crucial for overcoming these barriers and promoting effective AI regulation. But it is equally critical that the evaluation community champion this change and take decisive action, making its voice heard in the relevant institutional contexts.

The Future of Evaluator Competencies?

Nielsen concluded that AI-powered technologies will significantly affect evaluation. By applying a task-level analysis, he argued that AI will displace some and augment other human tasks (2025, Chapter 13). Elsewhere, Mason also took stock of how AI may affect evaluators' work, using the American Evaluation Association's (AEA) competencies framework (King & Stevahn, 2020). The AEA competency framework lists five different domains: professional practice, methodology, context, planning and management, and interpersonal, and 50 specific competencies that evaluators should possess. Mason's conclusion points toward competencies that are highly social, highly creative, and strategic are least at risk of replacement in the short term. Her conclusion is corroborated by broader reviews of what types of tasks are at risk of digital automation (cf. Eloundou et al., 2023).

Above, we listed a group of new competencies that we consider important if evaluators are to adapt to the fast-evolving world of AI. These are: digital literacy, data analysis, programming and coding basics, data ethics and privacy, and collaborative skills. As we see it, these skills must be integrated into existing evaluator competency frameworks.

To this end, Nielsen has called for actions to be taken by individual evaluators, evaluation providers, educational institutions (offering evaluation programs), and particularly Voluntary Organizations for Professional Evaluations (*VOPEs*) (2023; 2025, Chapter 13). VOPEs play a crucial strategic role in redefining evaluator competencies as we are facing a disruption of traditional evaluation practice. Furthermore, they must put into place a strategic upskilling program to ascertain that evaluators acquire the requisite competencies.

The Future of Evaluation Team Composition

Evaluation is rarely an individual endeavor. It is mostly done in teams of two or more people (Hwalek & Straub, 2018; Peck, 2018). Evaluation teams are traditionally comprised of a mix of evaluation methodology experts and subject matter experts (Nielsen, Lemire & Christie, 2018; Lemire, Nielsen, & Christie, 2018).

As AI will become part of both *what* and *how* we evaluate, we expect that evaluation team composition will change and data scientists will form integral members of evaluation teams in the future (Mazzeo Rinaldi et al., 2017). In the current volume, there are a number of examples of how such collaboration has already taken place (i.e., Næss et al., 2025, Chapter 6; Ziuli et al., 2025, Chapter 5). We are facing a future where many cognitive tasks previously managed by well-paid, highly educated staff will be handled by AI. This is a key finding, not just in evaluation but in all professional services.

This can be exemplified by a recent experimental study of how management consultants solved real-world tasks using Large Language Models (LLM) (in the case of ChatGPT 4.0). The study involved management consultants from Boston Consulting Group (BCG). BCG is known as a global top-tier strategy consulting firm. They recruit among the best and brightest from the cohorts of recent graduates from elite universities worldwide. Management consulting shares many similarities with evaluation. In fact, many providers and individuals who do evaluation work identify first and foremost as management consultants, not as evaluators. Management consulting is a professional service wherein you have to deliver advice (written and oral) based on analysis, expertise, contextual awareness, and relational skills in managing stakeholders. This is akin to evaluation practice. The study highlighted the significant benefits of using LLMs in terms of expedience, efficiency, and effectiveness (inside the technological frontier tasks). However, when tasks (deliberately) were more messy (with contradictory evidence), performance dropped as a higher proportion using LLM provided inaccurate answers compared to the control group (not using LLM) (outside the technological frontier tasks). The authors concluded that there are significant benefits of using LLM for most tasks but not for those outside the technological frontier. One must bear in mind that LLM is still an immature technology in its infancy. In such cases, the authors conclude: "On those tasks, this study highlights the importance of validating and interrogating AI and of continuing to exert cognitive effort and experts' judgment when working with AI" (Dell' Aqua et al., 2023:15).

The example highlights that we are facing a future where AI applications (such as LLMs) become part of the team. Nielsen (2025, Chapter 13) points out that we will see a future of collaborative intelligence, where professionals (and non-professionals) work alongside AI (Wilson and Daugherty 2018). This point has also been made by Leeuw, who foresees a future wherein AI and humans will need to collaborate in evaluation (2020; 2025, Chapter 12).

Artificial Intelligence's Influence on Evaluators' Professional Identity

As we described, AI technologies are upending how individuals work and relate to their work – their *professional identities.* Like many other professionals, evaluators may face challenges and changes in their roles with the advent of AI.

Different perspectives exist on the composition and evolution of an individual's professional identity. Self-determination theory (Deci & Ryan, 1985; Ryan & Deci, 2000) highlights three basic aspects: *competence* (valuing one's role and experience), *autonomy* (decision-making discretion), and *belonging* (meaningful connections). These elements must be addressed to manage AI's influence on evaluators' professional identity.

The need for *competence* is understood as the desire to interact effectively with one's environment. The extraordinary capabilities demonstrated by AI can significantly undermine the evaluator's perception of their own abilities to tackle specific tasks. Thus, as AI takes on tasks once done by humans, it may undermine evaluators' perception of their competence. In this evolving landscape, the question arises: what defines meaningful work for professionals such as evaluators? Organizations must pinpoint tasks that evaluators excel in to instill a fresh sense of purpose, setting them apart from AI's strengths. Leveraging AI for these unique tasks liberates people to engage in roles where they can surpass AI, strengthening their professional competence and adding value. AI can augment evaluators' work by automating routine tasks, data analysis, and report generation. This can free evaluators to focus on higher-level tasks such as interpreting results, designing evaluations, and making recommendations. In this scenario, the professional identity of evaluators may evolve, but they can still play a valuable role. Moreover, some evaluators may specialize in AI-related areas such as algorithm auditing or the development of AI-based evaluation tools. This specialization can help them maintain their professional identity and stay at the forefront of their field.

On the other hand, the need for *autonomy* refers to a sense of responsibility and active engagement in the task. If this does not happen, the worker may feel compelled to act under external forces, resulting in a decreased sense of responsibility and interest in the activity. AI possesses a unique capacity to make decisions or offer guidance on decision-making, which can be perceived as encroaching on professionals' autonomy (e.g., predictive and prescriptive AI algorithms). The increasing decision-making capabilities of generative AI intensify this concern. Organizations need to ensure that AI implementations include the option for human override. This reinstates a sense of control and establishes a positive feedback loop, increasing the likelihood of professionals, like evaluators, embracing AI. For instance, this approach can help mitigate the common human bias known as *algorithmic aversion*, where individuals distrust and reject AI. Allowing people to customize AI outputs can help overcome this aversion.

Finally, *belonging* to a professional group is one of the factors that, most of all, can influence an individual's motivation, as the integration aspects of belonging are among the most important for the individual's adaptation. AI can lead to changes in the professional community a person interacts with. For example, evaluators may need to collaborate more closely with data scientists or other specialists in AI and technology-related fields. This shift can alter their professional network and require them to adapt to new communities or collaborate with individuals from different backgrounds.

These changes can impact a person's sense of belonging to their original professional community as they navigate new relationships and collaborations. The field of evaluation, like many others, is constantly evolving. Evaluators have traditionally adapted to new methods, tools, and work contexts – and AI is *just* one more evolution. Continuous learning and adaptation are crucial to preserving a professional identity.

An Agenda for Research on the Evaluation of AI

Margaryan recently argued that to better understand the transformation of skills due to the spread of AI technologies, future research in this area needs to be guided by an integrated approach. That is, combining the perspectives of different actors and scientific disciplines using multimethod research designs. He identified four dimensions of integration central to prospective research in this area (Margaryan, 2023, p. 3). These may greatly interest the evaluation community when designing future evaluation teams and designs for assessing AI interventions and their consequences.

1. *Integrative exploration of work practices and skill requirements across the AI production chain* – Future research should investigate the diverse skill requirements of frontline actors in the AI production chain, which encompasses three stages: data production, AI design and development, and AI end-use.
2. *Integrating insights from social sciences, humanities, and computer science* – Research on AI and skills has traditionally occurred separately in computer science/engineering, social sciences, and humanities. Future research

should adopt a multidisciplinary approach to integrate findings, insights, and theoretical perspectives from diverse fields.

3. *Integrating quantitative and qualitative methodologies* – Skill analysis methods encompass both direct measures, such as expert ratings and self-report data, and indirect measures, like extrapolating skill data from wage and educational data or utilizing ML to identify AI skill gaps. The addition of ethnographic and interview-based research designs may effectively provide a holistic, contextualized understanding of AI's impact on skill requirements in AI-mediated workplaces.
4. *Integrating perspectives of different stakeholders* – Future research on AI and skills should adopt an integrative approach by analyzing the views of diverse stakeholders using participatory co-design methods.

Margaryan's call resonates strongly with the multi- and transdisciplinarity of evaluation.

Conclusion

In the era of AI, evaluators face complex challenges arising from the interdisciplinary nature, rapid evolution, and ethical considerations associated with AI systems. Collaboration across diverse fields is essential to understand AI's impact comprehensively. Keeping pace with technological advancements is a constant challenge, requiring evaluators to adapt to emerging innovations. Transparency and explainability are crucial but difficult to achieve, necessitating methods to address accountability and user trust concerns. Ethical challenges, including biases and societal implications, require evaluators to navigate complex dilemmas and contribute to ethical guidelines. Evolving regulations demand constant vigilance for compliance, highlighting the need for global collaboration. A collective effort from the evaluation community is crucial to address these challenges, foreseeing an evolution toward collaborative intelligence through integrated teams of data scientists and AI solutions. Future research should explore interdisciplinary collaboration, diverse stakeholder perspectives, and responsible AI development to benefit society in the digital era.

References

Anand, A., Batra, G., & Uitto, J. I. (2025). Harnessing Geospatial Approaches to Strengthen Evaluative Evidence. In S. B. Nielsen, F. Mazzeo Rinaldi and G. J. Petersson (eds.). *Artificial Intelligence and Evaluation. Emerging Technologies and Their Implications for Evaluation* (pp. 196–218). London: Routledge. https://doi.org/10.4324/9781003512493

Bruce, K., Gandhi, V., & Vandelanotte, J. (2025). Emerging Technology and Evaluation in International Development. In S. B. Nielsen, F. Mazzeo Rinaldi and G. J. Petersson (eds.). *Artificial Intelligence and Evaluation. Emerging Technologies and Their*

Implications for Evaluation (pp. 13–36). London: Routledge. https://doi.org/10.4324/9781003512493

Deci, E. L., & Ryan, R. M. (1985). *Intrinsic Motivation and Self-Determination in Human Behavior*. New York, NY: Plenum.

Dell'Acqua, F., McFowland, E., Mollick, E.R., Lifshitz-Assaf, H., Kellogg, K., Rajendran, S., Krayer, L., Candelon, F., & Lakhani, K. R. (2023). Navigating the Jagged Technological Frontier: Field Experimental Evidence of the Effects of AI on Knowledge Worker Productivity and Quality. *Harvard Business School Technology & Operations Mgt. Unit Working Paper.* 24-013.http://dx.doi.org/10.2139/ssrn.4573321

Eloundou, T., Manning, S. Mishkin, P., & Rock, D. (2023). GPTs are GPTs: An Early Look at the Labor Market Impact Potential of Large Language Models. *arXiv:2303.10130* [econ.GN*]*. https://doi.org/10.48550/arXiv.2303.10130

European Parliament. (2023). *Artificial Intelligence Act 2021/0106(COD)*. Retrieved from https://oeil.secure.europarl.europa.eu/oeil/popups/ficheprocedure.do?reference=2021/0106(COD)&l=en

Felten, E., Raj, M., & Seamans, R. (2023). *How Will Language Modelers Like Chatgpt Affect Occupations and Industries?* arXiv preprint arXiv:2303.01157

Gatto, L., & Bundi, P. (2025). The Use of Quantitative Text Analysis in Evaluations. In S. B. Nielsen, F. Mazzeo Rinaldi and G. J. Petersson (eds.). *Artificial Intelligence and Evaluation. Emerging Technologies and Their Implications for Evaluation* (pp. 144–167). London: Routledge. https://doi.org/10.4324/9781003512493

Green, B. (2022). The Flaws of Policies Requiring Human Oversight of Government Algorithms. *Computer Law & Security Review*, 45. https://doi.org/10.1016/j.clsr.2022.105681

Greenstein, N., & Cho, S.-W. (2025). Ethics & Equity in Data Science for Evaluators. In S. B. Nielsen, F. Mazzeo Rinaldi and G. J. Petersson (eds.). *Artificial Intelligence and Evaluation. Emerging Technologies and Their Implications for Evaluation* (pp. 56–77). London: Routledge. https://doi.org/10.4324/9781003512493

Hatzius J., Briggs J., Kodnani D., & Pierdomenico, G. (2023). *The Potentially Large Effects of Artificial Intelligence on Economic Growth.* Goldman Sachs, Global Economics Analyst. Retrieved from https://www.gspublishing.com/content/research/en/reports/2023/03/27/d64e052b-0f6e-45d7-967b-d7be35fabd16.html

Holm, J., Newman-Griffis, D., & Petersson, G. J. (2025). Big Data for Big Investments: Making Responsible and Effective Use of Data Science and AI in Research Councils. In S. B. Nielsen, F. Mazzeo Rinaldi and G. J. Petersson (eds.). *Artificial Intelligence and Evaluation. Emerging Technologies and Their Implications for Evaluation* (pp. 120–143). London: Routledge. https://doi.org/10.4324/9781003512493

Hwalek, M. A., & Straub, V. L. (2018). The Small Sellers of Program Evaluation Services in the United States. *New Directions for Evaluation, 160*, 125–143. https://doi.org/10.1002/ev.20340

King, J. A., & Stevahn, L. (2020). Presenting the 2018 AEA Evaluator Competencies. *New Directions for Evaluation, 168*, 49–61. https://doi.org/10.1002/ev.20435

Laux, J., Wachter, S., & Mittelstadt, B. (2024). Trustworthy Artificial Intelligence and the European Union AI Act: On the Conflation of Trustworthiness and the Acceptability of Risk. *Regulation & Governance, 18*(1), 13–32. https://doi.org/10.1111/rego.12512

Leeuw, F. L. (2020). Program Evaluation B: Evaluation, Big Data, and Artificial Intelligence: Two Sides of One Coin. In E. Vigoda-Gadot and D. R. Vashdi (eds.).

Handbook of Research Methods in Public Administration, Management and Policy (pp. 277–297). https://doi.org/10.4337/9781789903485.00024

Leeuw, F. L. (2025). The Algorithmization of Policy and Society: The Need for a Realist Evaluation Approach. In S. B. Nielsen, F. Mazzeo Rinaldi and G. J. Petersson (eds.). *Artificial Intelligence and Evaluation. Emerging Technologies and Their Implications for Evaluation* (pp. 242–265). London: Routledge. https://doi.org/10.4324/9781003512493

Lemire, S., Nielsen, S. B., & Christie, C. A. (2018). Toward Understanding the Evaluation Market and Its Industry — Advancing a Research Agenda. *New Directions for Evaluation, 160*, 145–163. https://doi.org/10.1002/ev.20339

Margaryan, A. (2023). Artificial Intelligence and Skills in the Workplace: An Integrative Research Agenda. *Big Data & Society, 10*(2). https://doi.org/10.1177/20539517231206804

Mason, S., & Montrosse-Moorhead, B. (2023). Editors' Notes. *New Directions for Evaluation, 178–179*, 7–10. https://doi.org/10.1002/ev.20563

Mazzeo Rinaldi, F., Celardi, E., Miracula, V., & Picone, A. (2025). Artificial Intelligence and Text Analysis in Evaluating Complex Social Phenomena. The Russia-Ukraine Conflict. In S. B. Nielsen, F. Mazzeo Rinaldi and G. J. Petersson (eds.). *Artificial Intelligence and Evaluation. Emerging Technologies and Their Implications for Evaluation* (pp. 168–195). London: Routledge. https://doi.org/10.4324/9781003512493

Mazzeo Rinaldi, F., Giuffrida, G., Nicotra, S., & Dispinseri F. (2024). A Classification Algorithm to Link Official Documents to Sustainable Development Goals. In I. K. Lindkvist., K. Forss & P. Øyvind Bastøe (eds.). *Towards Sustainable Futures - A role for evaluations?* (pp. 166–185). London: Routledge.

Mazzeo Rinaldi, F., Giuffrida, G., & Negrete, T. (2017). Real-time monitoring and evaluation - Emerging news as predictive process using Big Data based approach. In G. J. Petersson & J. D. Breul (eds.). *Cyber Society, Big Data and Evaluation* (pp. 191–214). New York, NY: Routledge.

McKinsey & Co. (2023). *The Economic Potential of Generative AI: The Next Productivity Frontier*. McKinsey. Retrieved from https://www.mckinsey.com/capabilities/mckinsey-digital/our-insights/the-economic-potential-of-generative-ai-the-next-productivity-frontier#introduction

Montrosse-Moorhead, B. (2023). Evaluation Criteria for Artificial Intelligence. *New Directions for Evaluation, 2023*, 123–134. https://doi.org/10.1002/ev.20566

Nielsen, S. B. (2025). The Evaluation Industry and Emerging Technologies. In S. B. Nielsen, F. Mazzeo Rinaldi and G. J. Petersson (eds.). *Artificial Intelligence and Evaluation. Emerging Technologies and Their Implications for Evaluation* (pp. 266 286). London: Routledge. https://doi.org/10.4324/9781003512493

Nielsen, S. B. (2023). Disrupting Evaluation? Emerging Technologies and Their Implications for the Evaluation Industry. *New Directions for Evaluation, 178–179*, 47–57. https://doi.org/10.1002/ev.20558

Nielsen, S. B., Mazzeo Rinaldi, F., & Petersson, G. J. (2025). Evaluation in the Era of Artificial Intelligence. In S. B. Nielsen, F. Mazzeo Rinaldi and G. J. Petersson (eds.). *Artificial Intelligence and Evaluation. Emerging Technologies and Their Implications for Evaluation* (pp. 1–12). London: Routledge. https://doi.org/10.4324/9781003512493

Nielsen, S. B., Lemire, S., & Christie, C. A. (2018). The Commercial Side of Evaluation: Evaluation as an Industry and as a Professional Service. In J. E. Furubo and N. Stame (eds.). *The Evaluation Enterprise: Evaluating Evaluation* (pp. 243–265). New York: Routledge.

Nielsen S. B., Ejler, N., & Schretzmann, M. (2017). Exploring Big (Data) opportunities: The Case of the Center for Innovation through Data Intelligence, New York City. In G. J. Petersson and J. D. Breul (eds). *Cyber Society, Big Data and Evaluation* (pp. 147–170). New York, NY: Routledge.

Næss, T., Prabhu, C., Mjaaland, M., Holtermann, H., & Engebretsen, L. S. (2025). Text Mining and Machine Learning in an Evaluation of Police Handling of Cybercrime in Norway. In S. B. Nielsen, F. Mazzeo Rinaldi and G. J. Petersson (eds.). *Artificial Intelligence and Evaluation. Emerging Technologies and Their Implications for Evaluation* (pp. 103–119). London: Routledge. https://doi.org/10.4324/9781003512493

Nohara, Y., Matsumoto, K., Soejima, H., & Nakashima, N. (2022). Explanation of Machine Learning Models Using Shapley Additive Explanation and Application for Real Data in Hospital. *Computer Methods and Programs in Biomedicine*, *214*, 106584. https://doi.org/10.1016/j.cmpb.2021.106584

Peck, L. R. (2018). The Big Evaluation Enterprises in the United States. *New Directions for Evaluation*, *160*, 97–124. https://doi.org/10.1002/ev.20341

Ryan, R. M., & Deci, E. L. (2000). Self-determination Theory and the Facilitation of Intrinsic Motivation, Social Development, and Well-being. *American Psychologist*, *55*(1), 68–78. https://doi.org/10.1037/0003-066X.55.1.68

Sabarre, N. R., Beckmann, B., Bhaskara, S., & Doll, K. (2023). Using AI to Disrupt Business as Usual in Small Evaluation Firms. *New Directions for Evaluation*, *178–179*, 59–71. https://doi.org/10.1002/ev.20562

Vedantam, K. (2023). *Is AI The Cause Of Job Cuts This Year?* Crunchbas. https://news.crunchbase.com/ai-robotics/artificial-intelligence-layoffs-job-market/ (Accessed July 2023)

Wanner, J., Herm, L. V., Heinrich, K., & Janiesch, C. (2022). The Effect of Transparency and Trust on Intelligent System Acceptance: Evidence from a User-based Study. *Electron Markets 32*, 2079–2102. https://doi.org/10.1007/s12525-022-00593-5

White House. (2022). *Blueprint for an AI Bill of Rights: Making Automated Systems Work for the American People*. Retrieved from https://www.whitehouse.gov/ostp/ai-bill-of-rights/

Wilson, J., & Daugherty, P. R. (2018). Collaborative intelligence: Humans and AI are joining forces. *Harvard Business Review*, July-August, 2018. Retrieved from https://hbr.org/2018/07/collaborative-intelligence-humans-and-ai-are-joining-forces

York, P. (2025). The Future of Evaluation Analytics: Case Studies of Structural Causal Modeling in Action. In S. B. Nielsen, F. Mazzeo Rinaldi and G. J. Petersson (eds.). *Artificial Intelligence and Evaluation. Emerging Technologies and Their Implications for Evaluation* (pp. 219–241). London: Routledge. https://doi.org/10.4324/9781003512493

York, P., & Bamberger, M. (2025). The Applications of Big Data to Strengthen Evaluation. In S. B. Nielsen, F. Mazzeo Rinaldi and G. J. Petersson (eds.). *Artificial Intelligence and Evaluation. Emerging Technologies and Their Implications for Evaluation* (pp. 37–55). London: Routledge. https://doi.org/10.4324/9781003512493

York, P., & Bamberger, M. (2020). *Measuring Results and Impact in the Age of Big Data: The Nexus of Evaluation, Analytics, and Digital Technology*. New York, NY: The Rockefeller Foundation. Retrieved from https://www.rockefellerfoundation.org/report/measuring-results-and-impact-in-the-age-of-big-data-the-nexus-of-evaluation-analytics-and-digital-technology/

Ziulu, V., Anuj, H., Hagh, A., Raimondo, E., & Vaessen, J. (2025). Extracting Meaning from Textual Data for Evaluation. Lessons from Recent Practice at the Independent Evaluation Group of the World Bank. In S. B. Nielsen, F. Mazzeo Rinaldi and G. J. Petersson (eds.). *Artificial Intelligence and Evaluation. Emerging Technologies and Their Implications for Evaluation* (pp. 78–102). London: Routledge. https://doi.org/10.4324/9781003512493

Index

Made in the USA
Las Vegas, NV
19 September 2026

eb5bb413-7fdf-47b5-ad4d-99b2e0067152R02